BIOMEDICAL WASTE MANAGEMENT

Recycling and Remediation Strategies

BIOMEDICAL WASTE MANAGEMENT

Recycling and Remediation Strategies

Younis Ahmad Hajam, PhD
Javid Ahmad Malik, PhD
Anu Pandita, MSc
Deepu Pandita, MPhil
Editors

AAP | APPLE ACADEMIC PRESS

First edition published 2025

Apple Academic Press Inc.
1265 Goldenrod Circle, NE,
Palm Bay, FL 32905 USA

760 Laurentian Drive, Unit 19,
Burlington, ON L7N 0A4, Canada

CRC Press
2385 NW Executive Center Drive,
Suite 320, Boca Raton FL 33431

4 Park Square, Milton Park,
Abingdon, Oxon, OX14 4RN UK

Library and Archives Canada Cataloguing in Publication

Title: Biomedical waste management : recycling and remediation strategies / Younis Ahmad Hajam, PhD, Javid Ahmad Malik, PhD, Anu Pandita, MSc, Deepu Pandita, MPhil, editors.
Names: Hajam, Younis Ahmad, dedicator | Malik, Javid Ahmad, 1987- editor. | Pandita, Anu, editor. | Pandita, Deepu, editor.
Description: First edition. | Includes bibliographical references and index.
Identifiers: Canadiana (print) 20250133008 | Canadiana (ebook) 20250133016 | ISBN 9781774919286 (hardcover) | ISBN 9781774919293 (softcover) | ISBN 9781003595434 (ebook)
Subjects: LCSH: Medical wastes—Management.
Classification: LCC RA567.7 .B56 2025 | DDC 363.72/88—dc23

Library of Congress Cataloging-in-Publication Data

CIP data on file with US Library of Congress

ISBN: 978-1-77491-928-6 (hbk)
ISBN: 978-1-77491-929-3 (pbk)
ISBN: 978-1-00359-543-4 (ebk)

About the Editors

Younis Ahmad Hajam, PhD
Assistant Professor, Department of Life Sciences and Allied Health Sciences, Sant Baba Bhag Singh University, Padhiana, Jalandhar, Punjab, India

Younis Ahmad Hajam, PhD, is currently working as an Assistant Professor in the Department of Life Sciences and Allied Health Sciences at Sant Baba Bhag Singh University, Padhiana, Jalandhar, Punjab, India. Formerly, Dr. Hajam worked as an Assistant Professor and Head of the Department of Zoology at Career Point University, Hamirpur, Himachal Pradesh, India. He teaches undergraduate, postgraduate, and doctorate students. He holds a doctorate in Zoology from Guru Ghasidas Vishwavidyalaya (a Central University) in Bilaspur, Chhattisgarh, India. Dr. Hajam has published over 40 research papers in leading national and international journals, over 45 book chapters, and four papers presented at national proceedings. He has published eight books as an editor with Springer, Apple Academic Press, Nova Science Publishers, Taylor & Francis, etc. He has guided five MSc students under the thesis mode and 33 MSc students under project mode. Dr. Hajam has completed one major research project as principal investigator and three projects under the HP Chief Minister Start-up scheme. He has also served as a Junior Research Fellow (JRF) in Zoology at Guru Ghasidas Central University, Bilaspur, Chhattisgarh, India. He has presented his research work at more than 35 national and international conferences and has attended more than 30 national and international conferences and several workshops. He has organized various conferences, seminars, workshops, and webinars at national and international levels. Dr. Ahmad has also delivered invited lectures at international conferences. He has received a national award for his contributions to biological science. He is an editorial member of various international journals and is also a member of the Asian Council of Science Editors.

Javid Ahmad Malik, PhD

Academician/Researcher, Department of Zoology, Guru Ghasidas Central University Bilaspur, Chhattisgarh, India

Javid Ahmad Malik, PhD, is an Academician/Researcher working in the Department of Zoology at Guru Ghasidas Central University, Bilaspur, Chhattisgarh, India. Dr. Malik completed his MSc and MPhil in Zoology from Jiwaji University, Gwalior, Madhya Pradesh, and his PhD from Guru Ghasidas Central University, Bilaspur, Chhattisgarh, India. He has qualified for many national and state-level competitive exams, such as the National Eligibility Test (CSIR-NET), Graduate Aptitude Test (GATE-XL), and Jammu and Kashmir State Level Eligibility Test (JKSET) in life sciences. He was awarded a prestigious International Young Scientist Award for his paper presentation in 2016. His areas of interest include toxicology, pharmacology, and the evaluation of phytotherapeutic approaches using different animal models. He has published various research papers and book chapters in national and international journals and has reviewed several book chapters. Dr. Malik has also presented his work at a number of national and international conferences.

Anu Pandita, MSc

Senior Dietician, Vatsalya Clinic, New Delhi, India

Anu Pandita, MSc is working as a Senior Dietician at Vatsalya Clinic, Krishna Nagar, New Delhi, India. Previously, she has worked as a Lecturer at Bee Enn College of Nursing, Talab Tillo, Jammu, India, and as a Dietician at Ahinsa Dham Bhagwan Mahavir Charitable Health Center, New Delhi, India. Anu Pandita completed her MSC Internship and a course in the Dietetics Department of PGI, Chandigarh, India. She has conducted a case study at the Pediatric Gastroenterology Ward at Nehru Hospital PGI, Chandigarh, India, on a patient suffering from chronic liver disease. She has also completed a certificate course in food and nutrition. She has a number of training sessions, refresher courses, and workshops to her credit. Anu Pandita is a lifetime member of the Indian Dietetic Association and the Indian Science Congress Association, Kolkata, India. She has presented her research work at both national and international conferences. She has published various book chapters in Springer and CRC Press and has a number of review and research articles in various journals of national and international repute, such as *Cells* (MDPI) (IF=7.666) and *Frontiers in Physiology* (IF=4.755). Anu Pandita has edited seven books to date and has a number of books currently in progress.

Deepu Pandita, MPhil

Senior Lecturer, Government Department of School Education, Jammu, Union Territory of Jammu and Kashmir, India

Deepu Pandita, MPhil, is working as a Senior Lecturer in the Government Department of School Education, Jammu, Union Territory of Jammu and Kashmir, India. Deepu Pandita has 20 years of teaching experience and has completed her master's in Botany (MSc) from the University of Kashmir, Jammu and Kashmir, India, and a Master of Philosophy (MPhil.) in Biotechnology from the University of Jammu, Jammu and Kashmir, India. She has a number of international and national courses to her credit and has qualified for fellowships such as JRF NET and SRF from the Council of Scientific and Industrial Research (CSIR), New Delhi, India, Biotechnology Fellowship from the Government Department of Science and Technology, Jammu and Kashmir, India, and the IASC-INSA-NASI Summer Research Teacher Fellowship, India. Deepu Pandita has presented her research work at both national and international conferences and was awarded the Best Oral Presentation Award at an International Conference on Biotechnology for a Better Tomorrow, the Women Researcher Award, and the Research Excellence Award by two professional associations in India. Deepu Pandita is a life member of various scientific societies. She acts as a reviewer for 18 journals as well as an associate editor and editor for a number of international journals. Deepu Pandita has published a number of editorials, over 40 book chapters (Springer, Elsevier, CRC, etc.), and over 22 review and research articles in various journals of national and international repute, such as *Cells* (MDPI) (IF=7.666), *Frontiers in Plant Sciences* (IF=6.627), *Journal of Fungi* (IF=5.724), *Journal of Plant Growth Regulation* (IF=4.8), and *Frontiers in Physiology* (IF=4.755). Deepu Pandita has edited seven books to date and has a number of books currently in progress.

Contents

Contributors

Yasmeen Bano
Department of Molecular Human Genetics, Jiwaji University, Gwalior, Madhya Pradesh, India

Monika Bhadauria
Department of Rural Technology and Social Development, Guru Ghasidas Vishwavidyalaya, Bilaspur, Chhattisgarh, India

Nishat Bhatia
Department of Chemistry, Career Point University, Bhoranj (Tikker–Kharwarian), Hamirpur, Himachal Pradesh, India; Center for Nano-Science and Technology, CPU, Hamirpur, Himachal Pradesh, India

Debraj Biswal
Department of Zoology, Government General Degree College, Mangalkote, Burdwan, West Bengal, India

Yash P. Dadarkar
Department of Pharmaceutics, St. John Institute of Pharmacy and Research, Palghar, Maharashtra, India

Ashish S. Dubey
Department of Pharmaceutics, St. John Institute of Pharmacy and Research, Palghar, Maharashtra, India

Amol D. Gholap
Department of Pharmaceutics, St. John Institute of Pharmacy and Research, Palghar, Maharashtra, India

Younis Ahmad Hajam
Department of Life Sciences and Allied Health Sciences, Sant Baba Bhag Singh University, Jalandhar, Punjab, India

Kounsar Jan
Government Degree College Thindim Kreeri, Jammu and Kashmir, India

Ritesh A. Khairnar
Department of Pharmaceutics, St. John Institute of Pharmacy and Research, Palghar, Maharashtra, India

Ajay Kumar
Department of Biosciences, Himachal Pradesh University, Summer Hill, Shimla, Himachal Pradesh, India

Rajesh Kumar
Department of Biosciences, Himachal Pradesh University, Summer Hill, Shimla, Himachal Pradesh, India

Suresh Kumar
Department of Biosciences, Himachal Pradesh University, Summer Hill, Shimla, Himachal Pradesh, India

Asha Kumari
Department of Chemistry, Career Point University, Bhoranj (Tikker–Kharwarian), Hamirpur, Himachal Pradesh, India; Center for Nano-Science and Technology, CPU, Hamirpur, Himachal Pradesh, India

Prem Lata
Department of Biosciences, Division of Zoology, Career Point University, Hamirpur, Himachal Pradesh, India

Javid Manzoor
Department of Environmental Science, JJT University, Churela, Jhunjhunu, Rajasthan, India

Ab Qayoom Mir
Department of Environmental Science, Government Degree College Hajin, Jammu and Kashmir, India

Neelam
Department of Life Sciences and Allied Health Sciences, Sant Baba Bhag Singh University, Jalandhar, Punjab, India

Satendra Kumar Nirala
Department of Zoology, Guru Ghasidas Vishwavidyalaya, Bilaspur, Chhattisgarh, India

Urooj Nissar
Government Medical College, Baramulla, Jammu and Kashmir, India

Apeksha Pandey
IPS Institute of Technology and Management, Indore; Amity Institute of Law, Gwalior, Madhya Pradesh, India

Tejal D. Parab
Department of Pharmaceutics, St. John Institute of Pharmacy and Research, Palghar, Maharashtra, India

Heena Qadir
Department of Sociology, Government Degree College Awantipora, Jammu and Kashmir, India

Sampada D. Sawant
Department of Pharmaceutics, St. John Institute of Pharmacy and Research, Palghar, Maharashtra, India

Rahul Sharma
Department of Chemistry, Career Point University, Bhoranj (Tikker–Kharwarian), Hamirpur, Himachal Pradesh, India; Center for Nano-Science and Technology, CPU, Hamirpur, Himachal Pradesh, India

Deepali Shukla
IPS Institute of Technology and Management, Indore; Amity Institute of Law, Gwalior, Madhya Pradesh, India

Piyush Shukla
Department of Molecular Human Genetics, Jiwaji University, Gwalior, Madhya Pradesh, India

Pramod Kumar Tiwari
Department of Molecular Human Genetics, Jiwaji University, Gwalior, Madhya Pradesh, India

Rafael Vargas-Bernal
Departamento de Ingeniería en Materiales, Instituto Tecnológico Superior de Irapuato, Irapuato, Guanajuato, Mexico

Khursheed Ahmad Wani
Department of Environmental Science, Government Degree College Thindim Kreeri, Jammu and Kashmir, India

Preface

Biomedical waste comprises any liquid or solid waste posing an infection risk to individuals. This encompasses non-liquid body tissues, organs, bodily fluids, blood, and blood derivatives from humans and other primates. Additionally, it includes secretions visibly tainted with blood and used absorbent materials saturated with blood and blood products. Secretions clearly tainted with blood, as well as absorbent materials dried with blood or blood products, also fall into this category. Hospitals, health clinics, nursing homes, medical research facilities, doctors', dentists', and veterinarians' offices, home healthcare facilities, and funeral homes are the primary sources of biomedical waste.

Improper handling throughout the creation, collection, storage, transport, and treatment stages constitutes inadequate hospital waste management. This encompasses insufficient use of personal protective equipment (PPE), substandard storage conditions (such as prolonged exposure to high temperatures before treatment), manual long-distance transport, and the utilization of uncovered containers instead of securely sealed plastic bags. To mitigate nosocomial or hospital-acquired infections, it is crucial to ensure the appropriate management of the diverse waste generated in hospitals. There is an urgent need for the development of scientific techniques to efficiently manage, dispose of, and possibly transform biomedical waste into usable goods because of the detrimental effects that biomedical waste has on both the environment and human health. To deal with this inevitable circumstance fully, such a strategy is necessary.

With an emphasis on preventing biomedical-induced illnesses, this book presents up-to-date research and a range of remediation strategies for the management of biomedical waste. In order to efficiently manage biomedical waste, we have carefully selected recent and current material to include in this book. These tactics include chemical, biochemical, microbiological, and biotechnological approaches

This book volume contains a total of 13 chapters related to different aspects of biomedical waste, such as general description, types, sources, risk and its assessment, and management strategies for recycling biomedical waste. These chapters have been contributed by researchers across the globe, from Europe to Asia, thus highlighting the importance of biomedical waste and its management.

CHAPTER 1

Introduction, Types, and Sources of Biomedical Waste

RAJESH KUMAR,* AJAY KUMAR, and SURESH KUMAR

Department of Biosciences, Himachal Pradesh University, Summer Hill, Shimla, Himachal Pradesh, India

*Corresponding author.

ABSTRACT

The world has entered into an era where conditions are changing at a very fast rate. Under the influence of global development, people are expressing their increasing concern for protecting human health and the quality of the environment. With increasing awareness, it has been stressed that all attempts should be made to preserve nature so that energy could be saved, and pollution could be minimized/controlled. A huge quantum of waste is produced in the present time due to rapid industrialization and open dumping and discharge of this waste in water bodies without proper treatment creates serious public and environmental hazards. Waste is getting generated at a very fast pace and has attracted the attention of environmentalists worldwide. In 2016 alone, 242 million tons of plastic waste was generated globally. It is estimated that by 2025 around 2.2 billion tons of waste will be generated annually across the globe. The waste generation rate in small towns in India is 0.41 kg/capita/day. The majority of this waste is organic, consisting of cattle dung and crop residues. The high quantum of waste production depends upon factors such as industrialization, urbanization, population explosion, and unsustainable economic development. The rise in waste generation is predicted to be highest in underdeveloped and developing countries. Unplanned waste management is the primary cause of methane emissions, as over 75% of waste is managed through the landfilling method globally. Biomedical waste

Biomedical Waste Management: Recycling and Remediation Strategies. Younis Ahmad Hajam, Javid Ahmad Malik, Anu Pandita, and Deepu Pandita (Eds.)

(BMW) is produced from healthcare activities like medical operations, surgeries, medications, wound repair, and so forth. Common places for the production of BMW are hospitals, health centers, medical research institutes, dental clinics, nursing homes, animal hospitals, home health care and mortuaries, and so forth. This waste may also be known as medical or clinical waste. BMW is different from normal waste, as it is composed of traces of living cells may be blood, tissues, or other similar kind of materials. Hence, medical waste requires special attention from environmentalists, researchers, and authorities so that it can be properly managed, and the least hazard may be caused to the environment as well as living beings.

1.1 INTRODUCTION

Waste is any worthless, defective, unusable substance that is discarded after its principal use. Almost every human activity produces waste and various types of waste generated through human activities are municipal, industrial, agricultural, biomedical, radioactive, and electronic waste. The proper management of biomedical waste (BMW) is a matter of concern to the environment and general public health as well. Healthcare facilities generate various types of toxins, hazardous, and contagious materials. The side effects of haphazard management of waste are well known. Improper management and keeping of BMW may become the reason for many communicable diseases. BMW is generated from the medical activities of health care and research facilities. These healthcare facilities are important for human health, life, and well-being. However, the hazardous and toxic part of this waste generated from health and research establishments comprising infectious, radioactive, toxic, and sharp materials poses a major risk to neighboring living creatures. Large-scale generation of BMW poses a big problem to human beings. In recent years due to the outbreak of COVID-19, the use of plastic disposables, expansion of health facilities and vaccination drives have resulted in the production of large amounts of BMW.

The waste coming out as the end product of any medical activity (may be small or large) is categorized as BMW. It includes discarded hand gloves, medicated masks, blood-soaked cotton, used syringes, drain bags, discarded blood, lancets, scalpels, blades, broken glasses, unused microbial cultures, and stocks, used bandages and dressings, food remnants, disinfectants, solvents, cytotoxic drugs, unused liquids from radiotherapy and contaminated glassware, and so forth. In a broad sense, anything that comes out during the treatment of patients falls under the ambit of BMW and refers to anything coming out of healthcare facilities. The proper management of BMW may involve its source identification, characterization, quantification, separation at source, proper storage, safe transport, and scientific treatment (Datta et al., 2018).

As per Biomedical Waste (Management and Handling) Rules (1998) of India "Any waste generated during the process of diagnosis and any activity leading to the formulation of biological products or testing" (Government of India, 1998). The World Health Organization (WHO) has defined "Healthcare waste as all waste produced by healthcare establishment, research facilities and laboratories conducting medical or pathological examination (such as dialysis and insulin injections)" (Shruti et al., 2020).

BMW may be any waste that is generated during medical activities, vaccination, animal surgeries, clinical trials, and left out from animal experiments in research institutions (Ministry of Environment and Forestry, 1998). Annually, it is estimated that 16,000 million injections are used in this world (Table 1.1).

TABLE 1.1 Medical Waste Generation

Country	Quantity (kg/bed/day)
United States of America	4.5
United Kingdom	2.5
Spain	3.0
France	2.5
India	1.5

1.2 CLASSIFICATION OF BIOMEDICAL WASTE (BMW)

Biomedical waste (BMW) is generally classified into two categories: hazardous and non-hazardous. Approximately 85% of the residue produced in health care activities is similar to domestic household waste which is non-hazardous to human life and the environment. The administrative and housekeeping functions of hospitals, nursing homes, clinics, and many research institutions generate this type of non-risky waste (Shikoska et al., 2016). The enduring 10–15% of medical waste as shown in Figure 1.1 is considered as hazardous waste and may produce several health and environmental hazards (WHO, 1999).

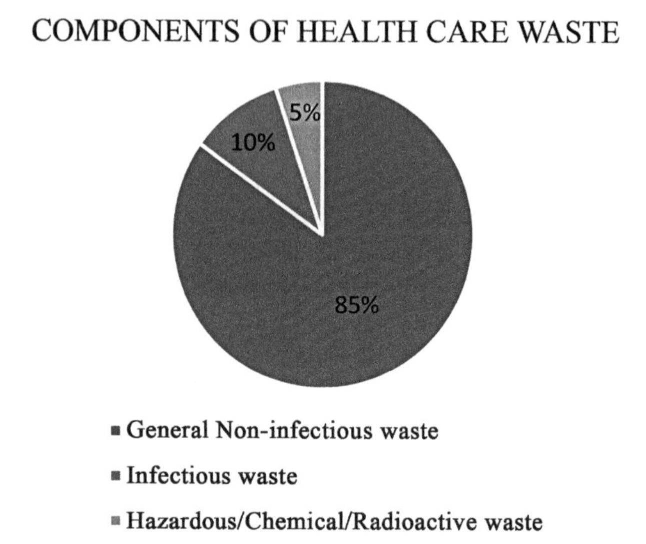

FIGURE 1.1 Components of healthcare waste.

Out of total BMW, about 10% to 25% of waste is supposed to be toxic and may have a negative impact on floral/faunal health. The remaining 75% to 90% of BMW is non-risky and similar to general domestic waste (Harshay et al., 2009).

1.2.1 NON-HAZARDOUS WASTE

In most healthcare and research establishments, about 85% of the generated waste is non-hazardous. It includes single-use paper towels; disposable covers of operating surfaces; unused/unutilized food and peels of fruits; wash water as well as cartons; paper mixing pads and materials used for packing.

1.2.2 HAZARDOUS WASTE

Approximately 15% of BMW is found to be toxic as it can disseminate infection among the masses including healthcare establishments as well as to waste handlers, rag pickers, and the general public as well. Almost all living beings are subjected to physical, chemical, and microbial risks by the hazardous part of BMW (Li and Jeng, 1993). These include infectious waste, pathological waste, sharps, chemical waste, cytotoxic waste, and contaminated and radioactive waste. Cytotoxic and chemical waste is mutagenic and/or teratogenic

(WHO, 1992). Singh et al. described that biomedical hazardous waste is of two types: (i) potentially contagious waste; and (ii) potentially toxic waste.

1.2.2.1 POTENTIALLY CONTAGIOUS WASTE

It is capable of causing infectious diseases. This type of waste has pathogens that cause disease in healthy humans. It constitutes about 10% of the total BMW and includes:

1. Clothes and wearing of patients treated for blood, pathogens, and other body fluids.
2. Discarded cultures and stocks of microbial infectious pathogens of diagnostic and research laboratory and unused vaccines.
3. Potentially infected material: It includes contaminated medical appliances, containers containing used gloves, body fluids, human pathological waste like excised tumors and organs of patients, placenta, and so forth.
4. Potentially infected animal carcasses.
5. Blood and blood products of patients and those used in research activities.

1.2.2.2 POTENTIALLY TOXIC WASTE

1. **Radioactive Waste:** Research and treatment activities and the use of radiotracers generate waste contaminated with radionuclides.
2. **Chemical Waste:** It Includes disinfectants, solvents, X-ray processing solutions, discarded laboratory reagents, monomers, and associated reagents, and base metal debris (dental amalgam in extracted teeth).
3. **Pharmaceutical Waste:** It includes creams, ointments, anesthetics, sedatives, vitamins, antibiotics, analgesics, supplements, and so forth (Hegde et al., 2007).

1.3 CLASSIFICATION GIVEN BY WORLD HEALTH ORGANIZATION (WHO)

The World Health Organization (WHO) has classified BMW into eight categories:

1. **Infectious Waste:** It includes waste that has disease-causing agents and may spread/transmit diseases, for example, lab materials,

discarded samples, used cultures, extracted tissues or equipment used for the treatment of diseased people, and so forth.

2. **Tissues or Fluids Waste:** It includes body parts and blood. Anything left out after minor or major surgery falls in the category of this type of waste as it has traces of fluid/tissue from the treated body.
3. **Sharp Waste:** It includes wastes from which there is a risk of injuries, for example, broken glass needles, disposable scalpels, hypodermic, and intra-venous needles, microscope slides, lancets, syringes pipettes, blades, knives, and infusion sets. Sharp waste can transmit diseases like HIV, Hepatitis B and C, and so forth.
4. **Pharmaceutical Waste:** Waste containing expired or discarded pharmaceuticals, such as antibiotics, anesthetics, analgesics, and sedatives, which are past their expiration date or contaminated by various chemicals.
5. **Genotoxic Waste:** This type of waste includes toxic drugs, hazardous mutagenic, and teratogenic substances; for example, cytotoxic drugs.
6. **Chemical Waste:** Waste having chemicals like reagents, dyes, stains, and discarded formalin disinfectants (hypochlorite, iodophors, phenolic derivatives, and alcohol-based disinfectants) are included in this category.
7. **Radioactive Waste:** Waste containing potentially radioactive materials; for example, anything that comes out from radiotherapy including liquid, glass, paper, cloth, excreta, and so forth.
8. **Heavy Metals Waste:** Various electrical equipment is used in healthcare facilities at almost all places, that is, in OPD, IPD, ICU, NICU, PICU, and so forth. Apart from this thermometers, blood pressure gauges, and many other quick test kits include electrical parts and sensors. All these are kept under the category of heavy metal waste and constitute about 1% of the total BMW (Rao, 2014).

1.4 RULES FRAMED IN INDIA

Various countries have formulated different guidelines for safe handling and disposal of medical waste. In India, the Ministry of Environment and Forest, GOI (1998) has notified BMW (Management and Handling) Rules (1998), according to which BMW can be categorized into 10 categories (Figure 1.2):

1. **Human Anatomical Waste:** This category comprises tissues, dissected organs, and surgically removed body parts.
2. **Animal Waste:** It is produced by institutions dealing with animals such as veterinary colleges, discharge from veterinary hospitals, and animal houses. It includes fluids, tissues, organs, body parts carcasses, blooding parts, fluid, blood, and dissected animals used in research.
3. **Microbial Waste:** This is a type of waste consisting of various samples, nutrient media, for culturing the samples, forensic samples from spots, micro-organisms used in laboratories, and cell culture used in industrial laboratories and research institutions.
4. **Waste Sharp:** It includes scissors, syringes, infusion sets, scalpels, hypodermic and intravenous needles, and so forth. Sharp waste may be broken, discarded, used, and unused sharps.
5. **Discarded Medicines and Cytotoxic Drugs:** Cytotoxic substances have the property of killing the cancerous cell, preventing the growth of cancerous cells, and are used in chemotherapy. Cytotoxic substances have mutagenic and teratogenic properties. It includes wastes comprising expired, contaminated, and discarded drugs.
6. **Soiled Waste:** It includes items used in treatment research that are filthy with body fluids.
7. **Solid Waste:** It consists of wastes from disposable solid items used in health care services other than the waste sharps.
8. **Liquid Waste:** It includes syrups, sanitizers, and solvents used in healthcare facilities.
9. **Incineration Ash:** It includes the ash that is generated during the scientific burning of BMW.
10. **Chemical Waste:** It includes reagents used in the production of healthcare items such as fluids, medicines, equipment, and so forth.

As per Biomedical Waste Management Rules (2016), waste generated from healthcare facilities can be classified as:

1. **Biomedical Waste (BMW):** Waste coming out from any medical activity/facility or testing laboratory is termed BMW.
2. **General Waste:** Anything that is generated along with BMW but not hazardous or infectious like newspapers, card boxes,

plastic water bottles, packaging material, food containers, and so forth, is kept in this category.

3. **Other Wastes:** These include electrical appliances used in health care facilities which are not included in BMW, however, may be hazardous to living beings. This type of waste may be managed as per the provisions laid down as under:

   i. E-waste (management) rules (2016);
   ii. Batteries (management and handling) rules (2001);
   iii. Rules/guidelines under the Atomic Energy Act (1962).

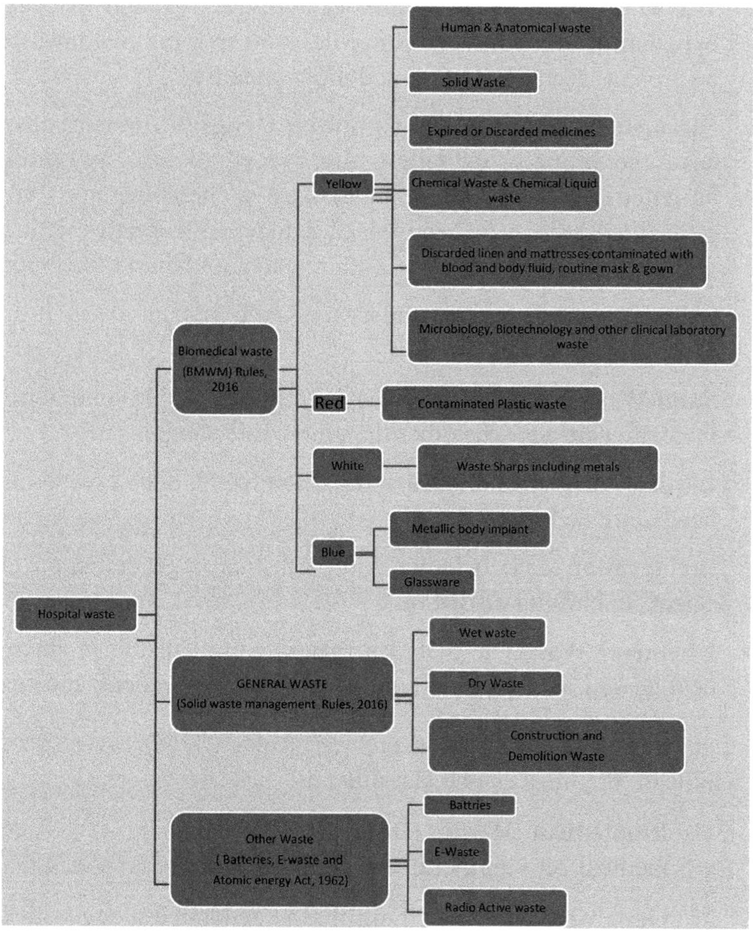

FIGURE 1.2 Waste categorization in health care facilities.

1.5 SOURCES OF BIOMEDICAL WASTE (BMW)

Biomedical waste (BMW) is generated primarily from healthcare facilities that may include hospitals, clinics, nursing homes, veterinary hospitals, blood banks, medical practitioners, animal houses, dispensaries, and research institutes (Babu et al., 2009). Some of the miscellaneous waste sources are as under:

1. Homes;
2. Pharm industries;
3. Research institutes;
4. Blood banks and testing labs.

Based on the quantification of BMW, it can be categorized into two types (Tiwari et al., 2013; Figure 1.3 and Table 1.2).

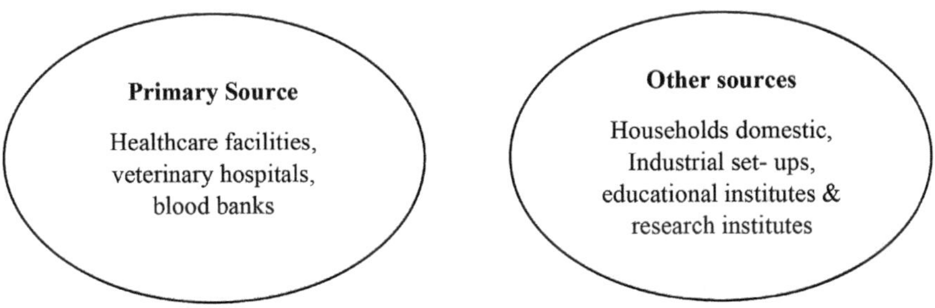

FIGURE 1.3 Sources of biomedical wastes.

TABLE 1.2 Sources of Biomedical Waste

Primary Sources of Biomedical Waste		Secondary Sources
Hospitals	Medical education colleges	Clinics
Nursing homes	Blood banks	Ambulances
Dialysis center	Nursing centers	Home treatments
Maternity centers	Animal research institutions	Animal slaughterhouses
Dispensaries	Immunization programs	Funeral services
Research laboratories	Industries	Educational institutes

1.6 MANAGEMENT OF BIOMEDICAL WASTE (BMW)

Management of waste is a major issue all over the globe. Developing countries are still far behind in managing BMW properly due to a lack of facilities/finances. The steps for waste management are shown in Figure 1.4.

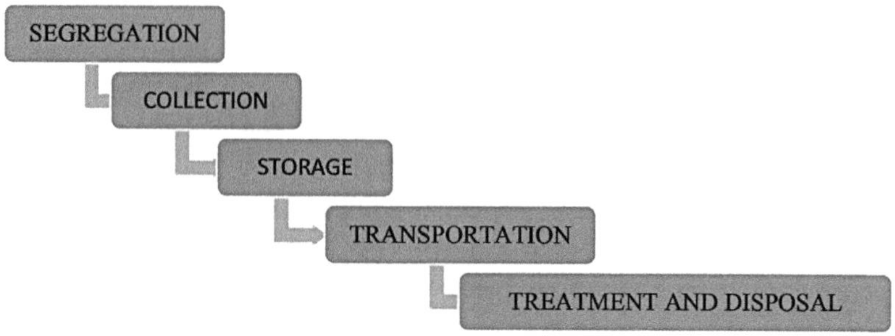

FIGURE 1.4 Steps for waste management.

1.6.1 SEGREGATION

Segregation involves separating the waste into appropriate containers with color codes. It involves the segregation of waste at the source of generation before its storage, transportation, treatment, and disposal thereby reducing the risks to human health and the environment. The cataloging of waste at source is a vital step in waste management. Biomedical waste (BMW) should not be allowed to get mixed with Municipal Solid Waste, as it may cause dreadful and infectious diseases like HIV, hepatitis B and C, tuberculosis, and other respiratory and skin ailments. The waste items sent to the incinerator/burial (Category 1, 2, 3, 6) should be properly packed in yellow bags for identification purposes. Red-colored bags should be used to store the waste for which chemical/microwave/autoclave treatment is required. Blue and white bags should be used to pack the waste that has to be sent for chemical treatment, is to be packed in Blue/White bag. All bags/containers must be labeled according to the rules (Schedule III of Biomedical Waste Rules, 1998). At the point of generation of waste containers different colored polythene bags should be placed. Proper segregation decreases the chances of infecting the other healthcare staff. Effective segregation helps in recycling certain components of BMW after thorough disinfection and prevents the mixing of harmful components of BMW with general municipal waste (Table 1.3).

1.6.2 COLLECTION

It involves gathering BMW by the use of different types of containers for operation theaters, wards, diagnostic laboratories, corridors, kitchens, and so forth. The collection of waste in the polyclinic is undertaken by a team of two members. One person pulls the cart and the other replaces the filled bags

TABLE 1.3 Biomedical Waste Packaging

Color Codes	Container	Waste Type	Provision
Yellow color	Plastic bags	Category 1, 2, 3, and 6	Deep burial in land/ Incineration
Red color	Disinfected container/plastic bags	Category 3, 6, and 7 (microbiological and biotechnological waste, solid waste)	Autoclaving/microwaving/ treatment by chemicals
Blue/white color	Plastic bags/ puncture-proof containers	Category 4 and 7 (waste sharps, solid plastic waste)	Autoclaving/microwaving/ treatment by chemical, shredding, and destruction
Black color	Plastic bags	Category 5 and 9	Disposed in sanitary landfills

Source: Biomedical Waste Management Rules (1998).

with empty Polythene bags. These workers must be trained and well aware of the associated hazards of the waste and take the requisite protective measures during waste handling bearing personal protective equipment during collection of the infectious-based segregation of various color-coded containers and transportation of the waste in the designated cart. All containers kept for the collection of hazardous waste were labeled with cytotoxic symbols as per the Biomedical Waste (Management and Handling) Rules (1998; Gupta et al., 2009).

1.6.3 STORAGE

After the collection, BMW should be at isolated places away from the reach of humans/animals. Temporarily stored waste should be immediately sent for disposal at a designated facility in specially designed vehicles. The storage facilities should be at a far distance from the patient wards.

1.6.4 TRANSPORTATION

Transportation of BMW for treatment should be made in covered trolleys specially designed for the purpose. Within the health care facilities, waste routes must be designated to avoid the passage of waste through patient care areas and separate time should be earmarked for transportation of BMW to decrease the probability of mixing BMW with general waste. An appropriate system for securing the weight during transport should be ensured (Singh et al., 2014).

1.6.5 TREATMENT AND DISPOSAL

Modification of waste before it gets to its ultimate resting place is termed treatment. It is the process of removing the toxic attributes of waste. Treatment of BMW also reduces environmental hazards. Glassware must be disinfected and cleaned. The basic principle involved in the disposal of BMW is the shredding and mutilation of waste of treatment with a 1% solution of hypochlorite so that unauthorized reuse of waste is prevented (Chakraborty et al., 2014). The various techniques used to dispose the BMW are given below:

1. **Incineration:** This converts the pathological and pharmaceutical waste into ash. Incineration is the burning of waste scientifically to transform it into inorganic/inert material. To ensure optimal combustion, incinerators have a primary chamber for solid-phase combustion and secondary burning chambers for gas-phase combustion. The temperature in incinerators ranges from 982°C to 1093°C. Incinerators have the advantage of a quick, easy disposal method. Specifications described in BMW rules (1998) for installation and operationalization of incinerators should be followed. Incineration reduces the volume of BMW and the resulting ash can be disposed of in landfills. It may decrease and reduce its effect by about 90% (Ferdowsi et al., 2013).
2. **Plasma Pyrolysis:** It is a new approach for converting waste into useful products for harnessing energy from it. Waste is a potential raw material to be used as fuel for the production of energy. In this context, pyrolysis is a type of thermal treatment involving high temperatures and a limited supply of oxygen. It is an environment-friendly technology that converts organic wastes into commercially useful by-products (Singh et al., 2014).
3. **Deep Burial:** A pit or trench of 2–3 m depth should be used for the burial of BMW. The pit may not be filled beyond its half limit and should be covered properly with lime and then the rest of the pit should be filled with the soil. Deep burial is only allowed for anatomical and soiled waste where proper facilities for disposal are not available (Chudasama et al., 2013).
4. **Autoclave Treatment:** This is a process of high-pressure steam treatment for sterilization of equipment for reuse and waste produced. It is comparatively cheaper and has not reported any health impact

(Pasupathi, 2011). An autoclave is a sophisticated machine used to deliver heat under pressure to a chamber to sterilize the contents of the chamber. BMW segregated in red bags is sent for autoclaving and then for shredding. After the treatment in autoclaves, the medical waste can usually be disposed of in landfills.

5. **Microwave Treatment:** In microwave treatment waste is shredded by cutting it into small pieces and then mixed with water to heat it internally and kill its harmful microbes. Waves of frequency 2450 MHz and wavelength 12.24 cm kill most of the microorganisms of BMW. The efficacy of microwave sterilization should be checked regularly with virological and bacteriological tests. Cytotoxic waste, radioactive waste, and body parts and metal items cannot be treated by microwaves.
6. **Shredding:** It is a process of cutting waste into smaller and simpler unrecognizable pieces. A shredder cuts the waste into small pieces. This technique requires skilled staff to run the machine. Shredding reduces the volume and helps in recycling plastics, but shredding does not disinfect the waste and cannot be used for heavy metal pieces.
7. **Encapsulation:** It involves filling a small number of hazardous materials in a mass of inert materials. The main objective of encapsulation is to prevent the contact of hazardous material with human beings and the environment. In this process, toxic and hazardous waste is packed in a polythene drum and then inert material like bituminous sand, clay plastic foam, and lime are put into it. The sealed containers are then disposed of in sanitary landfills. Encapsulation can only be used for small quantities of waste. The drawback of encapsulation is that it makes the waste bulky, and voluminous.
8. **Chemical Treatment:** It involves the use of various chemical agents to treat waste BMW (Pasupathi, 2011). Several techniques such as precipitation, oxidation, reduction, and pH neutralization are used in this method to transform waste into less harmful material.
9. **Irradiation:** These systems are currently being used to disinfect the waste by exposure to radiation (Odumosu, 2016).
10. **Secured Landfill:** Landfills are scientifically designed pits for the dumping of inert material. BMW category No. 9 is disposed of in a municipal landfill.

1.7 PERSONNEL PROTECTIVE EQUIPMENT

The use of the following safety devices should be made compulsories for all handling BMW:

1. **Gloves:** Good quality gloves should be used by waste handlers. The gloves should be bright yellow colored and should be washed with carbolic soap and detergents.
2. **Aprons, Gowns,** and Apparels: Aprons and gowns should be used by waste handlers to avoid direct contact with skin. It should be made of non-inflammable material.
3. **Masks:** Good quality masks, and eye and face shields should be worn by the personnel working with BMW.
4. **Boots:** Rubber-soled, anti-skid boots with a cover for the leg up to the ankle should be used by healthcare workers.

1.8 IMPACTS OF BIOMEDICAL WASTE (BMW)

Improper disposal of BMW poses harmful effects on human health and the environment as discussed in subsections.

1.8.1 BIOMEDICAL WASTE (BMW) AND HUMAN HEALTH

In many countries, a huge amount of healthcare waste is now posing serious health risks. Improper management of BMW by healthcare facilities in developing countries can lead to serious health hazards and environmental pollution. Inside hospitals, nursing homes, and other health care centers, the main groups of people subjected to risks due to BMW are doctors, surgeons, nursing staff, people engaged in housekeeping, maintenance, attendants, and any other person visiting medical facilities.

According to the WHO among the 35,000,000 healthcare workers worldwide it is estimated that every year about 3,00,000 receive hard exposures to blood-borne pathogens, 2,000,000 of those to Hepatitis B Virus, 0.9000000 to Hepatitis C Virus, and 1,70,000 to Human Immunodeficiency Virus.

BMW may contain harmful pathogens that can infect health workers, patients, and the general public. Pathogens may spread from healthcare facilities into the environment through BMW. The health hazards associated with BMW are discussed in subsections.

1.8.1.1 SHARPS-INFLICTED INJURIES

It is estimated that in each year 15 billion injections are administered world-wide. In low-income countries, all the used syringes and needles are not disposed of safely which may cause health and environmental hazards.

1.8.1.2 SPREAD OF DISEASES

Malpractices such as haphazard dumping of BMW in open areas, barren land, or water bodies without proper segregation and treatment can lead to the spread of diseases.

1.8.1.3 IN-HOME HEALTHCARE ACTIVITIES

Biomedical waste (BMW) generated during In-home healthcare activities is not properly disposed of. BMW generated by home nursing activities and from diabetics who are injecting themselves with insulin is not disposed of properly and gets mixed with domestic waste, which can cause health hazards.

1.8.1.4 TOXIC EXPOSURE

Cytotoxic drugs, mercury, and dioxins are toxic to human beings.

1.8.1.5 AIR POLLUTION

Incineration of medical waste results in the generation of dioxin which is a carcinogen. It binds with the DNA and is a potential cancer promoter.

1.8.1.6 RADIATION

Some diagnostic technologies use radiation, if these techniques are not contained and monitored properly, they can travel long distances to other areas for radiation burns.

1.8.2 BIOMEDICAL WASTE (BMW) AND ENVIRONMENT

Production of large amounts of BMW has started having a bad impact on the quality of water, air, and soil. Usually, BMW is mixed with domestic waste without rendering it harmless, and then it is disposed of anywhere without following BMW management rules. Improper management of BMW results in air, water, and soil pollution. Proper waste management is essential for the protection of the environment. The purpose of proper waste management is to keep the environment clean and avoid any risk to mankind. Improper disposal of BMW can deteriorate the quality of the environment including air, water, and soil as discussed in subsections.

1.8.2.1 EFFECT ON AIR

Storage of BMW in open air can release harmful gases such as methane and sulfide, which are polluting the atmosphere (Husain et al., 2011). Improper burning of BMW can introduce dangerous gases and particulate matter into the atmosphere such as polycyclic aromatic hydrocarbons (PAHs), chlorobenzenes, chlorophenols lead, cadmium, and mercury (Biswapriya et al., 2021). Moreover, carcinogens such as dioxins and polychlorinated biphenyls (PCBs) are released during incineration. Every hospital and healthcare facility should have proper Bio-medical treatment facilities so that waste is treated within 48 hours of its generation. Incineration and open burning are widely used and the most widely used methods of disposing majority of hazardous medical waste. However, open burning or burning of BMW in incinerators emits toxic gases such as black smoke, harmful flue gas, fly ash, and bad odors which result in atmospheric pollution producing respiratory, skin, and even dreadful cancer-like diseases. Open disposal and burning of BMW can mix different harmful pollutants in environment. Burning of BMW may release harmful pollutants that cause various types of respiratory problems among people. According to the U.S. Environmental Protection Agency, Biomedical waste is the third main source of dioxin emission and 10% of mercury emission.

1.8.2.2 EFFECT ON WATER

Leachate from the waste dumping sites may contaminate the surface and groundwater resulting in harmful effects on human and faunal health.

1.8.2.3 EFFECT ON SOIL QUALITY

Unscientific management of BMW may deteriorate the quality of soil near waste dumping sites. It has been studied the change in soil quality and decrease in floral diversity due to open dumping of BMW. A significant change has been observed in the physical and chemical properties of soil near dumping sites of BMW (Biswapriya et al., 2011).

1.8.3 IMPACT OF COVID-19 ON BIOMEDICAL WASTE (BMW) AND OCEAN PLASTIC POLLUTION

The COVID-19 pandemic was a disaster for humans and caused to production of large amounts of BMW. Each year approximately 8 million metric tons of plastic enter into oceans. According to WHO, during the COVID-19 pandemic huge quantum of medical waste was generated as people who were working day and night in healthcare facilities had to use personal protective equipment all the time and changing PPEs very frequently. This led to an increase in plastic pollution in the oceans by 10 times. Personal. The general public was also using masks and other protective shields to stay infected which resulted in an elevated quantum of biowaste. As per the WHO, more than 140 million test kits were manufactured all over the world resulting in the production of about 7.31 lakh liters of waste. In addition, the COVID-19 vaccine manufactured in different countries led to the generation of more than 1.44 lac tons of waste during the year 2020–2022.

1.9 BIOMEDICAL WASTE (BMW) REGULATIONS

1.9.1 UNITED STATES

Medical waste is regulated in the United States of America by the following legislation:

1. **US EPA Regulations:** The Environment Protection Agency of the U.S. government is responsible for setting and enforcement of pollution control standards.
2. **DOT Regulations:** Department of Transportation (DOT) regulations assign hospitals the responsibility to safely carry hazardous waste to disposal sites.
3. **Medical Waste Tracking Act:** The US Congress enforced the Medical Waste Tracking Act of 1988 to prevent potential health hazards from medical waste.
4. **OSHA Regulation:** These regulations were created by the US Congress under the Occupational Safety and Health Act of 1970.

1.9.2 BIOMEDICAL WASTE (BMW) REGULATIONS IN INDIA

In India, BMW management rules were framed under the Environment (Protection) Act (1986). These rules state that waste coming out of health-care facilities should be managed properly and disposal should be carried out as per regulations.

1.10 CONCLUSION

Biomedical waste (BMW) is generated by healthcare facilities; institutions carrying out research, and veterinary facilities. Improper management of BMW can result in health problems. BMW management includes collection, segregation at source, proper transportation, it's treatment, and disposal. The color coding of bins and the use of technologies such as incineration, autoclaving, and deep burial of BMW are useful for the environment. BMW management is a social responsibility as well as a legal necessity. BMW management should be included in the training curriculum of health care workers. In BMW management, the segregation of BMW at the initial stage is the most important step. Training should be imparted to healthcare staff through short-term courses and workshops. The main aim of BMW management is to reduce waste volume and ensure proper disposal. All those involved in the management of BMW should make an effort to keep this environment safe.

KEYWORDS

- **aprons**
- **biomedical waste**
- **environmental pollution**
- **healthcare centers**
- **health risks**
- **healthcare waste**

REFERENCES

Babu, B. R., Parande, A. K., Rajalakshmi, R., Suriyakala, P., & Volga, M. (2009). Management of biomedical waste in India and other countries: A review. *Journal of International Environmental Application & Science, 4*(1), 65–78.

Bakoğlu, M., Karademir, A., & Ayberk, S. (2004). An evaluation of the occupational health risks to workers in a hazardous waste incinerator. *Journal of Occupational Health, 46*(2), 156–164. doi: https://doi.org/10.1539/joh.46.156.

Biswapriya, J., Sanghamitra, P., & Nabnita, P. (2021). Impact of improper biomedical waste disposal on human health and environment during COVID-19 pandemic. *European Journal of Molecular & Clinical Medicine, 8*(3), 4137–4143.

Chakraborty, S., Veeregowda, B., Gowda, L., Sannegowda, S. N., Tiwari, R., Dhama, K. & Singh, S. V. (2014). Biomedical waste management. *Advances in Animal and Veterinary Sciences, 2*(2), 67–72.

Chudasama, R. K., Rangoonwala, M., Sheth, A., Mishra, S. K. C., & Kadri, A. M. (2013). Biomedical waste management: A study of knowledge, attitude, and practice among healthcare Personnel at tertiary care hospital in Rajkot. *Journal of Research in Medical and Dental Sciences, 1*(1), 17–22.

Datta, P., Mohi, G. K., & Chander, J. (2018). Biomedical waste management in India. Critical appraisal. *Journal of Laboratory Physicians, 10*, 6–14.

Ferdowsi, A., Ferdosi, M., & Mehrani M. J. (2013). Incineration or autoclave? A comparative study in Isfahan hospitals waste management system. *Materia Socio Medica, 25*(1), 48–51.

Government of India. (1998). *Biomedical Waste Management Rules, 1998* [Gazette notification]. Ministry of Environment and Forests, New Delhi, India.

Gupta, S., Boojh, R., & Mishra, A. (2009). Rules and management of biomedical waste at Vivekananda Polyclinic: A case study. *Waste Management, 29*, 812–819.

Harhay, M. O., Halpern, S. D., Harhay, J. S., & Olliaro, P. L. (2009). Health care waste management: A neglected and growing public health problem worldwide. *Tropical Medicine and International Health, 14*, 1414–1417.

Hegde, V., Kulkarni, R. D., & Ajantha, G. S. (2007). Biomedical waste management. *Journal of Oral and Maxillofacial Pathology, 11*(1), 5–9.

Li, C. S., & Jeng, F. T. (1993). Physical and chemical composition of hospital waste. *Infection Control and Hospital Epidemiology, 14*,145–150.

Ministry of Environment and Forests, Govt. of India. (1998). *Gazette Notification*, New Delhi.

Odumosu, B. T. (2016). *Biomedical Waste: Its Effects and Safe Disposal*. CRC Press, Taylor & Francis Group.

Pasupathi, P. (2011). Biomedical waste management for the health care industry. *International Journal of Biology and Medical Research, 2*(1), 472–486.

Rao, N. L. (2014). Biomedical waste management. *International Journal of Multidisciplinary Research and Development, 1*(2), 25–33.

Shikoska, B., Dimova, C., Schumanov, G., & Vankovski, V. (2016). Biomedical waste management. *Journal of the Macedonian Medical Association, 70*(1), 1–7.

Shruti, S., Babu, N. A., Behura, S. S., & Rajesh, E. (2020). Management of biomedical waste in hospital and health care organizations–A review. *Indian Journal of Forensic Medicine & Toxicology, 14*(4), 1256–1260.

Singh, H., Rehman, R., & Bumb, S. S. (2014). Management of biomedical waste: A review. *International Journal of Dental and Medical Research, 1*(1), 14–20.

Singh, H., Rehman, R., & Bumb, S. S. (2014). Management of biomedical waste: A review. *International Journal of Dental and Medical Research, 1*(1), 14–20.

Tiwari, A., & Kadu, P. A. (2013). Biomedical waste management practices in India–A review. *International Journal of Current Engineering and Technology, 3*(5): 2030–2033.

World Health Organization (WHO). (1992). *Managing Medical Wastes in Developing Countries*. Geneva, Switzerland.

World Health Organization (WHO). (1999). *Regional Workshop on Hospital Waste Management and Hospital Infection Control* (WHO Project: INDEHH001). Government Medical College and Hospital, Nagpur, India.

CHAPTER 2

Biomedical Waste: Sources, Characterization, Color-Coding, Segregation, Transportation, and Disposal Methods

NEELAM,[1] YOUNIS AHMAD HAJAM,[1*] and RAJESH KUMAR[2]

[1]*Department of Life Sciences and Allied Health Sciences, Sant Baba Bhag Singh University, Jalandhar, Punjab, India*

[2]*Department of Biosciences, Himachal Pradesh University, Shimla, Himachal Pradesh, India*

*Corresponding author.

ABSTRACT

Solid and liquid waste containing infectious or possibly contagious components that come from medical, laboratory, or research operations connected to disease diagnosis, prevention, and treatment are referred to as biomedical waste (BMW/BW). This kind of waste increases the danger of illness and harm to healthcare professionals, patients, and the neighborhood. Human anatomical waste, animal waste, trash from microbiology and biotechnology, sharps waste, unused medications, cytotoxic drugs, soiled waste, liquid waste, incinerator ash, and chemical wastes are all included in this category. Hospitals, clinics, medical and veterinary schools, blood banks, mortuaries, autopsy facilities, biotechnology organizations, research labs, home healthcare facilities, and funeral homes are common sources of biological waste. Despite being non-infectious, hazardous chemicals and radioactive waste require safe disposal. According to the World Health Organization (WHO), 5% of hospital wastes are harmful but non-infected, while 10% of hospital wastes are infectious. General, pathological, radioactive, chemical,

Biomedical Waste Management: Recycling and Remediation Strategies. Younis Ahmad Hajam, Javid Ahmad Malik, Anu Pandita, and Deepu Pandita (Eds.)

infectious, sharps, medicines, and pressurized wastes are the eight categories that WHO has created for medical waste. The Biomedical Waste (handling and Handling) Rules of 1998, as well as later changes, govern the handling of BMW in India. These regulations are divided into six schedules, which cover the classification of BMW, container types, color coding, labels for BMW containers or bags (which must be non-washable and visible), labels for transporting BMW containers or bags, standards for treatment and disposal, and schedules for waste treatment facilities like incinerators, autoclaves, and microwave systems. Operating requirements, such as combustion efficiency and pollution requirements, are spelled out. This review article focuses on fundamental issues, such as the definition, classification, difficulties associated with BMW, and treatment and disposal techniques within the context of BMW management.

2.1 INTRODUCTION

Biomedical waste (BMW), known as medical waste, encompasses waste generated during human or animal diagnosis, treatment, or immunization. This waste category includes items like syringes, needles, swabs, bandages, gloves, tissues, body fluids, expired medicines, and laboratory test chemicals (Rutala et al., 2015). BMW is considered hazardous due to its potential to spread infections and diseases. It may harbor infectious agents such as bacteria, viruses, and fungi, posing severe health risks if not handled and disposed of properly. Additionally, BMW can contain toxic substances like heavy metals, chemicals, and radioactive materials, posing threats to the environment and public health if not managed appropriately. Effective management of "biomedical waste" is crucial to prevent infection spread and safeguard "public health and the environment." It includes appropriate segregation, handling, storage, transport, and dumping of waste, minimizing exposure risks to hazardous materials. WHO has outlined rules for managing BMW, aiding healthcare facilities and regulatory authorities in ensuring safe and effective waste management practices (Figure 2.1).

BW is a comprehensive term encompassing all waste generated within healthcare facilities, spanning hospitals, clinics, dental clinics, nursing homes, and laboratories. This waste category comprises a diverse array of materials, including sharps (needles, syringes), surgical instruments, gloves, bandages, cultures, human or animal tissues, body fluids, expired or unused medicines, and laboratory test chemicals. In contrast, hospital waste is a more specific term denoting waste generated exclusively within hospital premises.

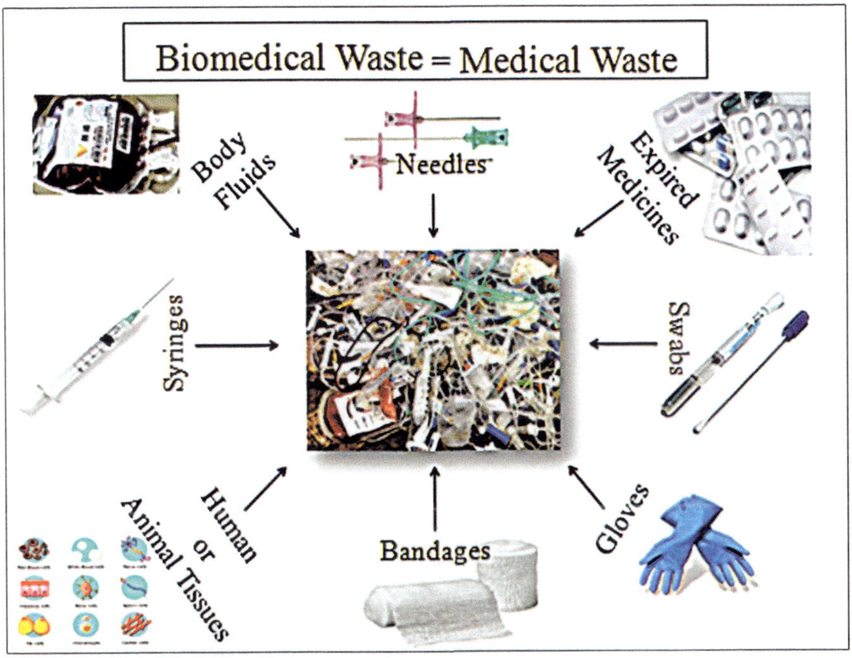

FIGURE 2.1 Different sources of biomedical waste.

Hospital waste not only includes BMW but also extends to non-BMW, such as kitchen waste, broken furniture, and electronic waste (*refer to* Figure 2.2).

Infectious waste, also referred to as hazardous waste, represents a specific category within BMW distinguished by its potential to transmit infections and diseases (Datta et al., 2018). It encompasses waste contaminated with "blood, body fluids, and other materials that" could harbor infections. Regardless of nomenclature, effective management of BMW, hospital waste, and infectious waste is paramount for community well-being and protection. The meticulous segregation, appropriate handling, secure storage, safe transportation, and proper disposal of these waste streams are imperative measures to curtail infection spread, mitigate environmental pollution, and cultivate a healthful and secure environment.

2.2 SOURCES OF BIOMEDICAL WASTE (BMW)

BW is produced from various sources within healthcare facilities, including hospitals, clinics, laboratories, and veterinary clinics (Sharma et al., 2002). Some of the common sources of BMW are discussed in subsections.

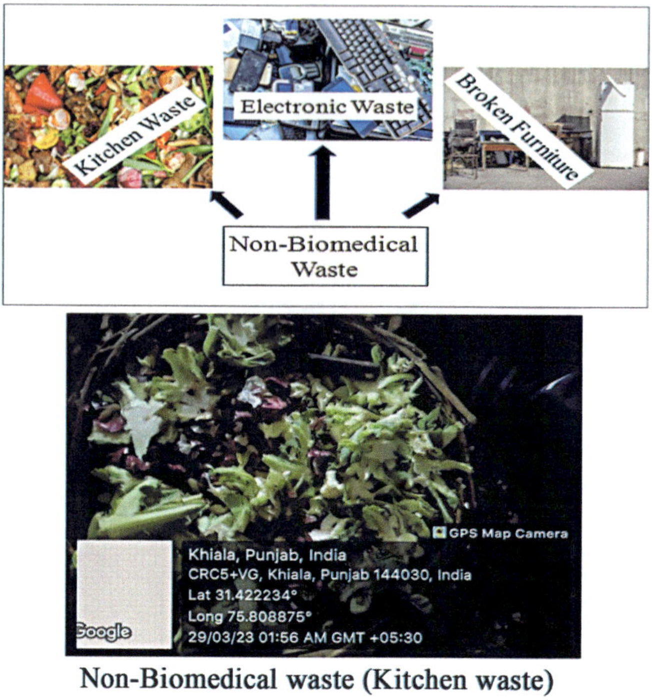

FIGURE 2.2 The segregation of non-biomedical waste (kitchen waste).

2.2.1 PATIENT CARE ACTIVITIES

This includes the "waste generated during diagnosis, treatment, and care of patients." Examples include disposable gloves, gowns, masks, bandages, and soiled dressings (Mohankumar and Kottaiveeran, 2007).

2.2.2 LABORATORY AND RESEARCH ACTIVITIES

This includes the waste generated from laboratory testing, research, and experimentation. Examples include "cultures, stocks, and specimens of microorganisms;" "contaminated animal carcasses, body parts, and bedding" and lab equipment such as pipettes, tubes, and slides (Rodríguez et al., 2007).

2.2.3 RADIOLOGY DEPARTMENTS

This includes the waste generated from diagnostic imaging procedures. Examples include used film, X-ray processing chemicals, and radioactive sources (Khunprasert et al., 2008).

2.2.4 DENTAL OFFICES

This includes the "solid waste consists of three main categories: infectious waste, non-infectious waste, and domestic-type waste" generated from dental procedures. Examples include extracted teeth, contaminated gloves, and disposable needles (Vieira et al., 2009).

2.2.5 MORTUARIES AND AUTOPSY CENTERS

This includes the waste generated from the preparation and handling of dead bodies. Examples include contaminated bedding, body fluids, and tissues (McGuone et al., 2020).

2.2.6 VETERINARY CLINICS AND ANIMAL RESEARCH CENTERS

This includes the waste produced from the care "and treatment of animals." Examples include animal carcasses, contaminated animal bedding, and animal tissue samples (McLean et al., 2007).

It is important to handle properly, "segregate, transport, and dispose of BMW to prevent the spread of infectious diseases and protect the environment" and public health. Healthcare facilities should have well-established procedures "and guidelines for the safe management of BMW" (Table 2.1).

TABLE 2.1 Sources of Biomedical Waste with Examples

Sl. No.	Sources of Biomedical Waste	Examples	References
1.	Patient care activities	Disposable gloves, gowns, masks, bandages, and soiled dressings.	Mohankumar and Kottaiveeran (2007)
2.	Laboratory and research activities	Cultures, stocks, and specimens of microorganisms; "contaminated animal carcasses, body parts, and bedding; and" lab equipment such as pipettes, tubes, and slides.	Rodríguez et al. (2007)
3.	Radiology departments	X-ray processing chemicals, and radioactive sources.	Khunprasert et al. (2008)
4.	Dental offices	Extracted teeth, contaminated gloves, and disposable needles.	Vieira et al. (2009)
5.	Mortuaries and autopsy centers	Contaminated bedding, body fluids, and tissues.	McGuone et al. (2020)
6.	Veterinary clinics and animal research centers	Include animal carcasses, contaminated animal bedding, and animal tissue samples.	McLean et al. (2007)

2.3 GENERATION OF BIOMEDICAL WASTE (BMW) IN HOSPITALS

Hospitals are significant producers of BW, encompassing waste produced during human or animal "diagnosis, treatment, or immunization." BMW is categorized into various types, including "infectious waste, pathological waste," sharps waste, chemical waste, and radioactive waste. Infectious waste consists of items "contaminated with blood, bodily fluids," or potentially infectious materials, like gloves, gowns, and disposable items used during medical procedures. Pathological waste comprises human or animal tissues, organs, and "body parts, such as amputated limbs," surgically removed organs and fetuses. Sharps waste includes needles, syringes, scalpels, and other sharp objects capable of puncturing the skin and potentially transmitting infectious diseases. Chemical waste involves expired drugs, solvents, disinfectants, and other chemical agents utilized in healthcare facilities. Radioactive waste includes materials contaminated with radioactive substances, as found in nuclear medicine.

The volume of BMW generated in hospitals can vary based on facility size, patient volume, and the types of medical procedures conducted. According to the World Health Organization (WHO), an average hospital bed produces approximately 0.5 kg of waste daily, with roughly 20–25% of that waste falling under the hazardous or BMW category. Consequently, a hospital with 100 beds may generate up to 50 kg of BMW each day (Floyd et al., 2006).

"Effective management of BMW" is paramount in curbing the transmission of infectious diseases and safeguarding healthcare workers and the broader public from potential hazards. Hospitals must implement well-defined protocols to "ensure the safe and" proper handling, segregation, storage, transportation, "and disposal of BMW" (*see* Figure 2.3).

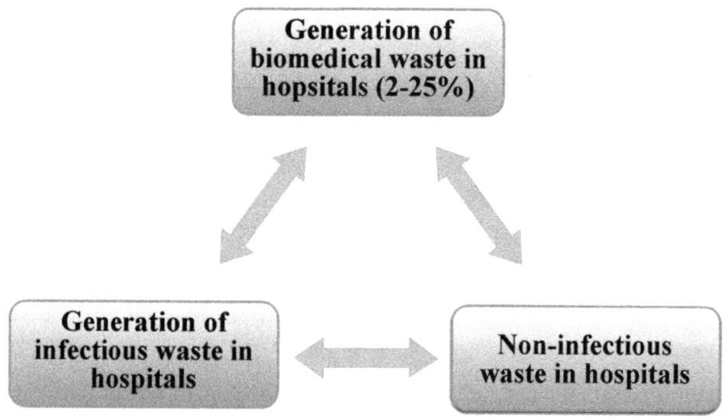

FIGURE 2.3 The infectious and non-infectious biomedical wastes.

2.3.1 GENERATION OF INFECTIOUS AND NON-INFECTIOUS WASTE IN HOSPITALS

The proportion of infectious and non-infectious BMW can fluctuate based on the healthcare facility type and its operational nature. Generally, infectious waste constitutes a smaller portion of the total BMW, with non-infectious waste making up the larger share. "As per the WHO," an average hospital bed generates approximately 0.5 kg of waste per day, with around 20–25% falling into the hazardous or BMW category. Within this hazardous waste, about 10–15% is infectious waste, and the remaining 85–90% is non-infectious waste. However, the percentage of infectious waste can vary based on the healthcare facility type, such as hospitals with infectious disease units or those providing specialized patient services (Floyd et al., 2006). "The Biomedical Waste (Management and Handling) Rules" were enacted in India in "1998 under the Environment (Protection) Act (1986)." These rules are designed to govern the proper generation, collection, handling, storage, transportation, treatment, and environmentally responsible "disposal of BMW."

2.4 CLASSIFICATION OF BIOMEDICAL WASTE (BMW)

The regulations categorize BMW into four distinct types: "human anatomical waste, animal waste," microbiological waste, and sharp waste. Responsibilities of healthcare facilities include ensuring safe "collection, segregation, and disposal" of BMW. It's mandatory for these facilities to designate a responsible individual for managing BMW and maintaining comprehensive records of waste generation and disposal (*refer to* Figure 2.4).

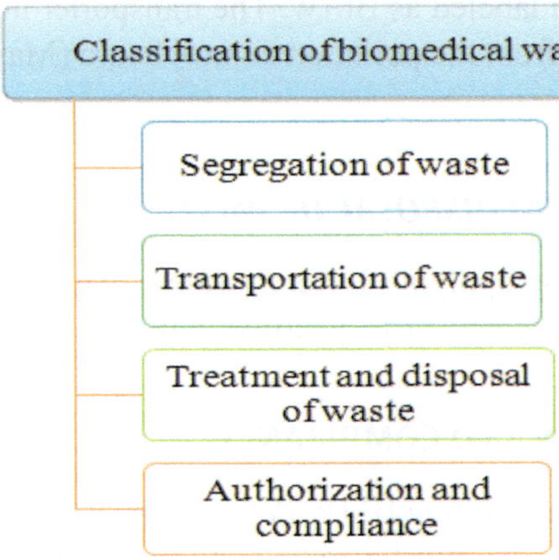

FIGURE 2.4 Classification of biomedical waste.

2.4.1 SEGREGATION OF WASTE

"Biomedical waste (BMW) must be segregated at the point of generation into different categories" as per the rules. The waste must be collected in color-coded bags or containers, as specified (Mandal and Dutta, 2009; Figure 2.5).

FIGURE 2.5 The segregation of different types of waste in separate containers.

2.4.2 TRANSPORTATION OF WASTE

Biomedical waste (BMW) must be transported in closed vehicles that are leak-proof, covered, and labeled as BMW. The transporter must possess the necessary authorization from the pollution control board (Mandal and Dutta, 2009).

2.4.3 TREATMENT AND DISPOSAL OF WASTE

BW must be managed by applying different rules. "The methods of treatment and disposal of waste are" also specified in the rules.

2.5 AUTHORIZATION AND COMPLIANCE

All healthcare facilities generating BMW must obtain authorization from the pollution control board. They must also "comply with the provisions of the rules," failing which they may be penalized. Overall, "the Biomedical Waste

(Management and Handling) Rules (1998/2000)" aim to ensure the safe and environmentally sound management of BMW generated in healthcare facilities. The rules are periodically updated to keep pace with emerging technologies and best practices in waste management (Sharma, 1998).

2.6 ELEVEN R CONCEPTS FOR THE MANAGEMENT OF BIOMEDICAL WASTE (BMW)

The 11R concept in BMW management outlines a comprehensive set of guidelines aimed at minimizing "the generation of BMW and ensuring its safe handling and disposal" (Upadhyaya et al., 2018). These guidelines emphasize the principles of "reduce, reuse and recycle" to mitigate the environmental impact associated with BMW.

1. **Refuse:** Healthcare facilities are encouraged to reject the utilization of unnecessary materials that contribute to BMW.
2. **Rethink:** They should critically evaluate the use of disposable items and actively seek out alternatives.
3. **Reduce:** Facilities should work towards reducing waste by minimizing the reliance on disposable items and implementing source reduction practices.
4. **Reuse:** Items that can be safely disinfected and reused, such as glassware and metal instruments, should be reused to the extent possible.
5. **Repair:** Healthcare facilities should prioritize the repair and maintenance of equipment to extend its lifespan and minimize the need for replacements.
6. **Refurbish:** Upgrading and refurbishing equipment to enhance its performance and longevity should be considered to reduce waste.
7. **Return:** Unused or expired pharmaceuticals should be returned to suppliers or manufacturers for proper disposal.
8. **Recycle:** Materials such as paper, cardboard, and plastic that can be safely recycled should undergo appropriate recycling processes.
9. **Recover:** Efforts should be made to recover materials like metals and chemicals that can be reused or repurposed.
10. **Reclaim:** Energy reclamation from waste through processes like incineration or anaerobic digestion should be explored.

11. **Dispose:** Lastly, any remaining waste should be disposed of responsibly, adhering to the stipulated guidelines and regulations for "biomedical waste management."

The 11R concept "of biomedical waste management" advocates for a sustainable and responsible approach to healthcare waste management, promoting environmental preservation and safeguarding public health.

2.7 CLASSIFICATION OF BIOMEDICAL WASTE (BMW) ACCORDING TO GUIDELINES OF WORLD HEALTH ORGANIZATION (WHO)

"The World Health Organization (WHO)" categorizes BMW into eight distinct categories, considering the potential risks it poses to both "human health and the environment" (Lee and Lee, 2022; Figure 2.6). These categories provide a structured classification system for BMW, aiding in its proper management and handling.

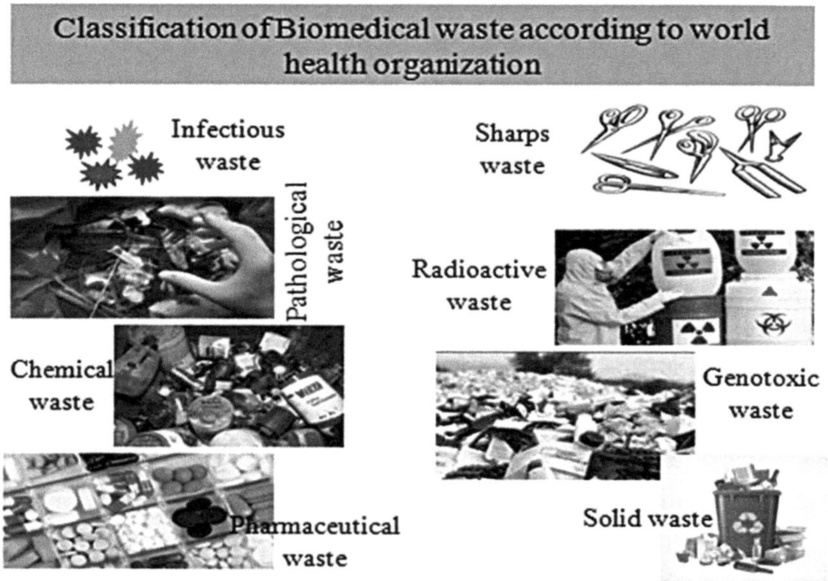

FIGURE 2.6 The classification is according to the guidelines of WHO.

1. **Infectious Waste:** This category encompasses waste containing pathogens like bacteria, viruses, and parasites. Examples include discarded culture dishes, instruments "contaminated with blood or other body fluids, and" animal carcasses.

2. **Pathological Waste:** This category includes waste containing human or animal tissues, organs, or body parts. Examples comprise amputated limbs, biopsy specimens, and tissues removed during surgery.
3. **Sharps Waste:** This category includes objects that can penetrate the skin, such as needles, syringes, scalpels, and broken glass.
4. **Chemical Waste:** "Waste containing chemical substances, like laboratory reagents, disinfectants," and heavy metals.
5. **Radioactive Waste:** "Waste containing radioactive substances," such as unused radioactive materials and contaminated items.
6. **Pharmaceutical Waste:** Waste encompassing expired, unused, or contaminated drugs, vaccines, and other pharmaceutical products.
7. **Genotoxic Waste:** This category involves waste with cytotoxic drugs capable of causing DNA damage.
8. **Solid Waste:** Waste that does not fall into the other categories, such as discarded medical equipment and other non-infectious waste.

Properly managing and disposing of BMW based on its classification is paramount to prevent infections from spreading, safeguard healthcare workers, and mitigate environmental damage. The WHO guidelines furnish a framework for the safe and responsible management of BMW.

2.8 MAJOR AND MINOR SOURCES OF BIOMEDICAL WASTE (BMW)

Healthcare Facilities, Primarily Hospitals: Hospitals are primary generators of BMW due to the nature of their operations. This includes surgical waste, sharps waste, pathological waste, and various other waste types stemming from patient care activities (Table 2.2).

1. **Clinics and Laboratories:** BMW is generated in clinics, diagnostic laboratories, and research facilities due to medical tests, research activities, and diagnostic procedures (Pasupathi et al., 2011).
2. **Blood Banks:** Waste is produced in blood banks through discarded blood bags, needles, and other blood collection and storage equipment (Abd El-Salam, 2010).

TABLE 2.2 The Major and Minor Sources of Biomedical Waste

Sl. No.	Sources	Examples	References
		Major Sources of Biomedical Waste	
1.	Clinics and laboratories	Medical tests, research activities, and diagnostic procedures.	Pasupathi et al. (2011)
2.	Blood banks	Discarded blood bags, needles, and other blood collection and storage equipment.	Abd El-Salam (2010)
3.	Veterinary clinics and animal research facilities.	Animal carcasses, surgical waste, and other waste generated during animal research activities.	McLean et al. (2007)
4.	Mortuaries and funeral homes	Human tissue, organs, and body fluids are generated during autopsy procedures and embalming.	Nolte et al. (2002)
5.	Home healthcare	Syringes, needles, and other medical supplies.	Abd El-Salam (2010)
		Minor Sources of Biomedical Waste	
6.	Outpatient clinics and health centers.	Syringes, bandages, gloves, and other medical supplies are used during patient care.	Abd El-Salam (2010)
7.	Laboratories	Microbiological cultures, discarded slides, and other materials that have been in contact with infectious agents.	Pasupathi et al. (2011)
8.	Blood banks	Blood bags, needles, and other materials are used during the blood donation and storage process.	Abd El-Salam (2010)
9.	Veterinary clinics	Animal tissue samples used needles, and other medical supplies used during animal care.	McLean et al. (2007)
10.	Mortuaries and funeral homes	Human tissue and fluids, needles, and other materials used during the embalming and burial process.	Nolte et al. (2002)

3. **Veterinary Clinics and Animal Research Facilities:** These facilities produce BMW through animal carcasses, surgical waste, and waste generated during animal research activities (McLean et al., 2007).
4. **Mortuaries and Funeral Homes:** BMW is generated in these facilities in the form of human tissue, organs, and body fluids resulting from autopsy procedures and embalming (Nolte et al., 2002).
5. **Home Healthcare:** Patients receiving home healthcare generate BMW in the form of used syringes, needles, and other medical supplies.

2.9 MINOR SOURCES OF BIOMEDICAL WASTE (BMW)

Minor sources of BW encompass various settings:

1. **Outpatient Clinics and Health Centers:** Waste generated includes used syringes, bandages, gloves, and other medical supplies used during patient care.
2. **Laboratories:** Waste generated comprises microbiological cultures, discarded slides, and other materials that have been in contact with infectious agents (Pasupathi et al., 2011).
3. **Blood Banks:** Waste includes used blood bags, needles, and other materials used during the blood donation and storage process.
4. **Veterinary Clinics:** Waste comprises animal tissue samples, used needles, and other medical supplies used during animal care (McLean et al., 2007).
5. **Mortuaries and Funeral Homes:** Waste involves human tissue and fluids, needles, and other materials used during the embalming and burial process (Nolte et al., 2002).

2.10 BIOMEDICAL WASTE (BMW) MANAGEMENT IS CRUCIAL IN HOSPITALS FOR SEVERAL REASONS

1. **Prevention of Infections:** Biomedical waste (BMW), especially infectious waste, can spread diseases and infections to hospital staff, "patients, and the general public. Proper management of BMW can prevent the spread of" these infections (Selvan et al., 2021).
2. **Compliance with Regulations:** Hospitals are required to comply with local and national regulations related to BMW "management. Failure to comply can result in" penalties and fines (Reinhardt, 2018).
3. **Protection of the Environment:** Biomedical waste (BMW) can have harmful effects on the environment if not properly managed. For example, if BMW is not disposed of properly, it can contaminate soil and water sources (Chakraborty et al., 2013).
4. **Reputation Management:** Hospitals that do not manage BMW properly can damage their reputation among patients and the public.

5. **Occupational Health and Safety:** Proper BMW management practices can protect hospital staff from exposure to hazardous waste and reduce the risk of workplace injuries.

Overall, proper BMW management is critical to ensuring a safe and healthy environment for hospital staff, patients, and the public (Walkinshaw, 2011).

2.11 BENEFITS OF BIOMEDICAL WASTE (BMW) MANAGEMENT

Proper "biomedical waste management has" several benefits, including:

1. **Protection of Human Health:** Proper "disposal of biomedical waste (BMW)" minimizes the risk of exposure to hazardous and infectious materials, "protecting the health of healthcare workers, patients, and the general public."
2. **Environmental Protection:** BMW contains hazardous substances that can harm the surroundings if not managed. Proper management of BW reduces the risk of environmental contamination and pollution.
3. **Compliance with Regulations:** BMW management is regulated by local and national regulations. Proper management of BMW ensures compliance with these regulations, avoiding fines and penalties.
4. **Reduced Risk of Disease Transmission:** Improper dumping of biomedical "waste can lead to the spread of infectious diseases." Proper management reduces the risk of disease transmission and helps to control outbreaks.
5. **Cost Savings:** Proper BMW management can reduce costs connected with the management, transport, and dumping of waste.
6. **Reputation Management:** Proper BMW management helps hospitals and healthcare facilities maintain a good reputation among patients, staff, and the general public.

2.12 SIX STEPS OF BIOMEDICAL WASTE (BMW) MANAGEMENT

The six steps of BW management are as follows (Mandal and Dutta, 2009; Figure 2.7):

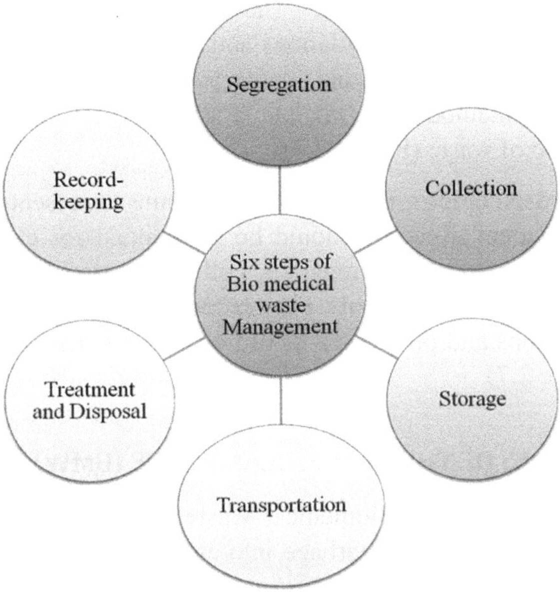

FIGURE 2.7 The management procedure of biomedical waste.

1. **Segregation:** The first step is to segregate BMW "at the point of generation into different categories such as" sharps, non-infectious waste, and pharmaceutical waste. This ensures that each "type of waste is disposed of in the" appropriate manner (Figure 2.7).
2. **Collection:** The segregated waste is then collected in specially marked and labeled containers that are leak-proof and puncture-resistant. These containers "are color-coded to indicate the type of waste" they contain (Figure 2.7).
3. **Storage:** The collected waste is then stored in a "designated area within the hospital" or healthcare facility until it is ready for disposal. The storage area should be secure, well-ventilated, and inaccessible to unauthorized personnel (Figure 2.7).
4. **Transportation:** BMW is transported from the healthcare facility to the disposal site using vehicles that are specially designed and labeled for this purpose. The vehicles should be equipped with the necessary safety features and should comply with local regulations (Figure 2.7).

5. **Treatment and Disposal:** "The BMW is treated and disposed of according to" local regulations and guidelines. Treatment methods may include autoclaving, incineration, or chemical treatment. Disposal methods may include landfill or recycling, depending on the type of waste (Figure 2.7).
6. **Record-Keeping:** Proper record-keeping is essential for BMW management. Records should be maintained for each step of the process, including segregation, collection, storage, transportation, treatment, and disposal. This helps to ensure compliance with regulations and provides a record of waste generation and disposal (Figure 2.7).

2.13 FEATURES OF THE BIOMEDICAL WASTE (BMW) RULES (2016)

The salient features of the Biomedical Waste Rules (2016) are as follows (Lee and Lee, 2022): sorting garbage into categories like infectious waste, sharps waste, and non-infectious waste at the place of generation; identification and labeling of BMW vehicles, bags, and containers to identify the waste kind and level of hazard; a provision for color-coding waste bags, containers, and vehicles based on the type of garbage; and establishing BMW treatment facilities inside hospitals or through shared facilities that are used by several hospitals. The licensing and registration of medical establishments that produce BMW are provided. Specified procedures for the processing, moving, and disposing of biological waste. There are specific guidelines for disposing of various kinds of biological waste, such as cytotoxic waste, pathological waste, and waste from sharps disposal. Healthcare facilities are required to keep records about the generation, collection, transportation, treatment, and disposal of biological waste. Penalties and fines for breaking the laws and guidelines governing the management of biological waste. Overall, with a focus on safeguarding public health and the environment, the Biological Waste Rules (2016) offer a thorough framework for the appropriate treatment of biological waste in India.

2.14 SPECIAL CATEGORY OF WASTE DISPOSAL

Sharps and plastic garbage must often be destroyed by a unique sort of waste disposal known as mutilation. Needles, syringes, and scalpels are only a few examples of objects that fall and are grouped as "sharps waste."

Plastic products such as IV bags, tubing, and connections are included in the category of plastic waste (Patil and Shekdar, 2001). These items can be rendered useless by being mutilated, which also lowers the possibility of harm or infection. The procedure usually starts with shredding or cutting the garbage into little pieces, then cleaning and sterilizing it. The resultant material can then be safely disposed of, usually through landfill or incineration. Sharps and plastics must be destroyed since doing so helps prevent their reuse. Reducing the amount of trash also makes transporting and getting rid of it easier and less expensive. Healthcare institutions should have sufficient processing and disposal procedures in place and should abide by all rules governing the management of biological waste if they generate a lot of sharps and plastic waste (Rao, 2015).

2.14.1 DISCARDED MEDICINES (CYTOTOXIC DRUGS AND ANTIBIOTICS)

If not properly disposed of, discarded medications, notably cytotoxic medications, and antibiotics, can have detrimental effects on the environment and well-being of mankind (Bashaar et al., 2017). By halting or reducing the proliferation of cancer cells, cytotoxic medications, commonly referred to as antineoplastic agents, are used to treat cancer. If not disposed of appropriately, these medications can be dangerous to both people and the environment. They can contaminate the environment if they are dumped down the drain or flushed down the toilet. This might get into the food chain and harm aquatic life. Drugs called antibiotics are used to treat bacterial infections. Antibiotic-resistant bacteria can grow as a result of improper antibiotic disposal, making it challenging to treat diseases in the future. Furthermore, if antibiotics are not properly disposed of, they may penetrate the environment and endanger plants and wildlife.

Follow the directions on the medication label or seek advice from your neighborhood waste management organization on how to properly dispose of used medications, especially antibiotics and cytotoxic medications. For the proper disposal of hazardous trash, including pharmaceuticals, many cities have detailed rules. In general, it is better to dispose of pharmaceuticals at a site approved for the disposal of hazardous waste or through a drug take-back program. Unless specifically told to do so by the drug label or your healthcare provider, never flush or pour medications down the drain.

2.14.2 MERCURY AND LEAD WASTE

Mercury and lead waste from biomedical "waste can be harmful to the environment and" human health if not disposed of properly (Henry, 2000). Here are some guidelines for their safe disposal:

1. **Segregate the Waste:** Mercury and lead waste should be segregated from other BMW and stored separately in a marked container.
2. **Use Appropriate Packaging:** Place the mercury or lead waste in a leak-proof container with a secure lid. If the waste is in liquid form, use an unbreakable container such as a plastic bottle with a screw-top lid.
3. **Label the Container:** Clearly label the container with the words "Mercury Waste" or "Lead Waste" and the date of accumulation.
4. **Contact Local Hazardous Waste Disposal Facilities:** Contact your local hazardous waste disposal facilities to find out if they accept mercury or lead waste. They may also have specific guidelines for packaging and transporting the waste.
5. **Do Not Dispose of in the Regular Trash:** Do not dispose of mercury or lead waste in the regular trash as it can contaminate the environment and harm wildlife and humans.
6. **Follow the Regulations:** Follow all regulations and guidelines set by your local, state, and federal authorities regarding the "disposal of hazardous waste."

Proper disposal of mercury and lead waste is important for protecting "the environment and human health." By following these guidelines, you can help ensure that these hazardous materials are disposed of safely and responsibly.

2.14.3 LIQUID CHEMICAL WASTE

"The management and disposal of" liquid chemical waste from BMW requires careful consideration to ensure the safety of individuals and the environment (Chitnis et al., 2005). Here are some guidelines to follow for proper management and disposal:

1. **Segregate the Waste:** Segregate liquid chemical waste from other BMW and store it separately in a clearly labeled, leak-proof container.

2. **Characterize the Waste:** Determine the chemical composition and properties of the liquid waste. This information is important for selecting the appropriate disposal method.
3. **Follow Regulatory Requirements:** Check with your local and state regulations for specific requirements related to the disposal of liquid chemical waste. Some chemicals may require special handling and disposal methods.
4. **Contact a Licensed Hazardous Waste Disposal Facility:** Contact a licensed hazardous waste disposal facility to determine if they accept liquid chemical waste. They may require specific documentation and labeling of the waste.
5. **Do not Dispose of in the Regular Trash:** Do not dispose of liquid chemical waste in the regular trash as it can contaminate the environment and harm wildlife and humans.
6. **Neutralize the Waste if Possible:** If the liquid chemical waste is not hazardous, it may be possible to neutralize it using a chemical treatment process. Consult with a licensed hazardous waste disposal facility for guidance on this process.
7. **Transport the Waste Safely:** Transport the liquid chemical waste in a secure container with appropriate labeling and documentation. Follow all regulatory requirements related to transport.

"Proper management and disposal of" liquid chemical waste from BMW is critical to protecting the environment and human health. By following these guidelines and consulting with licensed professionals, you can "ensure that the waste is handled and disposed" of safely and responsibly.

2.15 LABORATORY BIOMEDICAL WASTE (BMW)

Laboratory BMW is generated from various laboratory procedures such as microbiological cultures, chemical analyses, and other diagnostic procedures. This waste must be treated properly to stop the communication of infections and safeguard the environment. Here are some common types of laboratory BMW and their treatment methods (Chitnis et al., 2005).

1. **Microbiological Waste:** This type of waste includes cultures of microorganisms, used petri dishes, and contaminated laboratory

equipment. The waste should be autoclaved at a temperature of 121°C for 20–30 minutes or treated with a chemical disinfectant to kill the microorganisms.

2. **Chemical Waste:** This type of waste includes chemicals used in laboratory analyses such as acids, bases, and organic solvents. Chemical waste should be collected in separate, clearly labeled containers and treated using chemical neutralization or adsorption methods before disposal.
3. **Sharps Waste:** This type of waste includes needles, blades, "and other sharp instruments. Sharps waste should be collected in puncture-resistant containers and" sterilized using an autoclave or other suitable method (Lee and Lee, 2022).
4. **Radioactive Waste:** This type of waste includes materials contaminated with radioactive isotopes. Radioactive waste should be stored separately and disposed of according to specific regulations for radioactive waste.
5. **Pathological Waste:** This type of waste includes "human or animal tissues, organs," and body fluids. Pathological waste should be collected in biohazard bags and treated using incineration or other suitable methods.
6. **Liquid Waste:** This type of waste includes liquid materials such as blood, urine, and other bodily fluids. Liquid waste should be treated using chemical disinfectants or other appropriate methods before being disposed of according to regulations.

Proper treatment of laboratory BMW "is essential to prevent the spread of infections and protect the environment." By following specific guidelines for each type of waste and consulting with local regulations and guidelines, laboratory personnel can "ensure that waste is treated and disposed of" safely and responsibly.

2.16 MICROBIOLOGY, BIOTECHNOLOGY WASTE, AND INFECTIOUS WASTE

"Microbiology and biotechnology waste," as well as infectious waste, are generated in laboratories and healthcare facilities and require careful management and disposal to prevent the spread of infections and safeguard the environment. Here are some common types of microbiology, biotechnology, and infectious waste and their treatment methods:

1. **Microbiology and Biotechnology Waste:** This type of waste includes cultures of microorganisms, recombinant DNA, and genetically modified organisms (GMO). The waste should be sterilized using an autoclave or other suitable method before being disposed of in a biohazard waste container.
2. **Sharps Waste:** Waste of this nature encompasses needles, blades, and other sharp instruments. It is essential to collect sharps waste in puncture-resistant containers and sterilize it using an autoclave or another appropriate method.
3. **Infectious Waste:** This type of waste includes waste that is contaminated with potentially infectious materials "such as blood, body fluids," or tissues. Infectious waste should be collected in biohazard bags and sterilized using an autoclave or other suitable method.
4. **Chemical Waste:** Laboratory analyses generate various types of chemical waste, including acids, bases, and organic solvents. It is essential to collect this waste in separate, clearly labeled containers and treat it through chemical neutralization or adsorption methods prior to disposal (Figure 2.8).

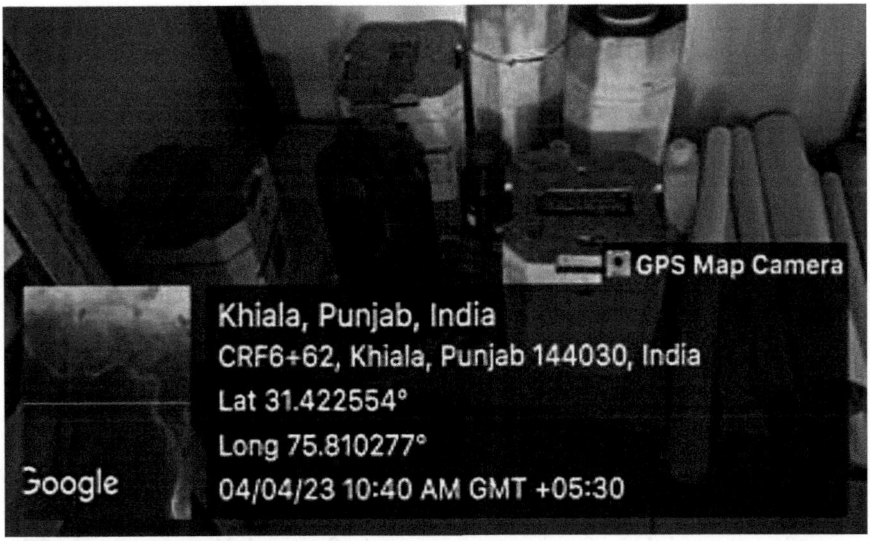

FIGURE 2.8 The disposal of liquid waste.

2.17 RADIOACTIVE WASTE

This type of waste includes materials contaminated with radioactive isotopes. Radioactive waste should be stored separately and disposed of according to specific regulations for radioactive waste (Khunprasert et al., 2008).

Proper management and disposal of microbiology and biotechnology waste, as well as infectious waste, are "essential to prevent the spread of infections and protect the environment." By following specific guidelines for each type of waste and consulting with local regulations and guidelines, laboratory personnel and healthcare professionals can "ensure that waste is treated and disposed of" safely and responsibly.

2.17.1 PACKAGING, LABELING, AND TRANSPORT TO COMMON WASTE SITE

Packaging, labeling, and transport to a common waste site are important steps in properly disposing of waste materials (Desole et al., 2022). Some key points to consider are discussed in subsections.

2.17.1.1 PACKAGING

Use appropriate containers for the type of waste being disposed of (e.g., hazardous, non-hazardous, liquid, solid, and so forth). Make sure containers are leak-proof and sealed tightly to prevent spills and contamination. Label containers with the type of waste, date, and any other relevant information. Store containers in a secure location until ready for transport.

2.17.1.2 LABELING

Label waste containers clearly with the type of waste, date, and any other relevant information. Follow any regulations or guidelines for labeling specific types of waste (e.g., hazardous waste). Make sure labels are visible and not easily removed or damaged.

2.17.1.3 TRANSPORT

Use a licensed waste transporter who is authorized to transport the type of waste being disposed of Ensure the transport vehicle is appropriate for the type of waste and that containers are securely fastened. Keep a record of the transport, including the name of the transporter, date and time of transport, and any other relevant information. Follow any regulations or guidelines for transporting specific types of waste (e.g., hazardous waste). By following these steps, waste materials can be disposed of safely and in compliance with regulations and guidelines (Mandal and Dutta, 2009).

2.18 TRANSPORTATION TO COMMON BIOMEDICAL WASTE (BMW) TREATMENT FACILITY

The transport of BMW and its management requires special precautions. Here are some key points to consider (Kumar et al., 2015).

Select a licensed waste transporter who is authorized to transport BMW. Use appropriate containers for the type of BMW being transported. These may include leak-proof bags or containers made of puncture-resistant materials. Label containers clearly with the biohazard symbol and any other required information such as the type of waste, date, and generator's name. Segregate and package different types of BMW separately to prevent cross-contamination. Make sure the transport vehicle is appropriate for the type of waste being transported and that containers are securely fastened and stored. Ensure that the driver of the transport vehicle is trained and equipped to handle BMW. Keep a record of the transport, including the name of the transporter, date and time of transport, and any other relevant information. Follow any regulations or guidelines for transporting BMW, including those related to storage, labeling, and handling. By following these guidelines, BMW can be transported safely and efficiently to a treatment facility for proper disposal. It's important to take all essential precautions to shield the environment, public health, and the safety of those involved in the transport and disposal process (Figure 2.9).

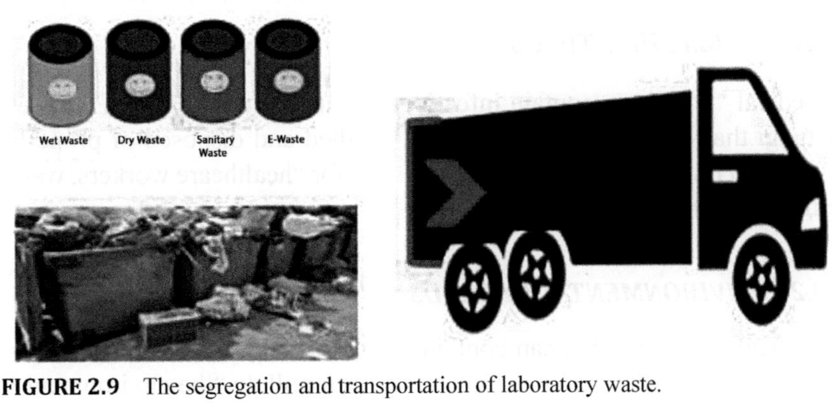

FIGURE 2.9 The segregation and transportation of laboratory waste.

2.19 PROBLEMS ASSOCIATED WITH BIOMEDICAL WASTE (BMW)

Biomedical waste (BMW), also known as medical waste or healthcare waste, can pose a range of problems (Figure 2.10), if not managed properly (Kalpana et al., 2016). Some of the key problems associated with BMW are explained in subsections.

FIGURE 2.10 Problems associated with improper management of biomedical waste.

2.19.1 *PUBLIC HEALTH RISKS*

Biomedical "waste can contain infectious agents such as" bacteria, viruses, and fungi that can spread disease if not handled and disposed of properly. This poses a risk to public health, particularly for "healthcare workers, waste handlers, and the general public."

2.19.2 *ENVIRONMENTAL HAZARDS*

Biomedical waste (BMW) can contain hazardous chemicals, such as disinfectants, solvents, and heavy metals, which can pollute "the environment if not disposed of properly." This "can contaminate soil, water, and air, leading to long-term environmental damage."

2.19.3 *LEGAL AND REGULATORY COMPLIANCE*

"Biomedical waste (BMW) is subject to strict regulations and guidelines" at the local, national, and international levels. "Failure to comply with these regulations can result in fines, legal action, and damage to an organization's reputation."

2.19.4 *OCCUPATIONAL HAZARDS*

Healthcare workers and waste handlers who handle biomedical "waste are at risk of exposure" to infectious agents, hazardous chemicals, and sharps injuries, which can cause serious health problems such as infections, injuries, and long-term illnesses.

2.19.5 *IMPROPER DISPOSAL*

Improper disposal of BMW can lead to it being scavenged by animals or informal waste pickers, leading to further health and environmental risks.

To mitigate these problems, it's important "to ensure that BMW is handled," transported, "and disposed of safely and" in compliance with regulations and guidelines. This includes using appropriate containers, labeling, and transport methods, and following proper procedures for treatment and disposal.

2.20 TREATMENT AND DISPOSAL TECHNIQUES FOR BIOMEDICAL WASTE (BMW)

Biomedical waste (BMW) requires specialized treatment and disposal techniques (Lee et al., 2004; Figure 2.11). Some of the common treatment and disposal techniques used for BMW are discussed in subsections.

2.20.1 *INCINERATION*

This is a common technique for treating BMW, particularly hazardous and infectious waste. "It involves burning the waste at high temperatures to destroy pathogens and reduce the volume of the waste." Incineration produces ash and emissions, which need to be treated before disposal.

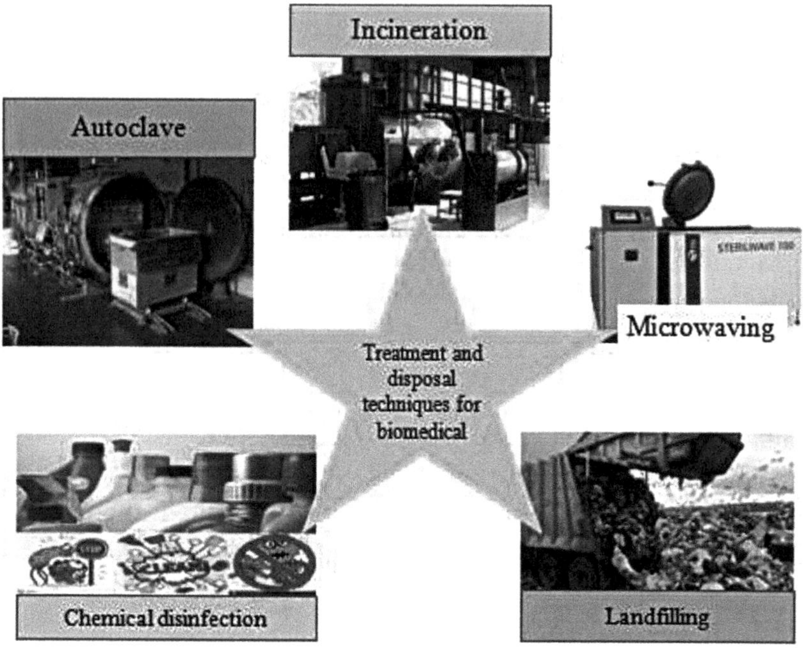

FIGURE 2.11 Treatment and disposal techniques for biomedical waste.

2.20.2 AUTOCLAVING

This is a process that uses steam and high pressure to sterilize BMW. The waste is placed in a sealed chamber and subjected to high temperatures and pressure, which destroys pathogens. The waste can then be disposed of in a landfill (Figure 2.12).

2.20.3 CHEMICAL DISINFECTION (CD)

This is a process that uses chemicals such as chlorine, hydrogen peroxide, or ozone to disinfect BMW. The waste is soaked in a disinfectant solution, which destroys pathogens.

2.20.4 MICROWAVING

This is a process that uses microwaves to heat the waste and destroy pathogens. The waste is placed in a microwave chamber and subjected to high temperatures, which kill the pathogens.

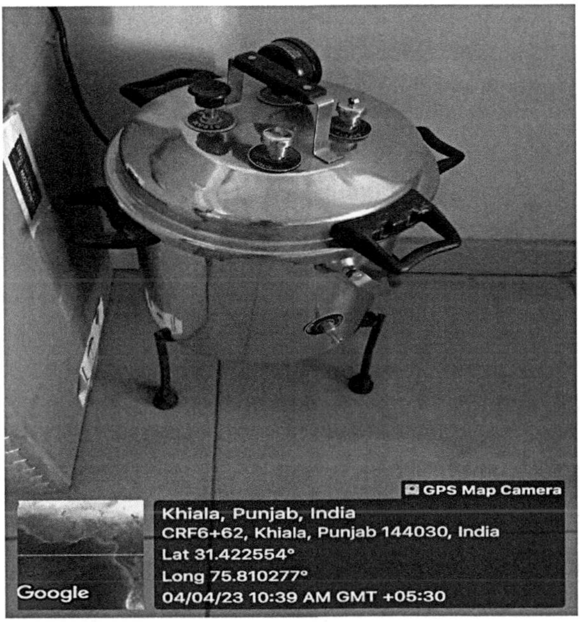

FIGURE 2.12 The autoclaving of biomedical waste.

2.20.5 *LANDFILLING*

This is a common disposal technique "for non-hazardous" BMW. The waste is placed in a designated landfill, which is lined to stop pollution of the soil and groundwater. The waste is then covered with soil and compacted to reduce its volume.

"It is important to note that the" appropriate treatment and disposal technique for BMW depends on the type of waste being generated, the local regulations and guidelines, and the available resources. It's important to follow proper procedures "to ensure that BMW is handled and" disposed of safely and in compliance with regulations and guidelines.

2.20.6 *THERMAL INACTIVATION*

Thermal inactivation is a general procedure for the treatment of BMW. This process involves subjecting the waste to high temperatures to kill any harmful microorganisms and pathogens present in the waste. The thermal inactivation process typically involves the use of specialized equipment such as autoclaves, incinerators, and microwaves. Autoclaves are the most commonly used equipment for thermal inactivation. They work by subjecting the waste to high temperatures and pressure in a closed container for a specific period of time. During the thermal inactivation process, the waste is heated to a temperature that is high enough to kill any pathogens present in the waste. The temperature and duration of the treatment depend on the type and quantity of waste being treated. For example, the temperature may range from 121 to 134°C, and the duration of the treatment may be between 20 and 90 minutes (Smelt and Brul, 2014). Once the thermal inactivation process is complete, the waste is considered to be safe for disposal. The treated waste can be disposed of in a landfill or incinerated in a specialized facility. Thermal inactivation is an effective method for the treatment of BMW. It ensures that harmful pathogens are killed, reducing the risk of spreading disease and protecting the environment from contamination.

2.20.7 GAS/VAPOR STERILIZATION

Gas/vapor sterilization is another method for the treatment of BMW. This process involves using specialized gases or vapors to kill any microorganisms present in the waste. The most commonly used gas for sterilization is ethylene oxide (ETO). ETO is a highly effective sterilant that can "kill all forms of microorganisms, including viruses, bacteria, fungi, and spores." Other gases that can be used for sterilization include hydrogen peroxide, ozone, and chlorine dioxide. Gas/vapor sterilization typically involves placing the waste in a sealed chamber and exposing it to the sterilizing agent for a specific period of time. The temperature, humidity, and concentration of the sterilizing agent are carefully controlled to ensure effective sterilization. One advantage of gas/vapor sterilization is that it can penetrate even small crevices and hard-to-reach areas, ensuring that all surfaces of the waste are sterilized. However, gas/vapor sterilization requires specialized equipment and is more expensive than other methods of waste treatment. Once the gas/vapor sterilization process is complete, the waste is considered to be safe for disposal. The sterilized "waste can be disposed of in a landfill" or incinerated in a specialized facility. Overall, gas/vapor sterilization is an effective method for the treatment of BMW. It ensures that harmful microorganisms are killed, reducing the risk of spreading disease and protecting the environment from contamination (Linton et al., 2005).

2.20.8 CHEMICAL DISINFECTION (CD)

"Chemical disinfection (CD) is another method for the treatment of" BMW (Chitnis et al., 2005). This process involves using chemical agents to kill or inactivate any microorganisms present in the waste. Chemical disinfection agents include chlorine, bleach, and other oxidizing agents. The concentration and duration of exposure to the disinfectant are carefully controlled to ensure effective disinfection. Chemical disinfection is commonly used for the treatment of surfaces and equipment in healthcare facilities. However, it can also be used for the treatment of BMW. One advantage of chemical disinfection is that it is a relatively simple and inexpensive method for waste treatment. However, it may not be as effective as other methods, such as thermal inactivation or gas/vapor sterilization, in killing all types of microorganisms, especially spores. Once the chemical disinfection process is complete, the waste is considered to be safe for disposal. The disinfected waste can be disposed of in a landfill or incinerated in a specialized facility. Overall, chemical disinfection is an effective method for the treatment of BMW, especially for waste that does not contain highly resistant microorganisms. It is a simple and relatively inexpensive method that can be easily implemented in healthcare facilities. However, it may not be as effective as other methods in killing all types of microorganisms.

2.20.9 STEAM TREATMENT TECHNOLOGIES

Steam treatment technologies can also be used for the treatment of BMW, which is generated by hospitals, clinics, and other healthcare facilities (Maamari et al., 2016; Figure 2.13). Some common steam treatment technologies used for BMW are explained in subsections.

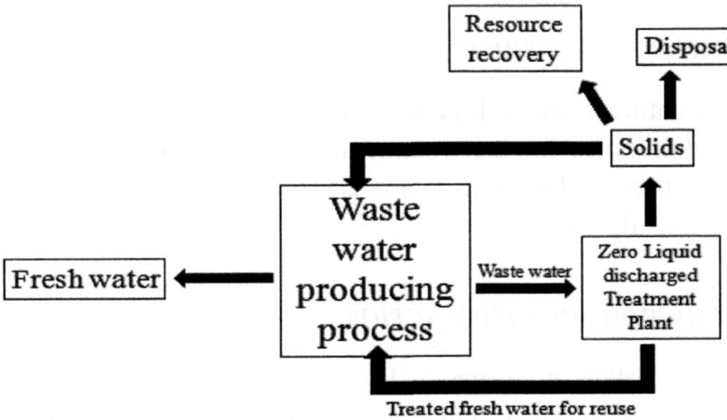

FIGURE 2.13 Steam treatment technologies.

2.20.9.1 AUTOCLAVES

As mentioned earlier, autoclaves are widely used for sterilization of medical equipment. Autoclaves can also be used for the treatment of BMW by exposing it to high-temperature steam for a specific period. The steam kills the pathogens present in the waste and makes it safe for disposal.

2.20.9.2 MICROWAVE DISINFECTION

This technology uses microwaves to generate heat that can effectively kill the pathogens present in BMW. The waste is placed in a microwave chamber, and the microwaves are directed toward it, heating it to a high temperature, and killing the pathogens.

2.20.9.3 HYDROCLAVE

A hydroclave is a type of autoclave that uses steam and pressure to sterilize and treat BMW. The waste is first shredded into smaller pieces and then exposed to high-temperature steam and pressure, which kills the pathogens present in the waste.

Steam disinfection systems: Steam disinfection systems are designed specifically for the treatment of BMW. These systems use high-temperature steam to disinfect and sterilize the waste, making it safe for disposal.

Overall, steam treatment technologies are an effective and environmentally friendly way of treating BMW. They are commonly used by hospitals, clinics, and other healthcare facilities to ensure the safe and proper disposal of BMW.

2.20.10 MICROWAVE TREATMENT TECHNOLOGIES

Microwave treatment technologies are an alternative to traditional steam treatment technologies for the treatment of BMW. Some common microwave treatment technologies used for BMW are explained in subsections (Giakoumakis et al., 2021).

2.20.10.1 MICROWAVE DISINFECTION

As mentioned earlier, microwave disinfection is a technology that uses microwaves to generate heat that can effectively kill the pathogens present in BMW. The waste is placed in a microwave chamber, and the microwaves are directed towards it, heating it to a high temperature, which kills the pathogens.

2.20.10.2 MICROWAVE-ASSISTED PYROLYSIS

Microwave-assisted pyrolysis is a technology that combines the use of microwaves with pyrolysis to treat BMW. Pyrolysis is a process that uses heat to decompose organic material. In this process, the waste is first shredded into smaller pieces, and then microwaves are used to heat the waste, which generates heat and breaks down the organic material into smaller molecules.

2.20.10.3 MICROWAVE HYDROLYSIS

Microwave hydrolysis is a process that uses microwaves to hydrolyze or break down the organic material present in BMW. The waste is placed in a microwave chamber with water, and the microwaves are directed toward it, heating it to a high temperature. The heat and pressure generated by the microwaves cause the organic material to break down into smaller molecules.

2.20.10.4 MICROWAVE PLASMA TREATMENT

Microwave plasma treatment is a technology that uses microwaves to generate plasma, which is a high-energy gas. The plasma is used to treat BMW, which is placed in a chamber with the plasma. The high-energy gas breaks down the organic material in the waste, making it safe for disposal.

Overall, microwave treatment technologies are an effective and efficient way of treating BMW. They are an alternative to traditional steam treatment technologies and offer several advantages, including faster treatment times, reduced energy consumption, and reduced greenhouse gas (GHG) emissions. However, they require careful design and operation to ensure that they are safe and effective.

2.20.11 DRY HEAT TREATMENT TECHNOLOGIES

Dry heat treatment technologies are another method for the treatment of BMW. Some common dry heat treatment technologies used for BMW are explained in subsections (Veronesi et al., 2005).

2.20.11.1 INCINERATION

Incineration is a process that uses high temperatures to burn BMW, reducing it to ash. The waste is placed in a specially designed incinerator, and the temperature is raised to between 800 and 1000°C. The heat generated by the incineration process destroys the pathogens present in the waste, making it safe for disposal.

2.20.11.2 HOT AIR STERILIZATION

Hot air sterilization is a process that uses dry heat to sterilize BMW. The waste is placed in a specially designed chamber, and hot air is circulated through the chamber, raising the temperature to between 160 and 180°C. The heat generated by the hot air destroys the pathogens present in the waste, making it safe for disposal.

2.20.12 MICROWAVE-ASSISTED THERMAL STERILIZATION

Microwave-assisted thermal sterilization is a process that uses microwaves to heat BMW to a high temperature. The waste is placed in a specially designed chamber, and microwaves are directed towards it, heating it to a high temperature of between 110 and 130°C. The heat generated by the microwaves destroys the pathogens present in the waste, making it safe for disposal.

Overall, dry heat treatment technologies are effective and efficient ways of treating BMW. They are commonly used by hospitals, clinics, and other healthcare facilities to ensure the safe and proper disposal of BMW. However, they can be energy-intensive and require careful design and operation to ensure that they are safe and effective.

2.20.13 SODIUM HYPOCHLORITE (NAOCL, 1–12%)

Sodium hypochlorite (NaOCl) is a commonly used disinfectant for the treatment of BMW. It is a strong oxidizing agent and can effectively kill a wide range of microorganisms, including bacteria, viruses, and fungi. NaOCl is also relatively inexpensive and readily available, making it a popular choice for use in healthcare settings. The recommended concentration of NaOCl for the treatment of BMW varies depending on the type and quantity of waste being treated. For example, the WHO recommends a concentration of 1% NaOCl for the disinfection of general healthcare waste, while a concentration of 5% is recommended for the disinfection of highly infectious waste. It is important to note that the use of NaOCl for the treatment of BMW should be done in a well-ventilated area and with appropriate personal protective equipment (PPE), including gloves and eye protection. Additionally, proper disposal of NaOCl-treated waste is essential to prevent contamination and potential harm to the environment and public health (Tsai and Lin, 1999).

2.20.14 CHEMICAL DISINFECTANTS

Chemical disinfectants are commonly used for the management of BMW. Some of the commonly used disinfectants for BMW include (Chitnis et al., 2005).

2.20.14.1 SODIUM HYPOCHLORITE (NAOCL)

As discussed earlier, NaOCl is a strong oxidizing agent that can effectively kill a wide range of microorganisms, including bacteria, viruses, and fungi. It is commonly used for the disinfection of general healthcare waste.

2.20.14.2 HYDROGEN PEROXIDE (H_2O_2)

Hydrogen peroxide is another commonly used disinfectant for BMW. It is a strong oxidizing agent that can kill a wide range of microorganisms. It is often used as a fumigant for the disinfection of large areas or for the decontamination of equipment.

2.20.14.3 GLUTARALDEHYDE

Glutaraldehyde is a powerful disinfectant that is effective against a wide range of microorganisms, including spores. It is commonly used for the disinfection of equipment and surfaces.

2.20.14.4 QUATERNARY AMMONIUM COMPOUNDS (QACS)

QACs are a group of disinfectants that are effective against a wide range of microorganisms. They are commonly used for the disinfection of surfaces and equipment. Peracetic acid (PAA): PAA is a strong oxidizing agent that is effective against a wide range of microorganisms, including spores. It is commonly used for the disinfection of medical instruments and endoscopes.

It is important to choose the appropriate disinfectant for the type and quantity of BMW being treated. Additionally, it is important to follow the manufacturer's instructions for use and to ensure that the disinfectant is used in a well-ventilated area and with appropriate personal protective equipment (PPE). Proper disposal of disinfectant-treated waste is also essential to prevent contamination and potential harm to the environment and public health (Gerba, 2015).

2.20.15 ENCAPSULATION AND INTERSTATION

Encapsulation and inertization are two treatment methods that can be used for the disposal of BMW. Encapsulation involves the use of materials such as cement, plastic, or resin to encase the BMW. This method is commonly used for the disposal of sharps and other types of BMW that may pose a risk of injury or infection. Encapsulation effectively immobilizes the waste, preventing any potential release of contaminants. Inertization involves the treatment of BMW with chemicals such as lime, sodium hydroxide, or other alkaline agents to raise the pH level to a level that effectively sterilizes the waste. The treated waste is then buried in a landfill or other suitable location. Inertization is a safe and effective method for the disposal of BMW that is not suitable for other treatment methods. Both encapsulation and inertization can be effective methods for the disposal of BMW. However, it is important to ensure that the waste is properly sorted and segregated before treatment and that the treatment is performed by trained personnel using appropriate equipment and procedures. Proper disposal of treated waste is also essential to prevent contamination and potential harm to the environment and public health (Alcázar Román, 2016).

2.21 HEALTH HAZARDS FROM BIOMEDICAL WASTE (BMW)

Biomedical waste (BMW) can pose significant health hazards to healthcare workers, patients, and the general public if it is not handled and disposed of properly. The following are some of the potential health hazards from BMW (Chitnis et al., 2005).

2.21.1 *INFECTIOUS DISEASES*

Biomedical waste (BMW) may contain infectious agents such as bacteria, viruses, and parasites, which can cause a range of diseases including hepatitis B and C, HIV/AIDS, tuberculosis, and other respiratory infections.

2.21.2 *SHARPS INJURIES*

Healthcare workers are at risk of needle-stick injuries and other sharps injuries, which can transmit blood-borne infections such as hepatitis B, hepatitis C, and HIV (Lee and Lee, 2022).

2.21.3 *CHEMICAL EXPOSURE*

Biomedical waste (BMW) may contain chemicals such as formaldehyde, mercury, and solvents, which can cause a range of health problems including respiratory and skin irritation, cancer, and neurological disorders.

2.21.4 *RADIATION EXPOSURE*

Medical waste generated from radiology departments and nuclear medicine facilities can contain radioactive materials that can cause radiation exposure and related health problems.

2.21.5 *ENVIRONMENTAL CONTAMINATION*

Improper disposal of BMW can lead to contamination of the environment, which can result in health hazards for the general public.

To minimize the health hazards of BMW, it is essential to properly segregate, handle, and dispose of the waste using appropriate methods and procedures. Healthcare workers should receive adequate training on the safe handling of BMW and should use appropriate personal protective equipment (PPE) to minimize their exposure to potential hazards. Additionally, proper disposal of treated waste is essential to prevent contamination and potential harm to the environment and public health.

2.22 CHALLENGES OF BIOMEDICAL WASTE (BMW) IN INDIA

India faces numerous challenges in managing BMW, which is generated in large quantities from hospitals, clinics, and other healthcare facilities. Some of the major challenges are explained in subsections.

2.22.1 LACK OF INFRASTRUCTURE

The lack of proper infrastructure, including collection, segregation, storage, transportation, and disposal facilities, is one of the biggest challenges in managing BMW in India. Many healthcare facilities do not have the necessary resources to segregate waste at the source, leading to the mixing of hazardous waste with regular waste.

2.22.2 INADEQUATE REGULATIONS

The existing regulations for managing BMW are often not enforced effectively, and the penalties for non-compliance are not stringent enough to act as a deterrent. As a result, many healthcare facilities do not adhere to proper disposal practices.

2.22.3 LOW AWARENESS

The level of awareness among healthcare workers and the general public about the dangers of BMW is often low. This can lead to unsafe practices, such as reusing syringes and needles, which can spread infections and diseases.

2.22.4 LACK OF TRAINED PERSONNEL

The proper management of BMW requires trained personnel who can handle and dispose of it safely. However, there is a shortage of such personnel in India, leading to unsafe disposal practices.

2.22.5 HIGH COST

The cost of managing BMW is high, and many healthcare facilities may not have the financial resources to implement safe disposal practices.

2.22.6 LIMITED ACCESS TO TECHNOLOGY

Many healthcare facilities in India lack access to modern technology for BMW management, such as autoclaves, shredders, and incinerators. This can lead to unsafe disposal practices and increase the risk of infection.

Addressing these challenges will require a concerted effort from healthcare facilities, regulatory authorities, and the general public to create a safer and healthier environment for all.

2.23 CONCLUSION

In conclusion, the management of biomedical waste (BMW) is essential to prevent potential health hazards to healthcare workers, patients, and the general public. Effective management of BMW requires the proper segregation, handling, treatment, and disposal of waste using appropriate methods and procedures. Segregation of BMW at the point of generation is the first step in effective waste management. Healthcare facilities should establish clear guidelines for the segregation of waste into different categories and provide adequate training to staff to ensure compliance.

Treatment of BMW can be done using various methods such as chemical disinfection, encapsulation, and inertization. The appropriate treatment method should be chosen based on the type and quantity of waste being treated. Proper disposal of treated waste is essential to prevent contamination and potential harm to the environment and public health. Healthcare facilities should follow the guidelines and regulations set by their local authorities for the disposal of BMW. Regular monitoring and auditing of the BMW management system can help identify any areas of improvement and ensure compliance with regulations and best practices. Overall, the proper management of BMW is critical to protecting the health and safety of healthcare workers, patients, and the general public, as well as minimizing the impact on the environment.

2.24 FUTURE PROSPECTUS

Biomedical waste (BMW) must be managed in accordance with local, state, and federal regulations. Generally, BMW is segregated by type (e.g., infectious, hazardous, and non-hazardous) and placed into separate containers that are properly labeled. Proper labeling includes an explanation of the type of waste, the date it was generated, and the name of the person generating the waste. Containers must be securely sealed and stored in a safe, secure location. Infectious waste should be autoclaved, incinerated, or treated with a chemical disinfectant before it is disposed of. Hazardous waste should be labeled and disposed of according to local, state, and federal regulations. Non-hazardous waste should be recycled or composted, if possible. All BMW should be collected and transported by a licensed waste management service.

KEYWORDS

- **biomedical waste**
- **color-coding**
- **dry heat treatment technologies**
- **greenhouse gas emissions**
- **microwave plasma treatment**
- **microwaves**
- **radiation exposure**
- **thermal sterilization**

REFERENCES

Abd El-Salam, M. M. (2010). Hospital waste management in El-Beheira governorate, Egypt. *Journal of Environmental Management*, *91*(3), 618–629.

Alcázar Román, Á. (2016). Synthesis of microcapsules containing extractant agents for the selective removal of heavy metals from wastewaters. https://pubmed.ncbi.nlm.nih.gov/26299426/.

Bashaar, M., Thawani, V., Hassali, M. A., & Saleem, F. (2017). Disposal practices of unused and expired pharmaceuticals among the general public in Kabul. *BMC Public Health*, *17*, 1–8.

Brindha, G., & Kanchan, K. (2013). Biomedical Waste Management. *Int J Health Sci Res*, *3*(8), 69–76.

Chakraborty, S., Veeregowda, B., Gowda, L., Sannegowda, S. N., Tiwari, R., Dhama, K., & Singh, S. V. (2013). Biomedical waste management. *Interaction*, *2013*, 12–02.

Chitnis, V., Vaidya, K., & Chitnis, D. S. (2005). Biomedical waste in laboratory medicine: Audit and management. *Indian Journal of Medical Microbiology*, *23*(1), 6–13.

Datta, P., Mohi, G. and Chander, J. (2018). Biomedical waste management in India: Critical appraisal. *Journal of Laboratory Physicians*, *10*(01), 006–014.

Desole, M. P., Aversa, C., Barletta, M., Gisario, A., & Vosooghnia, A. (2022). Life cycle assessment (LCA) of PET and PLA bottles for the packaging of fresh pasteurized milk: The role of the manufacturing process and the disposal scenario. *Packaging Technology and Science*, *35*(2), 135–152.

Floyd, K., Hutubessy, R., Samyshkin, Y., Korobitsyn, A., Fedorin, I., Volchenkov, G., Kazeonny, B., Coker, R., Drobniewski, F., Jakubowiak, W., & Shilova, M. (2006). Health-systems efficiency in the Russian Federation: Tuberculosis control. *Bulletin of the World Health Organization*, *84*(1), 43–51.

Gerba, C. P. (2015). Quaternary ammonium biocides: Efficacy in application. *Applied and Environmental Microbiology*, *81*(2), 464–469.

Giakoumakis, G., Politi, D., & Sidiras, D. (2021). Medical waste treatment technologies for energy, fuels, and materials production: A review. *Energies*, *14*(23), 8065.

Henry, J. R. (2000). *An Overview of the Phytoremediation of Lead and Mercury* (pp. 1–31). Washington, DC: US Environmental Protection Agency, Office of Solid Waste and Emergency Response, Technology Innovation Office.

Kalpana, V. N., Prabhu, D. S., Vinodhini, S., & Devirajeswari, V. (2016). Biomedical waste and its management. *Journal of Chemical and Pharmaceutical Research*, *8*(4), 670–676.

Khunprasert, P., Grisdanurak, N., Thaveesri, J., Danutra, V., & Puttitavorn, W. (2008). Radiographic film waste management in Thailand and cleaner technology for silver leaching. *Journal of Cleaner Production*, *16*(1), 28–36.

Kumar, A., Duggal, S., Gur, R., Rongpharpi, S. R., Sagar, S., Rani, M., Dhayal, D., & Khanijo, C. M. (2015). Safe transportation of biomedical waste in a health care institution. *Indian Journal of Medical Microbiology*, *33*(3), 383–386.

Lee, B. K., Ellenbecker, M. J., & Moure-Ersaso, R. (2004). Alternatives for treatment and disposal cost reduction of regulated medical wastes. *Waste Management*, *24*(2), 143–151.

Lee, S. M., & Lee, D. (2022). Effective Medical Waste Management for Sustainable Green Healthcare. *International Journal of Environmental Research and Public Health*, *19*(22), 14820.

Linton, R. H., Han, Y., Selby, T. L., & Nelson, P. E. (2005). Gas-/vapor-phase sanitation (decontamination) treatments. *Microbiology of Fruits and Vegetables*, 417–452.

Maamari, O., Mouaffak, L., Kamel, R., Brandam, C., Lteif, R., & Salameh, D. (2016). Comparison of steam sterilization conditions efficiency in the treatment of Infectious Health Care Waste. *Waste Management*, *49*, 462–468.

Mandal, S. K., & Dutta, J. (2009). Integrated biomedical waste management plan for Patna city. *Institute of Town Planners, India Journal*, *6*(2), 1–25.

McGuone, D., Sinard, J., Gill, J. R., Masters, A., Liu, C., Morotti, R., & Parkash, V. (2020). Autopsy services and emergency preparedness of a tertiary academic hospital mortuary for the COVID-19 public health emergency: The Yale plan. *Advances in Anatomic Pathology*, *27*(6), 355–362.

McLean, M., Watson, H. K., & Muswema, A. (2007). Veterinary waste disposal: Practice and policy in Durban, South Africa (2001–2003). *Waste Management*, *27*(7), 902–911.

Mohankumar, S., & Kottaiveeran, K. (2007). *Hospital Waste Management and Environmental Problems in India*. Framework.

Nolte, K. B., Taylor, D. G., & Richmond, J. Y. (2002). Biosafety considerations for autopsy. *The American Journal of Forensic Medicine and Pathology*, *23*(2), 107–122.

Pasupathi, P., Sindhu, S., Ponnusha, B. S., & Ambika, A. (2011). Biomedical waste management for the health care industry. *Int J Biol Med Res*, *2*(1), 472–486.

Patil, A. D., & Shekdar, A. V. (2001). Health-care waste management in India. *Journal of Environmental Management*, *63*(2), 211–220.

Rao, N. (2015). Biomedical waste management. *Environmental Science*, *10*(1).

Reinhardt, P. A. (2018). *Infectious and Medical Waste Management*. CRC Press.

Rodríguez, R., Sanchez, M., Ledesma, A., & Lloret, A. (2007). Experimental and numerical analysis of desiccation of mining waste. *Canadian Geotechnical Journal*, *44*(6), 644–658.

Rutala, W. A., & Weber, D. J. (2015). Disinfection, sterilization, and control of hospital waste. In Mandell, Douglas, and Bennett's *Principles and Practice of Infectious Diseases* (p. 3294).

Selvan Christyraj, J. R. S., Selvan Christyraj, J. D., Adhimoorthy, P., Rajagopalan, K., & Nimita Jebaranjitham, J. (2021). Impact of biomedical waste management system on infection control in the midst of COVID-19 pandemic. *The Impact of the COVID-19 Pandemic on Green Societies: Environmental Sustainability*, 235–262.

Sharma, A. K. (1998). Biomedical Waste (Management and Handling) Rules. *Bhopal: Suvidha Law House*, 50–70.

Sharma, M. (2002). *Hospital Waste Management and Its Monitoring*. Jaypee Brothers Publishers.

Smelt, J. P. P. M., & Brul, S. (2014). Thermal inactivation of microorganisms. *Critical Reviews in Food Science and Nutrition*, *54*(10), 1371–1385.

Tsai, C. T., & Lin, S. T. (1999). Disinfection of hospital waste sludge using hypochlorite and chlorine dioxide. *Journal of Applied Microbiology*, *86*(5), 827–833.

Upadhyaya, V., Sudhakaran, P., Shukla, A. K., Kumar, S., & Joshi, B. (2018). Application of small-scale renewable energy techniques in traditional buildings of Jaipur. *International Journal of Civil Engineering and Technology*, *9*(2), 627–643.

Veronesi, P., Leonelli, C., Moscato, U., Cappi, A., & Figurelli, O. (2005). Non-incineration microwave-assisted sterilization of medical waste. *Journal of Microwave Power and Electromagnetic Energy*, *40*(4), 211–218.

Vieira, C. D., de Carvalho, M. A. R., de Menezes Cussiol, N. A., Alvarez-Leite, M. E., Dos Santos, S. G., da Fonseca Gomes, R. M., Silva, M. X., & de Macêdo Farias, L. (2009). Composition analysis of dental solid waste in Brazil. *Waste Management*, *29*(4), 1388–1391.

Walkinshaw, E. (2011). Medical waste-management practices vary across Canada. *Canadian Medical Association Journal*, *183*(10), 1127–1128.

CHAPTER 3

Biomedical Wastes and Associated Risks

PREM LATA[1] and YOUNIS AHMAD HAJAM[2]

[1]*Department of Biosciences, Division Zoology, Career Point University, Hamirpur, Himachal Pradesh, India*

[2]*Department of Life Sciences and Allied Health Sciences, Sant Baba Bhag Singh University, Jalandhar, Punjab, India*

ABSTRACT

Biomedical waste (BMW) management is a critical aspect of healthcare services, ensuring the safe disposal and containment of hazardous materials generated during medical procedures, research, and diagnosis. This chapter provides a comprehensive analysis of BMWs and the associated risks they pose to public health, the environment, and healthcare workers. The improper handling and disposal of BMWs can result in severe consequences. The primary risk is the potential transmission of infectious diseases. Pathogens, such as bacteria, viruses, and prions, present in BMW, can lead to nosocomial infections and community outbreaks if not managed effectively. Moreover, antibiotic-resistant bacteria, a growing concern, can thrive in healthcare settings due to inadequate waste management practices, further exacerbating the global health crisis. Environmental contamination is another significant risk associated with BMW mismanagement. Chemicals, pharmaceuticals, and heavy metals found in medical waste can leach into soil and water bodies, polluting ecosystems and disrupting the delicate balance of nature. Health-care workers, who are at the forefront of handling BMW, face numerous risks. Accidental needle stick injuries and exposure to harmful substances pose immediate dangers, while long-term effects may include respiratory disorders due to inhalation of noxious fumes. Lack of proper training and

Biomedical Waste Management: Recycling and Remediation Strategies. Younis Ahmad Hajam, Javid Ahmad Malik, Anu Pandita, and Deepu Pandita (Eds.)

awareness regarding BMW handling can heighten these risks, compromising the safety and well-being of medical personnel. Effective BMW management strategies are essential to mitigate these risks. Implementation of stringent regulations, appropriate segregation, and proper disposal methods are crucial steps. Therefore, collective effort from healthcare institutions, governments, and the community is imperative to address this pressing global concern and safeguard the well-being of present and future generations.

3.1 INTRODUCTION

Among all the sources of contamination, medical waste is also a big contamination source of land and water. In the world, the medical sector growth increases and leads to generating a huge amount of waste of medical products. There are various causes of poor medical waste management, that is, odor, insect growth and development, pollution, growth of rodents, and worms as well. These all transmit various diseases like typhoid, cholera, and so forth, reported by different workers (Babanyara et al., 2013), and the generated waste can be varied according to the places or different regions, and from different hospitals. In India, biomedical waste (BMW) management has become a big problem in cities. The basic problem is the increasing population and urban development. Due to this, health issues arise, and healthcare facilities also increase, which leads to BMW generation in huge amounts. Therefore, lots of BMW is generated and has no proper treatment found for its disposal which is very non-hygienic for the environment and causes serious health issues.

Among different institutions, a hospital is one of the places where people walk every day in life. There is no distinction between the people like age, race religion, and so forth. In the hospital, hospital staff and patients are the normal inhabitants that produce waste in different ways according to the knowledge, and advancement in science, and technologies. There is the biggest risk to the patients and the person who handles all the processes around the patient that cause health hazards and environmental contamination. BMW is mainly produced at the time of treatment, diagnosis, and during research activities, or in different health care camps. There is some way to treat BMW, that is, its characterization, quantification, segregation, storage, transport, and treatment or disposal of, and also the 3R formula (reduce, recycle, and reuse). Different drugs are used by clinics, hospitals, and healthcare camps such as antibiotics, corrosive chemicals, and so forth, that become BMW as well.

Present day, it become a very serious issue in India and the whole world as well (Narang et al., 2012; Tambe et al., 2013). According to the World Health Organization (WHO), most of the waste about 15–25% generated from hospitals, which is very dangerous and causes health issues. There are some data which represent that every year 32%, 40%, and 5% of hepatitis B, hepatitis C, and HIV infections, respectively caused by contaminated sharps and syringes alone (Badi et al., 2019). Medical waste can be solid, liquid, or in gaseous form with organic or inorganic substances such as pathogenic radionuclides generated from *in vitro* analysis of body microorganisms, some tissues, and so forth, according to the WHO (1999) reported that among infectious, hazardous, and non-hazardous wastes, about 85% is non-hazardous, 10% is infectious and rest of is hazardous waste mostly methyl chloride and formaldehyde found in this (Glenn and Garwal, 1999). Nonhazardous waste causes health hazards to the workers, and the public as well (WHO, 1999).

In India, about 1 to 1.5% of municipal waste is BMW. This waste about 10–15% of waste is infectious and causes health hazards. On the other hand, in other countries mostly in developed countries, the mismanagement of waste is found so that about 5–6 kg per bed waste is generated in a day. The waste generated in different countries per day is different, in the UK 3.3 kg, in France 2.5 kg, in Norway 3.9, and in the USA 4.5 kg. This is in huge amounts in comparison to India and all developing countries. In India, the waste generated per bed per day is about 1–2 kg except for AIIMS and SKIMS. In New Delhi, there are about 40,000 beds, which generate about 60 metric tons of BMW in a day. In the present day, BMW is disposed of but not managed properly. It is either dumped on open land, or in slums, which causes environmental and health-related issues, or risks of diseases like AIDS, cholera, hepatitis, and many more.

3.2 TYPES OF WASTES

Wastes are of two types, that is, non-hazardous and hazardous waste.

3.2.1 NON-HAZARDOUS WASTE

The food remnants, paper cartons, packaging material, fruit peels, and so forth, are some non-hazardous which is about 85% known as non-hazardous waste.

3.2.2 HAZARDOUS WASTE

These are two types, that is, potentially contagious waste and potentially toxic waste shown in Table 3.1. Potentially contagious waste is infective medical waste, regulated or nonregulated medical wastes, and so forth, about 10% of waste includes, dresses that are infected with blood and other fluids, laboratory waste, infected material like organs, tumors, and so forth, the animals used in the research work, sharps like blades, needle, and so forth, blood and blood products. On the other hand, potentially toxic waste includes radioactive waste, chemical wastes like the solution used in X-rays, monomers, and associated reagents, and so forth, and pharmaceutical waste like antibiotics and sedatives, and so forth. The WHO has classified medical waste into eight categories: General waste, pathological, radioactive, chemical, infectious to potentially infectious waste, sharps, pharmaceuticals, and pressurized containers showed in Figure 3.1.

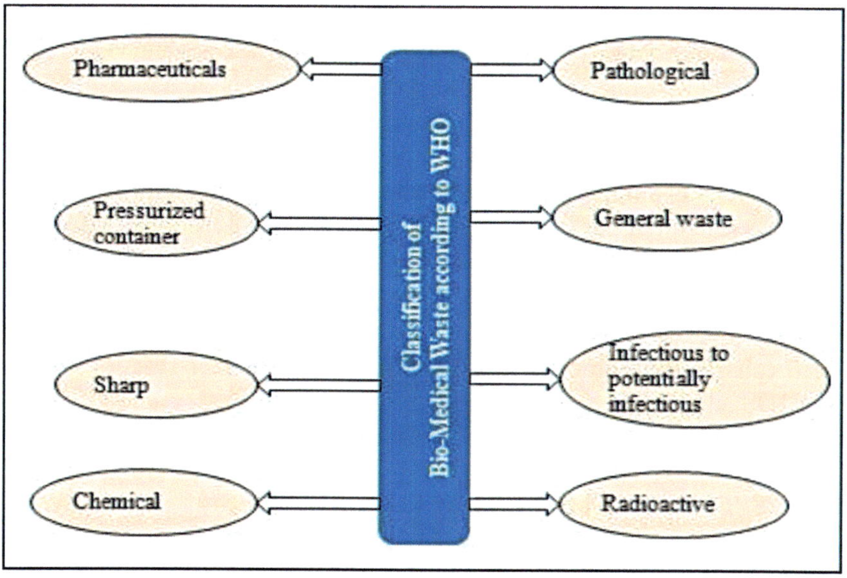

FIGURE 3.1 Classification of biomedical waste according to the World Health Organization (WHO).

Among the countries, in India, most of the waste is dumped in municipal waste sites and roads, or open landfills without complete burning of it. There are some techniques like incineration, autoclaving, hydroclaying, microwaving, shredding, mutilation, and pyrolysis that are used in waste

TABLE 3.1 Different Waste Content and Their Components

Sl. No.	Waste Content	Components
1.	Animal waste	Tissues, organs, bleeding parts, and so forth, from different sectors.
2.	Waste sharps	Needles, syringes, blades, glass, and so forth.
3.	Solid waste	Blood-contaminated cotton, dressings, soiled plaster casts, lines, and so forth.
4.	Liquid waste	Waste generated from laboratory and washing, cleaning, housekeeping, and disinfecting activities.
5.	Chemical waste	Chemicals used in the production of biological.
6.	Incineration ash	Ash from incineration of any biomedical waste.
7.	Medicine and drugs	Out-dated contaminated and discarded medicines.
8.	Laboratory (microbiology and biotechnology	Wastes from laboratory cultures, stocks, or specimens of micro-organisms used in research.
9.	Anatomical waste	Human tissues, organs, and body parts.

management according to the waste verities and based on color-coded categories. In our country India, the management ensured according to the Biomedical Waste Management Rule (2016). It is implemented on how to handle the waste, its segregation, disposal, and management according to the four colors-coded system, or well-trained health workers' availability. A healthy environment is essential for healthy living, and we know that the waste generated is linked with humans directly or indirectly if the management is not carried out according to the rules, then it also affects the environment's integrity. In India, different researchers reported that the management of BMW is the second highest problem in the world (Subrammani et al., 2014). In India, there are about 6 lakh hospitals, of which about 23,000 are basic units, 15,000 are small clinics or some private hospitals. About 4,20,461 kg of BMW is generated every day, and only 2,40,682 kg of waste can be treated every day.

3.3 EFFECTS OF BIOMEDICAL WASTE (BMW)

Sharma and workers reported that mostly the contaminated sharps cause an increase in risk associated with hospital waste, as well as with microbial waste handling. The high temperature while operating incinerators with toxic gases after the treatment of waste in the atmosphere causes health hazards (Sharma et al., 2013). This is caused by improper management of waste and inappropriate use of equipment without any knowledge (Chopra et al., 2017), which further creates an impact on the ecology of microbes that is very dangerous (Sharma et al., 2013; Sushma et al., 2010).

3.3.1 IMPACT ON WATER

Mismanagement of BMW causes a bad impact on the water when mixed with groundwater. Different researchers reported that different heavy metals found in water change water standards or quality (AlRaisi et al., 2014). Different elements have different concentrations, that is, aluminum 2.050, vanadium 0.9775, chromium 2.800, manganese 0.503, cobalt 0.128, nickel 0.773, beryllium 0.8575, lead 0.130, and iron 39.25 mg/L. In Lagos Nigeria, different workers reported the assessment of groundwater quality near it due to a waste dump site that caused a negative impact (Oluseyi et al., 2014). In Olusosun, high physical and chemical parameters were demonstrated at wells. According to WHO and NSDWQ, they limit the concentration of metals in both dumping sites, that is, 0.069 ± 0.075 mgL and 0.17 ± 0.086 mg/L for Oke-Afa and Olusosun dump sites, respectively. Researchers reported that in BMW there was a high concentration of heavy metals and PAHs as well that further contaminate and pollute groundwater and surface water as well (Heera and Rajor, 2014). So, the elimination of ash toxicity is important to be for landfill disposal (Table 3.2).

TABLE 3.2 Different Metals and Their Concentration in Water

Sl. No.	Different Elements	Concentration
1.	Aluminum	2.050 mg/L
2.	Vanadium	0.9775 mg/L
3.	Chromium	2.800 mg/L
4.	Manganese	0.503 mg/L
5.	Cobalt	0.128 mg/L
6.	Nickel	0.773 mg/L
7.	Beryllium	0.8575 mg/L
8.	Lead	0.130 mg/L
9.	Iron	39.25 mg/L

Water Pollution is caused by the liquid waste that is let into the water-bodies without treatment (Garg, 1994; Rao, 1995). It can alter pH, BOD, COD, and so forth, parameters. Near the incineration plants, dioxins have been reported in water bodies, entering through the air, as reported by different researchers (Ravikant et al., 2002). Also, radioactive waste originates from biological and chemical research like patient urine, and so forth, normally, the patient's urine and face are known as non-radioactive substances as long the monitoring of the room occurs everyday routine (Patil and Pokhrel, 2004).

3.3.2 IMPACT ON SOIL QUALITY

Biomedical waste (BMW) can change the soil quality near the dumping site. It occurs due to improper BMW management. There are many toxic substances that get mixed with the soil, due to which the soil's physico-chemical properties get changed. About five heavy metals reported by the researchers, that is, Cr, Ni, Zn, Pb, and Cu in soil (Abidemi and Theresa, 2015), and their concentration was 3.63 ± 2.46 mg/kg, 26.3 ± 51.1 mg/kg, 1133 ± 897 mg/kg, 137 ± 64 mg/kg, 110 ± 90 mg/kg, respectively. Differences win the heavy metal level in different sites, which washing her from background having Zinc (67), copper (18), and lead (20) factors. Furthermore, BMW can cause soil pollution with waste that is infectious, expired medicines, chemicals, and ash produced during treatment processes. All the heavy metals that were present in the waste will absorbed by the plants, due to which they enter the food chain. The nutrient elements in huge amounts can affect crops, humans, and animals as well.

3.3.3 IMPACT ON AIR QUALITY

At present air pollution become a problem for our county. Mostly, the burning of waste and other toxic substances leads to air pollution among which BMW burning is also a great threat, causing various health issues by releasing harmful pollutants and causing respiratory problems. Many researchers reported that dust, black carbon, ammonia, sulfate, and nitrate are the major species of PM10 and PM2.5 (Karthikeyan et al., 2006). Air is also contaminated with other hazardous compound such as diethyl phthalate, decane, dodecane, octane, nonane, methene amine, carbon disulfide, and acetone diperoxide. However, air pollution can be caused in both atmosphere, that is, indoor air pollution and outdoor air pollution.

In indoor, pathogens present in the waste. They remain in the air for a long time in the form of pathogens. So, the waste segregation, and its pretreatment help to inhibit its exposure, also the room sterilization helps in its checking reported by different workers (Askarian et al., 2004b; Baveja et al., 2000). Air pollution is also caused due to improper ventilation of chemicals as well that cause diseases like sick building syndrome. It can be handled by proper maintenance and good ventilation and overuse is also avoided (Bdour, 2004; Saurabh and Ram, 2006).

Outdoor air pollution is also caused by pathogens when pre-treated BMW is dumped outside near open areas or around institutions allowing the pathogen entry into the atmosphere. There are two major sources of outdoor air pollution, that is, burning of waste in open spaces and incineration which is very harmful, on inhalation it causes respiratory diseases because there are some organic gases that are carcinogenic like dioxin and furans (Burd, 2005).

There are some metals, acid gases, nitrogen oxide, sulfur, and xenobiotic substances produced by the BMW burning, causing cancer, respiratory symptoms, hormonal defects, increase in the sex ratio are its impact on burning. There are some environmental-related issues like global warming, acidification, and photochemical ozone formation caused by the burning of BMW.

There are some technologies that help to decrease the degradation of the environment such as hydroceles plasma pyrolysis which also neglects health impacts, waste handling is safe, maintenance cost is low microorganisms' reduction is effective, and very safe disposal as well. Research activities and radio-immunoassay activities generate radioactive gases as well in small amounts that are evacuated directly into the air. The utilization of all these can be maintained and reported by their searcher (Babu et al., 2009).

3.3.4 HEALTH HAZARDS

Recently, according to the WHO, life expectancy has increased every year, but other hand death is also increasing due the infectious diseases. WHO in 1996 reported that about 50,000 people die due to infectious diseases per day due to mismanagement of waste. There are different diseases and infections caused by it, that is, tuberculosis, pneumonia, diarrheal diseases, cough, and so forth (Figure 3.2), reported by various researchers (Chitins et al., 2002, 2003; Tudor et al., 2005).

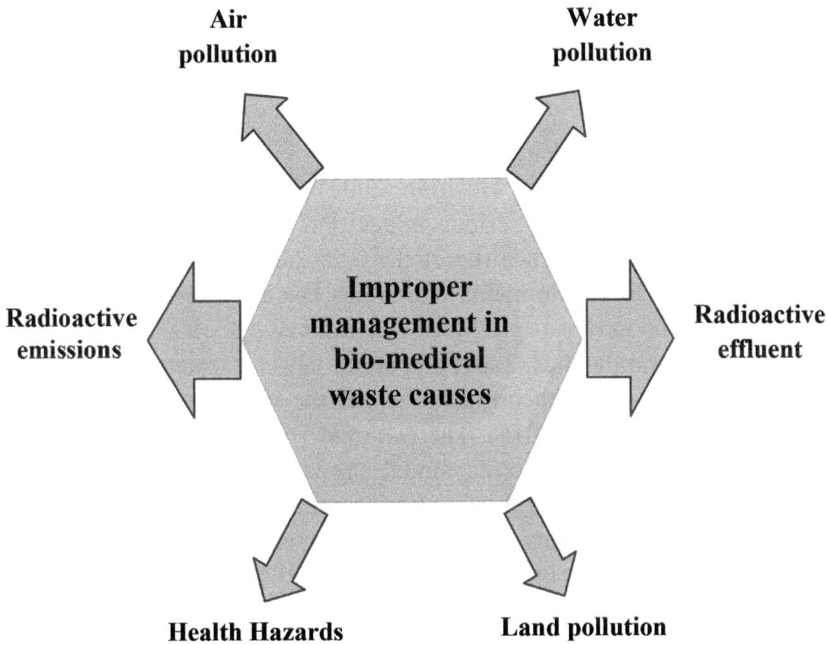

FIGURE 3.2 Causes of improper management of biomedical waste.

3.3.5 OCCUPATIONAL HEALTH HAZARDS

The hazards concerning laundry workers, janitorial, nurses, and so forth, that harmful chemical waste, sharps, and radioactive waste cause hazards to the employees in the institution generated BMW. However, it can be resolved by managing waste properly (Patil and Shekar, 2001).

3.3.6 HAZARDS TO THE GENERAL PUBLIC

Biomedical waste (BMW) also affects public health due to improper practices that are carried out by the workers, that is, dumping in open spaces, and water bodies that cause various health issues to the public. Also, the open burning of waste and incineration produces some harmful gases that cause cancer, and respiratory diseases reported by different researchers (DaSilva et al., 2005; Manohar et al., 1998). Animals are also affected by the plastic that is included in BMW; it chokes animals. On the other hand, animals get injured with the sharps, and chemicals like dioxin and furans cause health hazards to the animals as well as birds (Code and Christic, 1999).

The management of waste affects the economy living near the hospital, so the public concern is increased and can manage the waste systematically and in a strict manner. Minimization and recycling are the major stages of this. Different heavy metals found in huge amounts, that is, cadmium, zinc, lead, and copper reached the soil in deep and led to damage reported by the workers (Huiying and Huaqing, 2002). However, heavy metals are non-biodegradable, causing serious issues and accumulation in the environment. The different studies reported by various researchers showed the biggest challenge to development, and crises faced by developed countries and developing countries as well (Alloway and Ayres, 1997; Jensen, 1990; Kabata-Pendias and Pendias, 2001; Oyediran, 1994) researchers estimate the waste emission from healthcare centers, and the impact over public health (Eckelman and Sherman, 2016). The National Health Expenditure reported that over 10 years the economic model shows negative public health. In the world, it causes 12% acid rain, 10% greenhouse emission, 10% smog formation, 9% air pollutants, 1% stratospheric ozone depletion, and 1–2% carcinogenic and non-carcinogenic air toxic. About 44,000-98,000 people die in the US every year because of preventable medical errors in hospitals. If the environmental performance improved in the healthcare sector, then it could decrease waste reduction expenditures. In the US, the carbon footprint is estimated by researchers in the healthcare sector (Chung and Meltzer, 2009).

However, there was data that reported that about 8% of GHG emissions contributed by healthcare in the nation further caused 46% of direct purchase damage and about 54% to goods and services of healthcare centers. Similarly, about 3-4% of GHG emissions are contributed by the National Health Service. Various researchers reported that there is data in huge that helps to demonstrate the impacts of specific medical devices (Connor et al., 2011; Thiel et al., 2015). Since the effect of an Earth-wide temperature boost is wrecking, this can't be ignored. Medical care is not just addressing a picture of irritation, if waste can't be dealt with straightforwardly, yet additionally may become problematic when it emanates different GHG and is overburdened with various microorganisms that might harm human well-being and ecological administrations. A large portion of the examinations has likewise uncovered an elevated degree of suspended particulate matter in a space near emergency clinics (McMichael et al., 2006). Although medical clinics safeguard and save human existence, the results created extraordinarily harm the climate (Manyele and Mujuni, 2010).

Around 80% of the created squander is general and the rest is perilous material with irresistible specialists and poisonous radioactive squanders (WHO,

2011). The assessed 16,000 million infusions that are managed to the world are not arranged as expected a while later. The waste is completely stacked with hurtful microorganisms that contaminate medical clinic patients, medical services laborers, and the overall population. The irresistible squander, if not overseen as expected, prompts diseases, barrenness, genital deformations, hormonally set off malignant growths, mutagenicity, dermatitis, asthma, and neurological issues among individuals alongside kids, and typhoid, cholera, hepatitis, Helps, and other viral diseases from the sharps tainted with blood (Adedigba et al., 2010; Manyele and Mujuni, 2010).

Individuals who are in danger are the medical care laborers, patients, guests to medical care foundations laborers at garbage removal offices, hatchlings in the bellies of moms, individuals from people in general, and scroungers. The ill-advised treatment of BMW in medical clinics might expand the airborne pathogenic microscopic organisms, which could likewise unfavorably influence the clinic climate and local area at large. This genuinely affects the human climate alongside stylish issues (Ogbonna, 2011).

3.4 BIOMEDICAL WASTE (BMW) ADMINISTRATION CYCLE

The board of BMW is one of the greatest assignments of recent days because of the immediate effect on the soundness of people. Squanders created throughout the analysis, the raptor vaccination of persons or creatures, and the radioactive waste or waste produced from the research centers are termed bioclinical waste. The significant wellspring of waste is from the public authority and confidential emergency clinics. A few measures of waste are likewise produced from clinical labs, little facilities, dispensaries, nursing homes, morgues, and post-mortem focuses. Much significance ought to be given to the protected removal of waste. Treatment of clinical waste is a hazardous activity that requires an elevated expectation of planning (Manyele and Anicetus, 2006). The board, therapy, and removal of clinical waste and making mindfulness among the clinical consideration laborers and the overall population has turned into a developing concern. To accomplish this, the creation of information, mindfulness, and practice among the clinical consideration laborers is the need of great importance. Endeavors have been made by natural administrative experts for better administration of clinical waste to forestall the flare-up of infections. This can be accomplished provided that the medical services laborers collaborate in the waste administration process. Appropriate execution of clinical waste administration can prompt improved impacts for both the public and the climate. To manage healthcare waste, there is a huge network in India. Infectious and non-infectious waste, such as body parts, cotton, bandages, plaster, blood, organs, and so forth, is necessary for proper collection, segregation, storage, transport, treatment, and safe disposal to prevent infections acquired in hospitals.

3.4.1 WASTE SURVEY

A survey of waste is a very important component of management and the assessment of the type of waste (Ebenezer, 2021; Park, 2009; Singh et al., 2014). Enumerate the waste generated, the point of generation in different areas, and segregate waste properly. Demonstration of altitude of generation. There are six steps for the management of BMW (Figure 3.3), that is:

1. Collection of waste;
2. Waste segregation;
3. Waste transportation and storage;

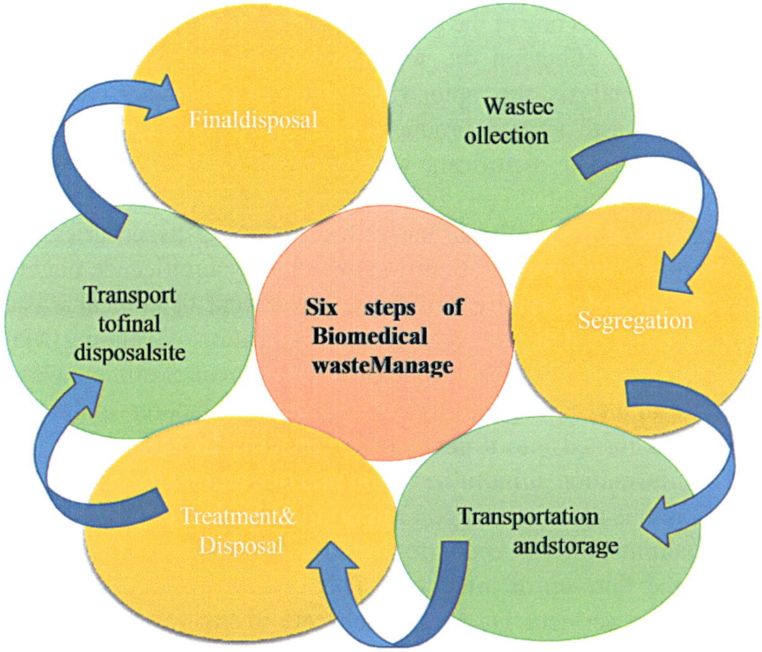

FIGURE 3.3 The management of biomedical waste.

4. Waste treatment and disposal;
5. Transport of waste to the disposal site; and
6. Final disposal.

Protecting the healthcare workforce against occupational health risks comes from waste management of hospital waste calls for effective infectious waste control measures. Worker's health can be protected from hospital waste, and this control measure helps in public health protection and the environment caused by hazards of hospital waste. So, the proper management at the time of waste generation can be handled by waste treatment and proper disposal of waste.

3.4.2 COLLECTION OF WASTE

Two types of waste are collected, that is, infectious and non-infectious waste by a group of two members, with one collecting waste and the other putting it into the polythene bags. These members are well trained and have knowledge about the waste hazards, so they follow the steps properly for handling waste. They have to wear personal equipment like masks, gloves, aprons, and so forth when they are collecting waste which is infectious. They used different color-coded containers to segregate waste and used designated carts for waste transportation. The waste collection process occurs four times a day at 8:00 a.m., at noon, at 4:00 p.m., and at 8:00 p.m. In polythene, 75% fill with waste is put into the polythene bags to avoid an overload of waste. Mostly the OT and CU waste are collected frequently to avoid infection depending upon the operation carried out on a particular day. A particular mark was carried on the containers for hazardous waste collection according to the Biomedical Waste Rule (1998).

3.4.3 SEGREGATION

After collection, the second step is waste segregation. The segregation of waste is done according to its characteristics. For the process, they used color codes and, best-quality polythene bags as prescribed in the II schedule of Biomedical Waste Rules (1998) for easy identification and as well as for waste segregation. Black color polythene bags are used for the collection of noninfectious and household-type waste. Red bags are used for the infectious waste collection. Sharps wastes are collected in blue polythene bags, and yellow color bags are utilized for the collection of cytotoxic waste. It decreases the cost of treatment and the bulk of the infectious waste. It reduces the chance of having an infection.

3.4.4 STORAGE AND ACCUMULATION OF BIOMEDICAL WASTE (BMW)

The storage of waste is done between the area of generation and waste treatment location and its disposal sites. The holding of waste near the generation point is temporary and, in small quantities refers to accumulation, and the holding of waste for a long time is known to storage in large amounts. The storage area is always near the treatment site of waste, but sometimes offside storage will be considered as per condition and requirement. As we discussed the different waste is collected in the different containers by which 100% of the waste gets collected. After collection, the waste is stored at a particular site. In big hospitals, the storage period is 8–10 hr. In nursing homes, the time of storage is about 24 hrs. Proper labeling is carried out on the containers for batter and safe storage of particular waste in particular sites and the sites should know with a caution sign. Polyclinics can have designed special storage areas having proper ventilation and fire protective facilities. In summer, the waste is stored for 24 hr, and in winter it was 48 hr (Table 3.3).

TABLE 3.3 Describes the Various Color Coding, Type of Container, and Waste Category

Category	Wastes	Type of Container	Color
1	Human anatomical waste, animal waste, solid (BMW), and laboratory wastes.	Plastic bags	Yellow
2	Chemical wastes, Incineration waste, cytotoxic waste, and expired medicine.	Plastic bags	Black
3	Solid (plastic), and waste sharp.	Plastic bag puncture-proof containers	White and blue

3.4.5 TRANSPORTATION OF WASTE

Now, after the segregation of waste, the next step is the transportation of waste and storage as well. Pre-planning is essential for the transportation of medical waste which is controlled by polyclinics, the specific corridors and elevators are required for the waste transportation of waste from one floor to another in the polyclinic basement. Carts and containers are used for the transportation of BMW. These carts and containers are not utilized for other purpose instead of waste transportation. Maintenance and cleaning should be done every day after work. Tagging on vehicles is best for the safe transportation option. The waste is secured properly during transportation to avoid biohazards and infections to the workers. Unprocessed BMW should not be stored more than 48 hr (Park, 2009).

3.4.6 TREATMENT OF WASTE

Treatment is important to stay away from well-being risks. Squander sanitizing and purification are expected to keep away from the disease because of pathogenic life forms. After the treatment cycle, some waste remains. These stays squander simply to deal with and shipped and put away effectively, for example:

1. Some broken things like glass, and surgical tool cutting edges are kept in isolated compartments with dye.
2. It can be moved in plastic boxes or cardboard boxes and fixed it appropriately.
3. Syringes spout and needles should be destroyed in needle destroyers.

China should be sanitized, disinfected, and cleaned. Culture plates with conceivable culture should be autoclaved; media are set in appropriate sacks and arranged off. The plates can be reused after sanitization.

1. Gloves ought to be cut/destroyed/ruined before removal.
2. Swabs ought to be artificially disinfected followed by burning. Assuming they have just a little amount of blood that doesn't dribble, they could be set in the trash.
3. Expendable articles ought to be plunged in newly pre-arranged 1% sodium hypochlorite for 30 to 60 minutes, after that mutilation is expected before removal ought to be the strategy taken on for such things.
4. Intensity ought to be utilized for the removal of combination, by no means. It will make mercury volatile and be released into the climate. Along these lines, teeth loaded up with mixture ought to be treated by drenching in undeniable level sanitizers (e.g., Glutaraldehyde) for 30 min.

5. Fluid waste delivered by the research facility is either obsessive or synthetic. Non-infectious waste ought to be best popped with reagents.
6. Fluid-defiled waste ought to be treated with a substance decontaminator and afterward stopped.

3.4.7 MINIMIZATION OF WASTE

Though customary strong and fluid waste requires no treatment before removal, essentially all infectious waste ought to initially be dealt with. The expense for the removal of irresistible waste might be multiple times more than the cost for the removal of common waste. Strategies that lessen the amount of irresistible burn-through created will simultaneously diminish the expense of irresistible waste disposal (Block, 2001).

Dispensable needles, blood, jugs, pee packs, careful gloves, catheters, and so on are a portion of the instances of plastic practice in medical services. Plastic has been connected with derivation in sperm count, genital irregularities, and an ascent in the event of bosom malignant growth. Copying plastics releases cancer-causing agents like dioxin and furan.

Practically every one of the nations on the planet is buckling down towards safe administration and removal of these medical care squanders. There is a ton of bewilderment with the troubles among the generators, administrators, chiefs, and the overall population about the protected administration of bio-clinical waste. The explanation might be an absence of mindfulness. In this way, asset material on the environmental elements for clinic managers, specialists, specialists, attendants, Paramedical staff, and waste retrievers, is the need of great importance (Almuneef and Memish, 2003). It has been emphasized (Summers, 1991) that for the appropriate removal of bio-clinical waste, the presentation of regulations isn't sufficiently adequate. The awareness of these regulations among the overall population as well as improvement of these strategies and severe execution that regard those regulations is essential. Different purposes behind the expansion in the clinical waste are the development of the populace with different sicknesses as well as expansion in the number of medical clinics. In any case, in many non-industrial nations, the board of these medical care squanders has become one of the serious issues because of the expansion in the utilization of expendable things. Besides, these nations need to handle issues of appropriate isolation and remove perilous medical services squandering. In this manner, handling what is going on with an advanced technique centering on the administration of bio-clinical squanders is likewise a need of great importance.

As of late, numerous endeavors have been made by ecological administrative organizations and waste generators to more readily deal with the loss from medical care offices. Jang et al. (2006) revealed that ongoing administration rehearses like age, organization, isolation, transportation, and removal of clinical squanders don't observe legitimate rules. Cremation is distinguished as the most favored removal technique and was the main accessible treatment choice until 2005. With expanded guidelines over poisonous air discharges (e.g., dioxins and furans), all the current little cremation offices that don't have air contamination control gadgets have halted their activity few years. There is an extraordinary potential to transmit air poisonous contaminations from such incinerators if inappropriately operated and maintained, because clinical waste normally contains various plastic materials, for example, polyvinyl chloride (PVC). They contended that squandering minimization and reusing alongside the control of harmful air outflows at clinical waste incinerators and other current choice strategies for cremation are significant difficulties later on.

3.5 DIFFERENT ADMINISTRATIVE REGULATIONS CONNECTED WITH BMW IN INDIA

Different administrative regulations connected with BMW in India are the following:

- The Water Act (1974; avoidance and control of contamination);
- The Air Act (1981; avoidance and control of contamination);
- The Climate Act (1986; security);
- The Perilous Waste Standards (1998; the executives and dealing with);
- The Biomedical Waste Principles (1998; the board and taking care of);
- City Strong Waste Guidelines (2000; the board and dealing with);
- The Biomedical Waste Standards Alteration (2000 and 2003; the executives and taking care of);
- The Biomedical Waste Standards (2011; management and dealing with).

The Service of Woodland and Climate, Government of India, on July 20, 1998, outlined strategies for legitimate BMW administration and taking care. These principles gave different codes for dealing with medical services squandering. This strategy was given to all exploration organizations, medical clinics, nursing homes, centers, dispensaries, creature houses, research and development, virology foundations, clinical demonstrative labs, veterinary establishments, and comparative establishments. This is material to the people who produce, gather, get, transport, store and discard any type of BMW (Government of India, 1998).

Overhauled rules in 2011 showed that each occupier who is creating BMW, regardless of the amount, will keep the BMW guidelines. Before the 1998 rules, it was referenced every one of the emergency clinics with 1000 beds should comply with this demonstration. It expresses the obligation of the occupier in medical services establishments and clinics to appropriately deal with the loss as per the means referenced and conventions given in the rules, and removal should be managed without making any antagonistic medical problems in the climate. To achieve this appropriate mindfulness, preparing is to be accommodated all the medical services laborers and the faculty responsible for this BMW taking care of and removal in the emergency clinic area. The administrators will report in Structure IV if any medical services organization isn't as expected in dealing with and discarding the medical care squander as per the rules recommended. Squander the executives pecking order includes decrease, reuse, reuse, recuperate, and decline. These are known as the Five R's administration. In reuse, non-biodegradable utilization of materials kept away that incorporate no plastic sacks to shoppers. In, lessen, decrease the bundle for customers, reusing materials over again for any planned reason. In reusing, it requires less energy than unrefined substance fabricating, and by separating the materials, new items are framed. Recuperate implies recuperating energy before the material is disposed of (Subramanian et al., 2021).

3.6 NEED OF BIOMEDICAL WASTE (BMW) ADMINISTRATION IN EMERGENCY CLINICS

1. Wounds from sharps prompting disease to all classifications of the clinic workforce and waste controller.
2. No so-comical diseases in patients from unfortunate contamination control practices and unfortunate waste management.

3. Chance of disease outside the medical clinic for burn through controllers and scroungers and at times overall population living near emergency clinics.
4. Risk related to dangerous synthetics, and medications to people taking care of squanders at all levels.
5. Dispensable being repacked and sold by deceitful components without being washed.
6. Drugs that have been discarded, repacked, and auctioned off to clueless purchasers.
7. Hazard of air, water, and soil contamination straightforwardly because of waste, or because of inadequate burning discharges and ash (Hem, 1999).

3.6.1 ADVANTAGES OF BIOMEDICAL WASTE (BMW) ADMINISTRATION

1. Cleaner and better environmental elements.
2. Decrease in the occurrence of emergency clinic procured and general contaminations.
3. Decrease in the expense of disease control inside the medical clinic.
4. Decrease in the chance of illness and demise due to reuse and repackaging of irresistible disposables.
5. Low occurrence of the local area and word-related well-being dangers.
6. Decrease in the expense of waste administration and age of income through fitting treatment and removal of waste.
7. Further developed picture of the medical services foundation and increment the personal satisfaction.

3.7 CONCLUSION

Biomedical waste (BMW) administration is pretty much as significant as a therapy for clinical experts. Mind fullness projects ought to be led as far as well-being might be concerned workforce and helper faculty of different medical services establishments to keep next to each other with the ongoing information on logical BMW administration framework and its significance and advantages to the patients, staff, and the local area as a whole. For severe and straightforward requirements of the warning of BMW (dealing with and the executives) rules (1998), individual medical care communities/units ought to develop and embrace a framework that suits them best. Each individual from the local area has the privilege to be educated about the destructive impacts of medical services squandering. Preparing programs needs to zero in on engaging the medical care experts on BMW administration with expansive extension and down-to-earth information from all perspectives. Preparing the staff with agenda as and standard.

KEYWORDS

- **biomedical wastes**
- **environmental hazards**
- **final disposal**
- **infectious disease**
- **personal protective equipment**
- **public health protection**
- **waste management**
- **waste treatment**

REFERENCES

Abidemi, O. O., & Theresa, O. C. (2015). Environmental fate of heavy metals in soil of Ido-Osun waste dump site, Osogbo, Osun, Nigeria. *American Journal of Environmental Protection, 3,* 1–4.

Adedigba, M. A., Nwhator, S. O., Afon, A., Abegunde, A. A., & Bamise, C. T. (2010). Assessment of dental waste management in a Nigerian tertiary hospital. *Waste Management Research, 28,* 769–777.

Al Raisi, S. A. H., Sulaiman, H., Suliman, F. E., & Abdallah, O. (2014). Assessment of heavy metals in leachate of an unlined landfill in the Sultanate of Oman. *International Journal of Environmental Science and Development, 5,* 60.

Alloway, B., & Ayres, D. C. (1997). *Chemical Principles of Environmental Pollution.* CRC press.

Almuneef, M., & Memish, Z. A. (2003). Effective medical waste management: It can be done. *American Journal of Infection Control, 31,* 188–192.

Askarian, M., Vakili, M., & Kabir, G. (2004). Hospital waste management status in university hospitals of the Fars province, Iran. *International Journal of Environmental Health Research, 14*(4), 295–305.

Babanyara, Y. Y., Ibrahim, D. B., Garba, T., Bogoro, A. G., & Abubakar, M. Y. (2013). Poor medical waste management (MWM) practices and its risks to human health and the environment: A literature review. *International Journal of Environmental Earth Science and Engineering, 11*, 1–8.

Babu, B. R., Parande, A. K., Rajalakshmi, R., Suriyakala, P., & Volga, M. (2009). Management of biomedical waste in India and other countries: A review. *Journal of International Environmental Application & Science, 4*, 65–78.

Badi, I., Shetwan, A., & Hemeda, A. (2019). A grey-based assessment model to evaluate health-care waste treatment alternatives in Libya. *Operations Research: Engineering Theory and Applications, 2*, 92–106.

Baveja, R., Buckshee, K., Das, K., Das, S. K., Hazra, M. N., Gopalan, S., & Saxena, B. N. (2000). Evaluating contraceptive choice through the method-mix approach: An Indian Council of Medical Research (ICMR) Task Force study. *Contraception, 61*, 113–119.

Bdour, A. N. (2004). *The Coupling of a Two-Dimensional Hydrodynamic/Sediment Routing Model with An Upland Watershed Erosion Model in A Mountain Watershed, Red River, Idaho* (Unpublished doctoral dissertation). Washington State University.

Block, S. S. (2001). *Disinfection, Sterilization, and Preservation.* Lippincott Williams & Wilkins.

Burd, M. (2005). Reducing the risks related to the handling and disposal of health-care waste. *Professional Nurse, 20*, 40–42.

Chitnis, V., Chitnis, D. S., Patil, S., & Chitnis, S. (2002). Hypochlorite (1%) is inefficient in decontaminating blood containing hypodermic needles. *Indian Journal of Medical Microbiology, 20*, 215–218.

Chitnis, V., Chitnis, S., Patil, S., & Chitnis, D. (2003). Treatment of discarded blood units: Disinfection with hypochlorite/formalin versus steam sterilization. *Indian Journal of Medical Microbiology, 21*, 265–267.

Chopra, R., Mathur, S., Dodwad, V., Sharma, N., & Tevatia, S. (2017). Awareness & attitude regarding biomedical waste disposal among post-graduate students, under-graduate students & auxiliary staff of a dental college–A questionnaire survey. *International Journal of Dental Research, 5*, 64.

Chung, J. W., & Meltzer, D. O. (2009). Estimate of the carbon footprint of the US health care sector. *JAMA, 302*, 1970–1972.

Coad, A., & Christen, J. (1999). *How Are We Managing Our Healthcare Wastes? Guidance On How to Investigate the Current Practices of Healthcare Waste Management in A Hospital or Other Healthcare Facility, With Some Information on Recent Practices Taken from Case Studies from Africa, Asia and the Middle East*. SKAT.

Connor, A., Lillywhite, R., & Cooke, M. W. (2011). The carbon footprints of home and in-center maintenance hemodialysis in the United Kingdom. *Hemodialysis International, 15*, 39–51.

Da Silva, C. E., Hoppe, A. E., Ravanello, M. M., & Mello, N. (2005). Medical wastes management in the south of Brazil. *Waste Management, 25*, 600–605.

David, N. O. (2011). Characteristics and waste management practices of medical wastes in healthcare institutions in Port Harcourt, Nigeria. *Journal of Soil Science and Environmental Management, 2*, 132–141.

Eckelman, M. J., & Sherman, J. (2016). Environmental impacts of the US health care system and effects on public health. *PLOS ONE, 11*, e0157014.

Glenn, M. R., & Garwal, R. (1999). Clinical waste in developing countries: An analysis with a case study of India, and a critique of the Basle TWG Guidelines.

Gupta, S., & Boojh, R. (2006). Biomedical waste management practices at Balrampur Hospital, Lucknow, India. *Waste Management Research, 24*, 584–591.

Heera, S., & Rajor, A. (2014). Bacterial treatment and metal characterization of biomedical waste ash. *Journal of Waste Management, 2014*(3), 1–6.

Hutajulu, S. M., Marsaulina, I., Siregar, F. A., Indirawati, S. M., & Utami, T. N. (2021). Environmental impact and cost analysis of solid medical waste treatment in hospitals. *Natural Volatiles and Essential Oils, 8*(4), 13018–13027.

Jang, Y. C., Lee, C., Yoon, O. S., & Kim, H. (2006). Medical waste management in Korea. *Journal of Environmental Management, 80*, 107–115.

Jeevan Rao, K., & Shantaram, M. V. (1995). Groundwater pollution from refuse dumps at Hyderabad. *Indian Journal of Environmental Health, 37*, 197–204.

Kabata-Pendias, A., & Pendias, H. (2001). *Trace Elements in Soils and Plants* (3rd ed.). CRC Press.

Karthikeyan, S., Balasubramanian, R., & Iouri, K. (2006). Particulate air pollution from bushfires: Human exposure and possible health effects. *Journal of Toxicology and Environmental Health Part A, 69*, 1895–1908.

Manohar, D., Rameshreddy, P., & Kotaiah, B. (1998). Characterization of solid waste of a super specialty hospital: A case study. *Indian Journal of Environmental Health, 40*, 319–326.

Manyele, S. V., & Anicetus, H. (2006). Management of medical waste in Tanzania hospitals. *Tanzania Journal of Health Research, 8*(3), 177–182.

Manyele, S. V., & Mujuni, C. M. (2010). Current status of sharps waste management in the lower-level health facilities in Tanzania. *Tanzania Journal of Health Research, 12*, 257–264.

McMichael, A. J., Woodruff, R. E., & Hales, S. (2006). Climate change and human health: Present and future risks. *The Lancet, 367*, 859–869.

Narang, R. S., Manchanda, A., Singh, S., Verma, N., & Padda, S. (2012). Awareness of biomedical waste management among dental professionals and auxiliary staff in Amritsar, India. *Oral Health & Dental Management, 11*, 162–168.

Oluseyi, T., Adetunde, O., & Amadi, E. (2014). Impact assessment of dumpsites on quality of nearby soil and underground water: A case study of an abandoned and a functional dumpsite in Lagos, Nigeria. *International Journal of Science, Environment and Technology, 3*, 1004–1015.

Oyediran, A. B. O. (1994). Waste generation and disposal in Nigeria: A keynote address presented at a workshop on waste generation and disposal in Nigeria NEST Annual conference 1994. *NEST Ibadan*, 95–100.

Park, K. (2009). Hospital waste management. In *Park's textbook of Preventive & Social Medicine* (pp. 795–800). Banarsidas.

Patil, A. D., & Shekdar, A. V. (2001). Health-care waste management in India. *Journal of Environmental Management, 63*, 211–220.

Patil, G. V., & Pokhrel, K. (2005). Biomedical solid waste management in an Indian hospital: A case study. *Waste Management, 25*, 592–599.

Rao, S. K., & Garg, R. K. (1994). Hospital waste disposal by incineration. *Journal of Academy of Hospital Administration, 6*, 43–47.

Ravikanth, J., Shah, D. D., Vijaya, R., Singh, B. P., & Shevgaonkar, R. K. (2002). Analysis of high-power EDFA operating in saturated regime at λ = 1530 nm and its performance evaluation in DWDM systems. *Microwave and Optical Technology Letters, 32*, 64–70.

Shang, H. (2002). Present situation and disposal of medical wastes in China. *Environmental Pollution Control, 24*, 312–313.

Sharma, R., Sharma, M., Sharma, R., & Sharma, V. (2013). The impact of incinerators on human health and environment. *Reviews on Environmental Health, 28*, 67–72.

Singh, H., Rehman, R., & Bumb, S. S. (2014). Management of biomedical waste: A review. *International Journal of Dental and Medical Research, 1*, 14–20.

Subramani, T., Anitha, P., & Sekar, S. (2014). Health-care waste management system. *International Journal of Engineering Research and Applications, 4*, 255–258.

Subramanian, A. K., Thayalan, D., Edwards, A. I., Almalki, A., & Venugopal, A. (2021). Biomedical waste management in dental practice and its significant environmental impact: A perspective. *Environmental Technology & Innovation, 24*, 101807.

Summers, J. (1991). Asset disposal: Follow company policies or follow the law? *Journal of Healthcare Material Management, 9*, 54–56.

Sushma, M. K., Bhat, S., Shetty, S. R., & Babu, S. G. (2010). Biomedical dental waste management and awareness of waste management policy among private dental practitioners in Mangalore city, India. *Tanzania Dental Journal, 16*, 39–43.

Tambe, V. H., Ali, F. M., & Kishor Patil, P. M. (2013). Awareness towards biomedical waste management: A review. *Critical Reviews in Pharmaceutical Sciences*, 2319–1082.

Thiel, C. L., Eckelman, M., Guido, R., Huddleston, M., Landis, A. E., Sherman, J., & Bilec, M. M. (2015). Environmental impacts of surgical procedures: Life cycle assessment of hysterectomy in the United States. *Environmental Science & Technology, 49*, 1779–1786.

Tudor, T. L., Noonan, C. L., & Jenkin, L. E. T. (2005). Healthcare waste management: A case study from the National Health Service in Cornwall, United Kingdom. *Waste Management, 25*, 606–615.

World Health Organization. (1999). *Guidelines for Safe Disposal of Unwanted Pharmaceuticals in and After Emergencies*, 31–31.

World Health Organization. (2011). *Report on the Burden of Endemic Health Care-Associated Infection Worldwide*

CHAPTER 4

Management of Biomedical Wastes

YOUNIS AHMAD HAJAM[1*] and PREM LATA[2]

[1]*Department of Life Sciences and Allied Health Sciences, Sant Baba Bhag Singh University, Jalandhar, Punjab, India*

[2]*Department of Biosciences, Division Zoology, Career Point University, Hamirpur, Himachal Pradesh, India*

*Corresponding author.

ABSTRACT

Biomedical wastes (BMWs), generated from healthcare facilities, research institutions, and laboratories, pose significant environmental and public health risks due to their hazardous nature. Proper management of these wastes is imperative to minimize potential adverse effects on human health and the environment. This abstract provides a comprehensive overview of BMW management practices. The abstract begins by highlighting the diverse sources of BMWs, encompassing infectious, sharps, pathological, pharmaceutical, and chemical wastes, emphasizing their distinct handling requirements. It delves into the potential consequences of improper disposal, such as the transmission of infectious diseases and the contamination of water bodies and soil. The chapter further explores various methods of BMW management, including segregation, collection, storage, transportation, treatment, and final disposal. Emphasis is placed on the importance of proper waste segregation at the point of generation to optimize the subsequent steps of the waste management process. Additionally, advancements in waste treatment technologies, such as autoclaving, incineration, and microwave irradiation, are discussed, along with their respective advantages and limitations. Furthermore, the abstract addresses the regulatory framework surrounding BMW management and the vital role of healthcare professionals and support

Biomedical Waste Management: Recycling and Remediation Strategies. Younis Ahmad Hajam, Javid Ahmad Malik, Anu Pandita, and Deepu Pandita (Eds.)

staff in adhering to these guidelines. Public awareness and education are emphasized as crucial factors to ensure the success of BMW management programs. In conclusion, effective management of BMWs necessitates a multifaceted approach that involves meticulous waste segregation, innovative treatment technologies, and strict adherence to regulatory standards. By implementing comprehensive waste management strategies, society can mitigate the potential hazards associated with BMWs, safeguarding both human health and the environment.

4.1 INTRODUCTION

The expression "clinical waste" is utilized in numerous nations, like the US, South Korea, what's more, China, while the European Association and World Wellbeing Association allude to it as a "medical care squander" (Yoon et al., 2022). The World Wellbeing Association characterizes medical care squander as any waste or results from emergency clinics and medical services offices for people and creatures utilized for analysis, cure, or inoculation, such as utilized needles, needles, metal sharps, dressings, blood tests, body parts, drug, compound, radioactive materials, and gadgets (Ananth et al., 2010). For the most part, nations with high income produce up to 0.5 kg/ emergency clinic bed of unsafe clinical solitary. The medical services area's waste broadly influences the climate and general well-being, demonstrating exorbitant. Also, the assembling and disposing of clinical and medical care areas squander lead to expanded levels of GHG outflows and contamination (Ordway et al., 2018). As a rule, the majority of the working room's waste can be considered non-unsafe on the grounds that it is produced even before the patient shows up and isn't polluted or tainted (Harding et al., 2021). Facial coverings make up a significant piece of the clinical waste (MW), particularly after the gigantic expansion being used because of Coronavirus and compulsory facial covering-wearing guidelines. As indicated by the discoveries of a new report that included seven emergency clinics and clinical focuses in the province of Massachusetts, USA, alongside three veterinary emergency clinics (Lee et al., 2002), plastic waste represented ~30% of the all-out squanders delivered by emergency clinics. Likewise, gloves contribute to climate contamination when improperly thrown because they are constructed of non-recyclable and non-biodegradable materials (Silva et al., 2020, Talvitie et al., 2017). A review featured that 15% of the all-out worldwide carbon spending plan is credited to the ozone-depleting substance (GHG) discharges coming about because of the existence pattern of plastics (Zheng and Suh, 2019). Consequently, unfortunate administration and removal of plastics compromise the capacity of the worldwide local area to meet fossil fuel byproduct targets and battle environmental change (Cf, O.D.D.S., 2015).

Sharma et al. (2013) explored the negative effects of burning MW caused by the cinders and vaporous emanations. Fly cinders as particulate matter, carbon monoxide, heavy metals, corrosive gases, and natural mixtures are just a few of the many toxins released. Likewise, leachable natural mixture's structure base buildups and cinders contain weighty metals and dioxins. Likewise, there is the carbon impression of transportation, autoclave disinfecting, and warm treatment, in addition to the fossil fuel byproducts delivered during reusing (Rizan et al., 2021). A concentrate in the UK has tracked down an impressive variety of various techniques for MW treatment. It was assessed that the carbon impression of MW treatment by high-temperature burning was 1074 kg CO_2e/t, pursuing the decision of waste treatment strategy affect the carbon impression of up to 50-crease (Rizan et al., 2021). An examination concentrates on that remembered three medical clinics in the Netherlands figured out that medical clinics figured out how to acquire in excess of 39,000 in just a half year from reusing and reusing restored instruments and materials gathered from these medical clinics. This benefit can support the clinics and the medical care area to carry out reusing rehearses and make the waste treatment processes more round than direct (van Straten et al., 2021). The particular goals of this survey are as per the following: (i) to order and arrange the various sorts of clinical squanders produced from medical care offices; (ii) to frame the means and cycles included in the removal, isolation, and ministration of MW; and (iii) to think about the practices for the board and ministration of MW in four nations from various regions of the planet furthermore, characterize the most un-malicious strategies and, in this way, help chiefs in the wellbeing area and industry to pursue better decisions, and, at last, to exhibit the effect of the Coronavirus flare-up on the measures of waste and the outcomes behind it.

The BMW representation, measurement, isolation, capacity, conveyance, and therapy are all adhered to under the support to grave technique. The 3 R's, or reduction, reuse, and reuse, are the cornerstone of great BMW practice. The best BMW the board (BMWM) methods put less emphasis on planning and more emphasis on preventing waste or recovering as much garbage as possible. In order of attractiveness, the many methods for removing a BMW are thus: avoid, lessen, reuse, reuse, recover, treat, and ultimately organize.

Hence, rather than managing waste at the end of the line, it should be handled from the source. Chartier (2014) officially paraphrases Any office where BMW treatment, removal, or procedures associated with such treatment and removal are conducted is referred to as a BMW treatment and removal office. Chartier (2014) official paraphrase Just approximately 10%–25% of BMW is harmful, while 75%–95% of it is nonhazardous. Everyone, including healthcare professionals involved in waste management, treatment, and disposal, is in danger from hazardous waste due to its physical, chemical, or possibly microbiological nature (Li and Jenq, 1993).

The World Health Organization (WHO) issued key recommendations for attaining secure and sustainable administration of healthcare waste during a meeting in Geneva in June 2007. By properly allocating resources and holding everyone accountable, it was intended to lessen the harmful impacts that healthcare waste has on people and the environment. The cost of BMW's legal administration should be borne by all parties involved in sponsoring and supporting healthcare exercises since they are morally and legally required to protect other people's safety. To ensure its safe removal, the company must also offer clinical tools that are favorable to the environment. WHO recommended that the government allocate a percentage of its budget to the creation, funding, and encouragement of efficient healthcare waste. These include original and creative techniques/mechanisms for lowering the quantity and toxicity of medical waste. Nongovernmental Organizations ought to assist the initiatives and projects that encourage this motivation (Li and Jenq, 1993). The WHO's "The Blue Book," or guidance to the safe management of waste from medical procedures, was first released in 1999. The updated processes for safe BMW removal, extra environmental contamination control measures, and location methodology are all included in the second edition of "The Blue Book," which was published in 2014. The second edition also included fresh material on subjects like healthcare waste in emergencies, emerging pandemics, drug-resistant microscopic organisms, and environmental changes (Chartier, 2014).

4.2 CHARACTERIZATION OF MEDICAL WASTE

As indicated by gauges by the WHO, 15 to 20% of clinical squanders can be delegated risky materials because of their infectivity, harmfulness, and, now and again, radioactivity (Tsai, 2021). In any case, clinical waste administration rehearses are not consistent or normalized altogether nations since this categorization isn't extremely clear or conclusive. Clinical waste alludes straightforwardly or in a roundabout way to irresistible, poisonous, or generally risky squander (HMW). Clinical foundations produce this kind of waste during clinical or safeguard care and related exercises, explicitly irresistible, obsessive, harmful, drug, and substance squandering (Yoon et al., 2022). Then again, non-perilous clinical waste (NHMW) incorporates all unique standard non-irresistible parts of waste, like metropolitan strong waste. HMW is as a rule debased with microbes. Subsequently, it can cause a great many diseases and sicknesses on account of abuse or unfortunate dealing with and disposing of. Adding to that, it can cause ecological tainting on account of unfortunate administration, and contamination of land, water, plants, creatures, and air, prompting the spread of sicknesses.

4.3 MEDICAL WASTE THE BOARD INTERACTION

Clinical waste management is a series of processes in which the MW generated is handled from the point of generation until it can be safely eliminated. The waste management process results in reducing the amount of waste sent for disposal and achieving a circular economy in which the materials used within the healthcare framework are used to their full potential, resulting in nearly zero waste.

The clinical waste produced is of different characterizations, as referenced in the past area. The basic part of this step is how much waste is created and the way things are taken care of to forestall perils to the faculty in touch with it. The waste produced by clinical establishments can be limited to lessen squandering gathering. The minimization can be drawn closer from various headings, like a decrease in squander at source, reusing, what's more, stock administration. The decrease in squandering at the source can accomplished by reusing materials that won't hurt the clients, like launderable decorative spreads, flatware, and top-off compartments for cleaning supplies (Schwendimann, 2011). Reusing can support limiting waste by reusing plastics and metals and treating the soil food squander. At long last, stock administration will help set a coordinated arrangement of the medications stock to forestall duplication and acquisition of pointless items that could lapse, consequently decreasing potential waste (Schwendimann, 2011).

4.3.1 SQUANDER ISOLATION

Isolation recognizes the different kinds of waste and how they can be gathered independently. Isolation is predominantly accomplished by isolating various classes of MW in different varieties of receptacles or packs indicated for every class (Schwendimann, 2011). Thus, this causes an issue while gathering from different sources because of the absence of normalization of the colors related to each waste classification (Windfeld and Brooks, 2015), in this manner expanding the required time and monetary expense of work and gear to isolate the waste and direct it to the legitimate squander stream, bringing about an issue. The isolation ought to for the most part happen at the point source with the exception of the waste going through a similar treatment, which could be isolated in the treatment office. The sharp articles ought to continuously be isolated at the source. The isolation is completed by clinical staff, which expects preparing to securely discard waste to stay away from diseases (Windfeld and Brooks, 2015). In the event that a misstep happens while isolating waste, it shouldn't be amended to forestall the tainting of the other waste (Schwendimann, 2011). Clinical waste ought to be put away securely to keep away from unapproved human contact, which can cause diseases (Windfeld and Brooks, 2015).

4.3.2 SQUANDER ASSORTMENT AND TRANSPORTATION

The recurrence of MW assortment ought to be pretty much as high as once each day to stay away from the aggregation of waste, which can spread contaminations. Moreover, the staff mindful of assortment ought to be furnished with well-being stuff to forestall defilements and contaminations that ought to be securely discarded after (Schwendimann, 2011). The waste is gathered from the medical services substance and moved involving auxiliary transportation to the treatment office for removal, reusing, and treatment processes. Treatment offices are either situated inside the well-being care office or off-site in a different area (Windfeld and Brooks, 2015).

4.3.3 SQUANDER TREATMENT

Clinical waste treatment is an interaction done before the removal of MW to limit the dangerous impacts of this sort of waste on the climate and well-being. The absence of appropriate treatment can have a few effects, as follows (Akter, 2000): (i) harming from poisonous components; (ii) bacterial and parasitic diseases; (iii) arrival of poisons into the air; (iv) filtering to the dirt and hidden springs; (v) bioaccumulation; (vi) leaving an impression on the climate; and (vii) annihilation of territories.

In the creation period of any clinical gear, the effect of these kinds of gear should be considered by playing out a day-to-day existence cycle examination (LCA) and rehearsing appropriate treatment strategies. Be that as it may, the strategies and methods for treatment have negligible effect as far as fossil fuel byproducts delivered into the surrounding air. For instance, a single intravitreal infusion causes the arrival of 0.05 kg CO_2e during the removal stage. Notwithstanding these difficulties, these techniques can diminish the risks referenced before. The medicines right now in the field are cremation, autoclave sanitization, microwave sterilization, and mechanical and synthetic sterilization (Akter, 2000).

4.3.4 INCINERATOR AND ITS PERILS

The major method of BMW elimination involved eating the trash. There was a spike in the installation of incinerators after the primary BMW regulation was put into effect in India in the late 1990s. It is predicated on high temperatures, which destroy the material in which microorganisms reside while also killing them (Mattiello et al., 2013). However, other poisons are produced during its operation, including dioxins and PIC (the result of insufficient ignition). During combustion and post-combustion cooling, waste components separate and recombine to create new hazardous particles known as PIC. Instead of being eliminated, metals are spread into the environment, raising serious health issues. Unexpected byproducts of waste ignition caused by incinerator operations include dioxins. These 75 artificial substances are related to the furans, a different class of poisons. These poisons often travel up the food chain and build up in oily tissues. One of the main environmental dioxin generators is the use of medical equipment made of polyvinyl chloride (PVC; Vilavert et al., 2015). Subramanian et al. (2007) The Public Organization for Multidisciplinary Research and Innovation and the Chamber of Logical and Modern Exploration have joined forces to conduct a review to determine whether dioxins are present in Thiruvananthapuram. This is being done in recognition of the importance of dioxins in the climate. Moreover, garbage from incinerators should be assessed for toxicity before being dumped in a landfill because it is dangerous. The majority of nations are choosing natural peaceful methods for banning BMW in light of these factors. Both Denmark and the Philippines have outlawed the expansion of incinerators. Burning produces fly debris and outflows, which include dioxins, furans, and mercury (Bano et al., 2017). According to Bano et al.

(2017), dioxins and furans are suspected of causing cancer, have a half-life of 7 to 11 years, and have a long-lasting effect on the environment. If the garbage is burnt, dioxin fluxes can be decreased (Bano et al., 2017). Specific non-synergist decrease (SNCR) can also be used to address the dioxins produced (Wielgosiski et al., 2020). This invention depends on the creation of free nitrogen through the interaction of nitric oxide with smelling salts, and the gas produced as a result is thought to be highly adequate and inexpensive (Wielgosiski et al., 2020). Mercury filters out 3–9% of the cremation emanations, which have an impact on the nervous system and general well-being. Fly debris is a significant accumulation following burning that is rich in heavy metals (Ababneh et al., 2020). In order to reuse fly debris, it must first undergo synthetic pre-treatment using sodium sulfide or ethylene diamine tetra acid corrosive disodium (EDTA), which removes the heavy metals from the fly debris (Ababneh et al., 2020). Burning MW contributes to dangerous atmospheric depletion by providing significant amounts of ozone-harming chemicals (GHS), primarily CO_2 (Wyssusek et al., 2019). About 3 kg of CO_2 is produced from consuming 1 kg of medical waste.

4.3.5 AUTOCLAVE SANITIZATION

A treatment method known as autoclave sanitization uses steam and temperature to continuously kill bacteria (Windfeld and Brooks, 2015). It uses strain and steam impact to produce sanitization while operating at a lower temperature than burning. To guarantee thorough waste sanitization, the working conditions are 60 minutes at 121°C and 1 bar, followed by 60 minutes at 134°C. Temperature (121–134°C), steam infiltration, waste load, treatment cycle length, and chamber air expulsion are the variables that determine how the autoclave operates. Due to the autoclaving process's inability to completely kill germs due to its low operating temperatures, pre-treatment of the waste is required before it can be disposed of in landfills (Windfeld and Brooks, 2015).

4.3.6 MICROWAVE SANITIZATION

In order to prevent the polymerization and corruption of natural materials and bacteria, microwave sanitization uses strong microwaves at low temperatures (Giakoumakis et al., 2021). The waves cause sub-atomic security vibrations that prevent emanations and save energy, making the technique even more innocuous to the ecology. The sterilizing is carried out at temperatures between 177 and 540°C using electromagnetic floods with frequencies between 1 mm and 1 m and recurrence ranges between 300 and 3000 MHz (Giakoumakis et al., 2021). Microwave sanitization has significant expenses and can be joined with cremation and autoclave. The following viewpoints oversee the activity of this strategy (Akter, 2000): Squander qualities, dampness content, microwave source strength, openness time, and level of waste blending.

4.3.7 COMPOUND STERILIZATION

Compound sterilization is utilized to kill microorganisms and fend off microbes by utilizing synthetic compounds. It is principally utilized for treating fluid-irresistible squanders like blood, pee, excrement, or clinic sewage. The synthetic sanitizers that are generally utilized are a blanch arrangement (1%) or a weakened dynamic chlorine arrangement (0.5%). Likewise, other sanitizers, for example, lime, and peracetic corrosive can be utilized. This treatment strategy straightforwardly influences those accountable for the treatment because of the inward breath of unpredictable synthetic substances or disturbances to the skin and eyes. The accompanying angles administer the adequacy of this strategy (Akter, 2000): pH, contact time, waste and substance blending, Distribution versus stream. The buildups of this treatment are fluid and strong deposits. The fluid buildups are discarded in the sewer framework, and strong buildups are discarded in the landfill. Deciding a satisfactory treatment technique includes characterizing the waste attributes, type, and ideal result, which ought to be checked against every treatment technique to show the benefits and deficiencies of each interaction (Akter, 2000).

4.3.8 SQUANDER REUSING

Squander reusing is utilizing created squander or results for something very similar or unique purposes. The majority of clinical waste is non-hazardous waste that is primarily disposed of in landfills. Reusing used products such as plastics, glass, metals, silver, and so forth, used in visual processing can reduce the use of waste dumps and landfills. Manure can be made from food and natural waste. Fly debris from cremation can be used in significant mixes and as building blocks following treatment (Ababneh et al., 2020). The intensity created from the cremation could be utilized to warm water for a brought-together warming framework.

4.3.9 *GARBAGE REMOVAL*

Rejects of the past advances are moved to a clean landfill for removal. Notwithstanding, landfills are not the ideal answer for dealing with MW due to their ecological impacts. These impacts are soil and water contamination brought about by leachate and gas discharges very high because of waste corruption (Giakoumakis et al., 2021). Hence, the waste being discarded ought to be limited to the most, and accomplishing a round economy ensures that. Long haul deterioration of waste is the essential cycle liable for garbage removal in landfilling (Giakoumakis et al., 2021). Preventive measures ought to be taken to guarantee the protected removal of MW, which are: Quick front of waste, Covering it under the old metropolitan misuse of least internment of 90 days, Waterproof base, least 2 m over the water table, and No removal of synthetic compounds.

4.4 ELECTIVE INNOVATION FOR BIOMEDICAL GARBAGE REMOVAL

The different new advances for BMW removal are classified into four gatherings – warm, compound processes, irradiative cycles, and organic cycles. The vast majority of these are still under research (Stringer et al., 2010). Warm cycles are assembled into three – low, medium, furthermore, high.

1. In autoclaves, steam is utilized as a cleaning agent. According to how air pickets are removed, autoclaves are divided into two categories: gravity/descending removal and pre-vacuum/high vacuum. With autoclaves with a pre-vacuum or different hoover cycles, air departure is more successful, hence these are desirable. A shredder or processor should be used if the volume of waste needs to be decreased, or the waste must be rendered unrecognizable. The evacuated air is sterilized before being released into the environment since it may contain pathogens by passing through a high-efficiency particulate air (HEPA) channel. Sharps and medical equipment are autoclaves cleaned along with all irresistible waste counting societies, human waste, lab waste, sensitive waste, and garbage. Synthetic materials and dangerous trash cannot be autoclaved because they release risky byproducts. Moreover, an autoclave cannot be used to clean heavy detritus like sheets, heat-resistant compartments, or other bulky items (Stringer et al., 2010). Microwaves use the moist intensity

and steam produced by the microwave energy to sanitize. Mercury, radioactive trash, and unpredictable, semi-volatile natural mixtures shouldn't be microwaved. There are fewer outflows when removing BMWs thanks to the microwave. Its drawbacks include a high setup cost in terms of capital, a smell problem around the machine, and the potential for a microwave energy gap.

2. Opposing polymerization and heated depolymerization are examples of medium-intensity advances that work between 177°C and 540°C. In order to break down organic molecules, BMW is using high-energy microwaves in nitrogen air. The trash's internal energy rises as a result of absorbing microwave energy and, anatomic-level compound decay takes place as a result. Burning is prevented by the oxygen-free atmosphere that nitrogen generates. After that, shredders are used to shred the trash (Stringer et al., 2010).
3. In pyrolysis oxidation, the organic waste is broken down into inert debris, glass, and metal shards after being heated to a high temperature (about 594°C) inside the pyrolysis chamber. The second stage involves supplying clean exhaust steam after the fumes are ignited in a chamber at a temperature between 982°C and 1093°C. To create plasma energy, plasma pyrolysis uses plasma lamps. IV. Ionized gases can conduct electricity in plasma, but because of their high resistance, the electricity is completely transformed into hot energy. The accumulation includes metallic buildups, carbon black, and totally vitrified glass. A variety of waste materials, including low-level radioactive waste, inedible waste, sharps, plastics, dialysis waste, hazardous waste, and chemotherapeutic waste, may be destroyed using plasma-based technology. A few benefits of this structure include a low outflow rate, a buildup of idle and sterile waste, or a pleasant environment. Moreover, there is a reduction in both mass (by 80%–90%) and volume (by 95%). High capital expenses, high activity costs, high energy consumption, a brief plasma light lifespan, and the potential for dioxin formation in an inadequately constructed system are some of its downsides.

4.4.1 CHEMICAL-BASED INNOVATION

Numerous substances utilized for BMW removal are right now being worked on. Waste treated by chemical innovation includes societies, human waste, lab trash, and delicate waste (dressing, gauze, and outfits). Chemicals should not be used to treat unstable, semi-volatile natural mixes, mercury, or radiological waste. BMW must be destroyed in chemical-based innovation. The advantages of chemical-based invention are a completely mechanized approach, ease of use, ease of release of fluid flowing into sewage, and no byproducts of burning framed. The obstacles include the generation of toxic byproducts from the widespread usage of chlorine and hypochlorite, synthetic hazards, and usually offensive odors. Chlorine- and non-chlorine-based frameworks can be distinguished in the chemical-based innovation. Sodium hypochlorite or chlorine dioxide are both used in the chlorine-based framework. Maybe the first chemical sanitizers used to cure BMW were sodium hypochlorite. Recent studies have demonstrated that poisons including dioxins, radiation acidic acids, as well as chlorinated sweet-smelling chemicals, are supplied in areas where sodium hypochlorite is used. As chlorine dioxide is unsteady, it is produced and used locally. That is one of a's strong points. When used, it produces less toxins, decomposes into salt, and doesn't react with ammonium or alcohol. Innovations based on nonchlorine treat BMW using gas, liquid, or on the other hand, a dry substance. Many of these frameworks have entered the market, but just a few are discussed here. The Steri Ecocycle 104 utilizes a convenient chamber for collecting waste into which a disinfectant vial made of per acetic acid is added. After 10–12 minutes, the acetic acid cleans the trash and uses a HEPA filter to prevent sprayed germs from escaping. The fluid that is emitted is discharged into the sewer, and the waste is treated like regular wastewater (Stringer et al., 2010). Waste reduction (WR) technology converts microbial and human waste into a neutral watery arrangement by using soluble hydrolysis at a high temperature. It is used for body fluids, degradable packs, and human/tissue waste. Moreover, it can handle chemotherapeutic waste. The soluble base also destroys fixative sin tissues and other dangerous synthetics, such as glutaraldehyde and formaldehyde. According to Stringer et al. (2010), this robotized framework has a steam-jacketed hardened steel compartment with a retainer for biodegradable debris like bone and teeth. Ozone is another innovation from Lynntech that is used to clean BMWs. Ozone destroys organisms and transforms them into atomic oxygen, which are areas of strength. Ionizing radiation harms DNA and delivering free extremists brings about additional harm to proteins furthermore, proteins of irresistible particles. The electron pillar innovation produces ionizing radiation as a light emission electron moves rapidly to strike the objective. The losses to be dealt with are irresistible squandering including human waste, research center waste, delicate squandering (dressing, gauze, and outfits), and sharps. The different benefits of this strategy are that it doesn't create harmful emanation, no fluid profluent, no ionizing radiation after the machine is turned off, completely computerized, and low functional expense. The burden incorporates insurance from radiation openness, concrete safeguard a few feet thick around the framework, expulsion of ozone gas, and there is no abatement in squander volume and need shredders or grinders. Natural strategies for the removal of BMW incorporate an arising framework called "Bio-converter." Using biodegradable plastics is another method for environmentally friendly BMW removal. Many biomedical inserts made of biodegradable polymers are naturally contaminated by microbial extracellular proteins. These biodegradable polymers are used as a substrate by these bacteria when they are starved and when there is no other viable option available. For the purpose of large-scale commercial production of biodegradable plastics, more research should be completed (Bano et al., 2017).

4.5 BIOMEDICAL WASTE (BMW) ADMINISTRATION OUTSIDE INDIA

The review covered written searches, distribution surveys, paper pieces, and various data sources. The status in each nation was evaluated on five principal areas of BMW, specifically, the executives, preparing, strategy and administrative structure, advances executed, and monetary assets. In the field of the executives, preparing, and arrangements with respect to BMWM, all West Pacific nations fared well with the exception of Micronesia, Nauru, and Kiribati. Just Japan and the Republic of Korea use BAT (the most ideal that anyone could hope to find advancements) for BMW coordinated factors and treatment, which were well-maintained and consistently tried. The greater part of the nations had no or very less monetary assets for BMWM. Subsequently, HCWM is still distant from ideal in the vast majority of West Pacific nations, and extra support for the extension of HCWM frameworks in nations is imperative to guarantee that in the following 10 years, safe HCWM frameworks are applied (Datta et al., 2018). In Canada, there is a variety found in the clinical squander – the executives rehearse across various regions. Not all areas have guidelines administering the dealing with and removal of clinical waste. Notwithstanding, Canada's clinic seems to getting away from on-site incinerators toward concentrated commonplace offices for BMW sterilization (Walkinshaw, 2011).

4.6 BIOMEDICAL WASTE (BMW) STATUS IN INDIA

In July 1998, the Indian government, along with the previous Forest Service and Climate Service, recommended the first set of BMW rules (Rules, 2009). The BMW issue was made worse by two factors: first, the reuse of needles without sufficient sterilization, and second, the existence of foragers in India who collect open, unprotected medical waste without gloves, veils, or shoes for reuse (WHO, 2004). Worldwide Clinical, 20 states across India between 2002 and 2004, the study of disease transmission network looked at the current BMW practices, procedures, and systems in basic, intermediate, and tertiary health care facilities (HCFs). They determined that a viable BMWM framework was missing from 82% of primary, 60% of secondary, and 54% of tertiary HCFs in India. As a result of reusing unsterilized syringes, about 240 individuals in Gujarat, India, contracted hepatitis B in 2009 (Seetharam, 2009). These and a host of other studies show that, even though India was among the first countries to start procedures for the safe removal of BMWs, there is an urgent need to take action to reinforce the current framework limit and enhance funding and responsibility for the safe removal of BMWs. The restrictions established by BMW in 1998 underwent revisions in the years 2000, 2003, and 2011. Due to the need for classification and standardization, the draft of the BMW rules for 2011 stayed in its original form and was not revised (Sharma, 1998). The Service of Climate, Forestry, and Environmental Change updated the BMWM regulations in Walk 2016. These new standards have boosted inclusiveness, sped up order and approval procedures, and enhanced measures for isolation, transportation, and removal to lessen natural contamination. It has four plans, five structures, and 18 standards, according to Panday et al. (2016). It has been asserted that the nation's BMW will be successfully managed by the new biological waste management laws. The application of these provisions has been clarified and the verb "taking care of" has been added. Also, stringent rules have been implemented to guarantee that no recyclables are stolen, handled inappropriately, distributed improperly, or spilled by animals when being transported from the HCFs to the conventional BMW treatment facility. Improvements are being made to garbage transit, isolation, and removal. The function of incinerators in escalating natural air contamination has also been investigated by establishing new rules for them and improving their operations (Pandey et al., 2016). Data from the Administration of India website show that the nation produces 484 TPD (tons per day) of total BMW from 1,68,869 HCFs. However, only 447 TPDs are treated, and another 37 TPDs are not. There are presently 198

CBMWTFs in use, and another 28 are being developed. About 21,870 of the 1,31,837 HCFs that use CBMWTFs have on-site treatment facilities (Pandey et al., 2016). According to the BMW Rules (1998), as amended, any HCF or CBWTF administrator who needs to use more inventive and sophisticated innovations than those detailed in Schedule I of the Guidelines must go to the Focal Contamination Control Board (CPCB) to have the standards established before the recommended power can consider granting approval. The PIWS 3000 innovation, the sharp blaster (also known as a needle blaster), the dry intensity sanitization and exemplification of waste sharps, and plasma pyrolysis are among them.

4.7 REMARKABLE HIGHLIGHTS OF BIOMEDICAL WASTE RULES (2016)

1. The requirements now cover a wider range of wellness camps, including immunization, blood donation, and caution camps (Pandey et al., 2016).
2. There have been changes to the requirements for HCF occupants. The individual who directly influences the HCF that creates BMW is an occupier (Walkinshaw, 2011).

   Before evacuation, either on-site or at the CBMWTF, the research facility, biohazardous waste, and blood packs must be pretreated. Developing the sanitization/sterilization process should be done in accordance with the Public Guides Control Association (NACO) or the WHO.

3. Cleaning and sterilization procedures. The prerequisites for HCF occupants have changed. The occupier is the one who directly controls the HCF that makes BMW (Walkinshaw, 2011).

   All significant mishaps, such as those brought on by fire hazards, collisions, handling a BMW, and medical interventions initiated by the advised power, ought to be documented.

   The current incinerator needs to be upgraded or replaced in a significant amount of time following the date of this warning in order to comply with the new standard. The BMW removal registration on the website needs to be updated every day and updated every month.

4. The administrator of a typical biological waste therapy and removal office (CBMWTF) now has extra responsibilities to carry out (Pandey et al., 2016). They ought to help put up HCW where trash is being collected. For the treatment of BMW in a year or less, barcoding and a global positioning system should be built as well. Keep a five-year record of all cremation, hydroclaving, and autoclaving activity data.
5. BMW has advanced its isolation, bundling, transportation, and capacity. Based on different waste codes, waste types, and treatment options, BMW has been categorized into four categories. Moreover, untreated biotechnology trash, animal physical waste, dirty waste, and human physical waste should not be stored for longer than 48 hours. If storage is required for longer than 48 hours, the occupier must take all reasonable precautions to ensure that the waste doesn't adversely affect human health or the environment (no permission is required).
6. If the CBMWTF arrangement is available far away from 75 kilometers, no HCF will set up an on-site BMW treatment and removal office. By receiving prior consent from the suggested power, the occupier will set up necessary BMW treatment facilities such as an autoclave, microwave, or shredder in the event that a CBMWTF is not available. After confirming autoclaving or microwave treatment of plastics and dishware, followed by mutilation or destruction, these recyclables should be sent to authorized recyclers.
7. Just one approval for BMW removal for non-bedded HCFs is given to the occupier. The legitimacy of permission and the legitimacy of assent orders for had contacts with HCFs will be synchronized.
8. Incinerator output standards have been modified to be more environmentally friendly. These have acceptable breaking points of 50 mg/nm^3 for SPM, two seconds for home time in the incinerator's auxiliary office, and 0.1 ng TEQ/nm^3 for dioxin and furans. Every year, the Service of Climate, Woodland, and Environmental Change will monitor how the rules are being followed. Establishing district-level councils with the Regional Authority, Regional Justice, or Extra Area Judge as the chairperson will fulfill each state's requirement to monitor for consistency. Also, this advisory body will deliver its findings to the State Contamination Control Board on a monthly basis.

4.7.1 *ADVANTAGES OF THE NEW BIOMEDICAL WASTE (BMW) PRINCIPLES*

The new specifications are strict and intricate, and they ought to transform the way BMW is run in India. The inclusion criteria have been expanded in accordance with the amended laws to cover a variety of health-care-related camps, including immunization camps, blood donation camps, and careful camps. Separation, packing, transport, and capacity are further factors that define BMW trash. To make isolation simpler, four classes have been introduced as opposed to eight. One of the basic guidelines for removing a BMW is that isolation must be finished when the trash begins to deteriorate. To overcome the confusion brought on by the abundance of classes, this has been condensed to be useful and inexpensive for all HCWs. Now, each colored bag or container (yellow, red, white, and blue) denotes a certain type of trash and a particular disposal technique. Synthetic toxic waste and cytotoxic waste, for instance, must be disposed of in a yellow pack that is sent for burning, plasma pyrolysis, or deep burial. Also, in order to lower the danger of diseases being transmitted to HCWs handling waste during the treatment stage, the HCF is required to pretreat different lab waste and blood packs in compliance with WHO and NACO requirements. After around two years, plastic bags, gloves, and blood packs must be discarded in order to remove dioxin and furan emissions that were discharged into the atmosphere. According to the new guideline, all bags and compartments used for BMW cleaning and removal must be labeled consistently. This action will help with quality assurance additional quality control monitoring and pack identification during inspection. Only an approved recycler will receive the BMW in the red/blue bag or holder, which is meant for reuse. As a result, the industry will continue to have control over numerous federal organizations. More focus than ever before has been placed on recycling trash in an effort to save resources and lessen pollution. The 2016 regulations are more precise in terms of who will donate land to build CBMWTF and how dependent HCFs must be on it. If a CBMWTF is situated within 75 kilometers, neither the state government nor the UT government may install an on-site treatment and removal facility; instead, they must grant land for its construction. This has a few benefits because it requires a substantial initial investment to establish a private BMW treatment facility and hire an autonomous, dedicated, and talented staff force. Also, this expedites the supervision and inspection processes at various waste disposal facilities. Also, maintaining logbooks and records will make documentation easier. The incinerator's outflow standards have been tightened up. This will result in the production of water and carbon dioxide while reducing the release of dioxins and furans, which are released at temperatures more than 600°C. The mechanism for obtaining clearance has been made incredibly simple by the new guidelines, which provide new criteria for HCF approval. Slept-in medical clinics will receive program approval, and HCFs without beds will receive a single approval. The checking area is another advancement in the new concepts. State well-being secretaries and the SPCB will conduct an annual survey of HCFs on behalf of the MoEF (Service of Climate, Timberland, and Environmental Change; State Contamination Control Board). Additionally, the warning board on BMWM is currently required to meet on a regular basis in accordance with the new regulations.

4.7.2 CHALLENGES IN THE EXECUTION OF NEW BIOMEDICAL WASTE (2016) PRINCIPLES

The major obstacle to the public authority emergency clinics and small HCFs implementing the BMW 2016 recommendations will probably be a lack of resources. The expense of replacing blood sacks, gloves, and packs/holders made of chlorinated plastic as well as creating a standardized identification framework will be significant. Furthermore, a time frame of two years is far too short for this. There are already 198 CBMWTFs operating in India, and another 28 are being built (Pandey et al., 2016). To meet the need for treatment and removal of every BMW made in India, there is a tremendous necessity for the rapid development of a great deal more CBMWTF. In addition, incineration debris may release toxic air contaminants.

4.8 CONCLUSION

The implementation of BMWM should be a team effort with unwavering support from the government, outstanding BMW practices followed by healthcare professionals and HCFs, regular auditing of BMW practices, and areas of strength. A clean and safe environment is what we look for in a home. The cornerstones of BMWM are WR and waste isolation at the source. The new BMWM 2016 recommendations reflect an improvement over earlier standards in terms of developing isolation, transportation, and removal techniques to lessen natural contamination and guarantee the safety of the personnel, patients, and general public. Also, it is important to promote the creation of newer, more creative, and environmentally friendly frameworks for the removal of BMW as well as the expansion of the usage of non-PVC medical equipment. Every BMWM member should make a commitment to uphold a cleaner, greener.

KEYWORDS

- **autoclaving**
- **biohazardous waste**
- **biomedical waste**
- **clinical waste**
- **compound sterilization**
- **hydroclaving**
- **waste management**

REFERENCES

Ababneh, A., Al-Rousan, R., Gharaibeh, W., & Abu-Dalo, M. (2020). Recycling of pre-treated medical waste fly ash in mortar mixtures. *Journal of Material Cycles and Waste Management*, 1–14.

Akter, N. (2000). Medical waste management: A review. *Journal of Environmental Health, 62*(5), 26–35.

Ananth, A. P., Prashanthini, V., & Visvanathan, C. (2010). Healthcare waste management in Asia. *Waste Management, 30*(1), 154–161.

Bano, K., Kuddus, M., Zaheer, M. R., Zia, Q., Khan, M. F., Gupta, A., & Aliev, G. (2017). Microbial enzymatic degradation of biodegradable plastics. *Current Pharmaceutical Biotechnology, 18*(5), 429–440.

Chartier, Y. (2014). *Safe Management of Wastes from Health-Care Activities*. World Health Organization.

Datta, P., Mohi, G., & Chander, J. (2018). Biomedical waste management in India: Critical appraisal. *Journal of Laboratory Physicians, 10*(1), 6–14.

Giakoumakis, G., Politi, D., & Sidiras, D. (2021). Medical waste treatment technologies for energy, fuels, and materials production: A review. *Energies, 14*(23), 8065.

Harding, C., Van Loon, J., Moons, I., De Win, G., & Du Bois, E. (2021). Design opportunities to reduce waste in operating rooms. *Sustainability, 13*(4), 2207.

Lee, B. K., Ellenbecker, M. J., & Moure-Eraso, R. (2002). Analyses of the recycling potential of medical plastic wastes. *Waste Management, 22*(5), 461–470.

Li, C. S., & Jenq, F. T. (1993). Physical and chemical composition of hospital waste. *Infection Control & Hospital Epidemiology, 14*(3), 145–150.

Mattiello, A., Chiodini, P., Bianco, E., Forgione, N., Flammia, I., Gallo, C., Pizzuti, R., & Panico, S. (2013). Health effects associated with the disposal of solid waste in landfills and incinerators in populations living in surrounding areas: A systematic review. *International Journal of Public Health, 58*, 725–735.

Nema, S. K., & Ganeshprasad, K. S. (2002). Plasma pyrolysis of medical waste. *Current Science, 82*(3), 271–278.

Ordway, A., Pitonyak, J. S., & Johnson, K. L. (2018). Durable medical equipment reuse and recycling: Uncovering hidden opportunities for reducing medical waste. *Disability and Rehabilitation: Assistive Technology, 15*(9), 1–8.

Pandey, A., Ahuja, S., Madan, M., & Asthana, A. K. (2016). Biomedical waste management in a tertiary care hospital: An overview. *Journal of Clinical and Diagnostic Research: JCDR, 10*(11), DC01–DC03.

Rizan, C., Bhutta, M. F., Reed, M., & Lillywhite, R. (2021). The carbon footprint of waste streams in a UK hospital. *Journal of Cleaner Production, 286*, 125446.

Rules, S. A. (2009). *Ministry of Environment and Forests Notification*. New Delhi, India.

Schwendimann, F. (2011). The legal framework of humanitarian access in armed conflict. *International Review of the Red Cross, 93*(884), 993–1008.

Seetharam, S. (2009). Hepatitis B outbreak in Gujarat: A wake-up call. *Indian Journal of Medical Ethics, 6*(3), 120–121.

Sharma, A. K. (1998). *Biomedical Waste (Management and Handling) Rules*. Suvidha Law House.

Sharma, R., Sharma, M., Sharma, R., & Sharma, V. (2013). The impact of incinerators on human health and the environment. *Reviews on Environmental Health, 28*(1), 67–72.

Silva, A. L. P., Prata, J. C., Walker, T. R., Campos, D., Duarte, A. C., Soares, A. M., Barceló, D., & Rocha-Santos, T. (2020). Rethinking and optimizing plastic waste management under COVID-19 pandemic: Policy solutions based on redesign and reduction of single-use plastics and personal protective equipment. *Science of the Total Environment, 742*, 140565.

Stringer, R., Kiama, J., Emmanuel, J., Chenya, E., Katima, J., & Magoma, F. (2010). Non-incineration medical waste treatment pilot project at Bagamoyo District Hospital, Tanzania. *Public Health Care Without Harm, 37*.

Subramanian, A., Ohtake, M., Kunisue, T., & Tanabe, S. (2007). High levels of organochlorines in mothers' milk from Chennai (Madras) city, India. *Chemosphere, 68*(5), 928–939.

Talvitie, J., Mikola, A., Koistinen, A., & Setälä, O. (2017). Solutions to microplastic pollution–Removal of microplastics from wastewater effluent with advanced wastewater treatment technologies. *Water Research, 123*, 401–407.

Tsai, W. T. (2021). Analysis of medical waste management and impact analysis of COVID-19 on its generation in Taiwan. *Waste Management & Research, 39*(1_suppl), 27–33.

United Nations. (2015). *Transforming Our World: The 2030 Agenda for Sustainable Development*. New York, NY: United Nations.

van Straten, B., Dankelman, J., Van der Eijk, A., & Horeman, T. (2021). A circular healthcare economy: A feasibility study to reduce surgical stainless-steel waste. *Sustainable Production and Consumption, 27*, 169–175.

Vilavert, L., Nadal, M., Schuhmacher, M., & Domingo, J. L. (2015). Two decades of environmental surveillance in the vicinity of a waste incinerator: Human health risks associated with metals and PCDD/Fs. *Archives of Environmental Contamination and Toxicology, 69*, 241–253.

Walkinshaw, E. (2011). Medical waste-management practices vary across Canada. *Canadian Medical Association Journal, 183*(10), E677–E678.

Wielgosiński, G., Czerwińska, J., Szymańska, O., & Bujak, J. (2020). Simultaneous NOx and dioxin removal in the SNCR process. *Sustainability, 12*(14), 5766.

Windfeld, E. S., & Brooks, M. S. L. (2015). Medical waste management: A review. *Journal of Environmental Management, 163*, 98–108.

World Health Organization. (2004). *Guidelines for Drinking-Water Quality* (Vol. 1). World Health Organization.

Wyssusek, K. H., Keys, M. T., & van Zundert, A. A. (2019). Operating room greening initiatives: The old, the new, and the way forward: A narrative review. *Waste Management & Research, 37*(1), 3–19.

Yoon, C. W., Kim, M. J., Park, Y. S., Jeon, T. W., & Lee, M. Y. (2022). A review of medical waste management systems in the Republic of Korea for hospital and medical waste generated from the COVID-19 pandemic. *Sustainability, 14*(6), 3678.

Zheng, J., & Suh, S. (2019). Strategies to reduce the global carbon footprint of plastics. *Nature Climate Change, 9*(5), 374–378.

CHAPTER 5

Risks Associated with Hazardous Biomedical Wastes to Employees

KHURSHEED AHMAD WANI,[1*] KOUNSAR JAN,[2] JAVID MANZOOR,[3] and AB QAYOOM MIR[4]

[1]*Department of Environmental Science, Government Degree College Thindim Kreeri, Jammu and Kashmir, India*

[2]*Government Degree College Thindim Kreeri, Jammu and Kashmir, India*

[3]*Department of Environmental Science, JJT University, Rajasthan, India*

[4]*Department of Environmental Science, Government Degree College Hajin, Jammu and Kashmir, India*

[*]Corresponding author.

ABSTRACT

Hazardous biomedical wastes (BMWs) can pose a significant risk to employees who handle them without taking proper precautions. Some potential risks include exposure to infectious agents, chemical exposure, physical injuries, and psychological impact. To protect employees from the risks associated with hazardous BMW, it is essential to implement proper waste management procedures, such as the use of personal protective equipment (PPE), proper labeling, and safe disposal. Training employees on these procedures and providing them with the necessary PPE is also important to minimize the risk of exposure and ensure their safety. This chapter provides a general account of the risks associated with hazardous BMW to employees who are directly and indirectly related to it.

Biomedical Waste Management: Recycling and Remediation Strategies. Younis Ahmad Hajam, Javid Ahmad Malik, Anu Pandita, and Deepu Pandita (Eds.)

5.1 INTRODUCTION

Biomedical waste (BMW), also known as medical waste, refers to any materials that are generated during the diagnosis, treatment, or immunization of human beings or animals. These materials can include sharps (needles and syringes), laboratory waste, pathological waste, and pharmaceutical waste. BMW can pose significant risks to human health and the environment if it is not properly managed. One of the main risks of BMW is the potential for the spread of infectious diseases. Sharps, for example, can carry blood-borne pathogens such as HIV and hepatitis B and C. If these sharps are not properly disposed of, they can be accidentally stuck by a person and cause infection. Similarly, pathological waste, such as tissues and organs, can also carry infectious agents. Another risk of BMW is the potential for environmental contamination.

These risks include the spread of infectious diseases, exposure to hazardous chemicals, and injury from sharps (such as needles). Improper disposal of BMW can also lead to environmental pollution and harm to wildlife. It is important for hospitals and other healthcare facilities to properly handle and dispose of BMW to minimize these risks. In addition, individuals should take precautions when handling BMW, such as wearing gloves and protective clothing, and properly disposing of sharps in designated containers.

BMW can contain hazardous chemicals and materials, such as heavy metals and radioactive substances, which can harm the environment if they are not properly contained and disposed of. For example, if BMW is not properly incinerated, the ash and other by-products can contain harmful pollutants that can contaminate the air, water, and soil. Additionally, BMW can also pose a risk to wildlife and other animals. Sharps and other hazardous materials can cause injury or death if they are ingested or come into contact with animals.

To minimize the risks of BMW, it is essential that proper management and disposal practices are followed. This includes segregation of different types of BMW, such as sharps and pathological waste, and proper labeling and packaging of the waste to ensure that it is handled and disposed of correctly. Additionally, it is important that healthcare facilities, laboratories, and other institutions that generate BMW have a plan in place for the safe and responsible disposal of this waste.

Knowledge, attitude, practice, and safe disposal of BMW are important factors in ensuring that healthcare facilities and practitioners handle and dispose of BMW properly and safely. Assessing these areas can help identify areas for improvement and ensure that proper protocols are being followed to protect the health and safety of healthcare workers, patients, and the wider community. The assessment can include evaluating the facility's policies and procedures, conducting employee training, and inspecting the facility for compliance with regulations. Safe disposal of BMW is crucial as it helps to reduce the risk of infection and contamination and protect the environment.

BMW can have a significant impact on human health if not properly handled and disposed of. Improper disposal of BMW can lead to the spread of infectious diseases, as well as pollution of air and water sources. Additionally, exposure to BMW can lead to injuries from sharp objects, such as needles and scalpels. It is important for hospitals and other facilities that generate BMW to follow proper protocols for handling and disposing of the waste to minimize the potential health risks to both workers and the general public.

There have been several studies conducted on the infectious risks associated with BMW. One study found that certain types of medical waste, such as sharps and infectious waste, can pose a high risk for the spread of bloodborne pathogens, such as HIV and hepatitis B and C. Another study found that improper disposal of BMW in low- and middle-income countries can lead to the spread of infectious diseases, particularly among children and healthcare workers.

Another study published in the Journal of Infection and Public Health found that the risk of infection from BMW is higher in developing countries due to a lack of proper waste management infrastructure and training of healthcare workers. It is crucial to follow proper procedures for the handling, transportation, and disposal of BMW to reduce the risk of infection. It is also important to note that the COVID-19 pandemic has put a spotlight on the importance of proper BMW management, as infectious waste such as used personal protective equipment (PPE), and swabs needs to be handled properly to prevent the spread of the virus.

There are several knowledge gaps in the understanding of BMW health hazards. One gap is in the understanding of the long-term health effects of exposure to BMW. While some studies have shown that exposure to certain types of BMW, such as sharps, can lead to immediate injury, there is limited research on the long-term health effects of exposure, such as cancer and chronic illness.

Another gap is in the understanding of the specific health hazards associated with different types of BMW. While some research has been conducted on the infectious risks associated with certain types of BMW, such as sharps and infectious waste, there is a need for more research to understand the specific health hazards associated with other types of BMW, such as pathological waste and chemical waste. Another gap is in the understanding of how BMW is handled and disposed of in different settings, such as in low- and middle-income countries, and how this affects the health risks to workers and the community. Finally, more research is needed to understand the impacts of the COVID-19 pandemic on BMW management and the risks of infection from BMW during the pandemic. Overall, these knowledge gaps highlight the need for ongoing research to better understand the health hazards associated with BMW and to develop effective strategies for reducing these risks.

In summary, BMW can pose significant risks to human health and the environment if it is not properly managed. Proper segregation, labeling, packaging, and disposal of BMW is essential in order to minimize these risks and protect public health.

5.1.1 TYPES OF BIOMEDICAL WASTE (BMW)

Biomedical waste (BMW) refers to any waste generated during the diagnosis, treatment, or immunization of human beings or animals (Figure 5.1). There are several types of BMW which are explained in subsections.

5.1.1.1 INFECTIOUS WASTE

This includes waste that is contaminated with blood, bodily fluids, or other potentially infectious materials. Examples include used needles, sharps, and cultures.

5.1.1.2 PATHOLOGICAL WASTE

This includes waste generated from human or animal tissue, organs, and body parts. Examples include amputated limbs and surgical specimens.

5.1.1.3 CHEMICAL WASTE

This includes waste generated from the use of chemicals in the healthcare setting. Examples include discarded medication, expired chemicals, and used batteries.

5.1.1.4 RADIOACTIVE WASTE

This includes waste generated from the use of radioactive materials in the healthcare setting. Examples include discarded radioactive sources and contaminated protective clothing.

5.1.1.5 PHARMACEUTICAL WASTE

This includes expired or unused medications, as well as empty medication containers.

5.1.1.6 SHARPS WASTE

This includes waste generated from the use of needles, syringes, and other sharp instruments.

5.1.1.7 GENERAL WASTE

This includes waste generated from the healthcare setting that is not infectious or hazardous. Examples include office paper, soiled linens, and plastic bottles.

It is important to note that BMW should be handled and disposed of separately from other types of waste in order to minimize the risk of harm to human health and the environment.

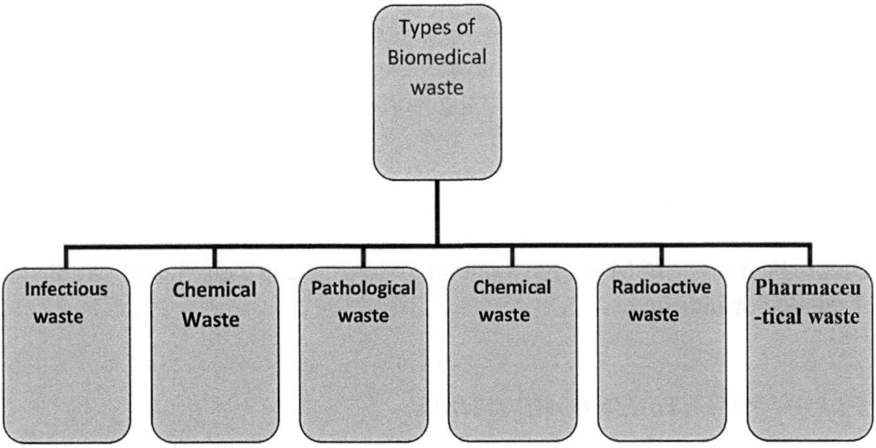

FIGURE 5.1 Types of BMW.

5.1.2 CHARACTERISTIC OF BIOMEDICAL WASTE (BMW)

Biomedical waste (BMW) has several characteristics that distinguish it from other types of waste. Some of these characteristics are explained in subsections.

5.1.2.1 POTENTIALLY INFECTIOUS

Biomedical waste (BMW) may contain infectious agents such as bacteria, viruses, and bloodborne pathogens that can cause disease in humans and animals.

5.1.2.2 POTENTIALLY HAZARDOUS

Biomedical waste (BMW) may contain chemicals, radioactive materials, and other hazardous substances that can cause injury or harm to human health and the environment.

5.1.2.3 POTENTIALLY SHARP

Biomedical waste (BMW) may contain sharp objects such as needles, scalpels, and glass that can cause injury if not handled properly.

5.1.2.4 POTENTIALLY ODOROUS

Biomedical waste (BMW) may have unpleasant odors that can be offensive and may cause health problems to the workers who handle it.

5.1.2.5 POTENTIALLY TOXIC

Biomedical waste (BMW) may contain toxins that can cause health problems if inhaled or ingested.

5.1.2.6 POTENTIALLY COMBUSTIBLE

Biomedical waste (BMW) may contain materials that can catch fire or explode if not handled properly.

5.1.2.7 POTENTIALLY VOLUMINOUS

Biomedical waste (BMW) can be voluminous and may require special handling and disposal methods.

5.2.1.8 POTENTIALLY HARMFUL TO THE ENVIRONMENT

Biomedical waste (BMW) can be harmful to the environment if not properly handled and disposed of. It is important to identify and classify BMW properly to handle and dispose of it safely and effectively, protecting human health and the environment from the potential hazards it poses.

5.2 OCCUPATIONAL HEALTH RISKS

Petersen's Accident/Incident and the human factors theories provide background for Reinach and Viale's (2006) discussion of the hazards faced by biological waste handlers on the job. The main components outlined by the theories are expanded upon by Muniafu and Otiato (2010), who provide a detailed explanation of the reasons for the ongoing issues of workplace dangers that are characteristic of County garbage management. Wickens et al. (2013) argue that the human factors theory accurately explains the origins of problems in BMW handling on the basis of overload, inappropriate responses, and inappropriate execution of waste collection, transportation, and disposal activities, despite some significant differences in perspectives on the application of the theories. Overload elements in human factors theory include things like workplace tension and imprecise directions or a lack of familiarity with risk factors that contribute to incorrect trash categorization. Examples of inappropriate reactions include waste producers and biological waste managers who discover refuse but fail to separate it properly. Careless use of color codes to designate refuse-gathering receptacles is just one example of a failure to adhere to health and safety regulations and the removal of safety signs. Inadequate biological waste handling skills and bad assessment of waste handling hazards were hypothesized to be at the root of the problem the research sought to illuminate. Overload in the context of worries at the workplace, use of alcohol and other drug abuse substances, lack of motivation, fatigue, and working under pressure to deliver are all sources of occupational safety and health hazards among BMW handlers, as reported by Reinach and Viale (2006), who cite Petersen's Accident/Incident theory. Ergonomic pitfalls also include conflicting standards, workplace problems, force, reach, sensation, and a lack of employees (Othigo, 2014). Petersen's Accident/Incident theory illustrates how mistakes in judgment can arise from insufficient refuse categorization, illogical choices, and an implicit wish to blunder (Reinach and Viale, 2006). Poor policies, laws, and rule applications; accountability; inspection; training; adjustments;

inspection; and a dearth of standards all contribute to system breakdowns. Errors made by humans are a major contributor to job mishaps, accidents, and damages. The purpose of incorporating both theories into the research is to outline and explain the conditions under which biological waste workers are more likely to be exposed to workplace health risks (Othigo, 2014). Perceptions and sensitivity of garbage workers and external variables are part of the propensity traits. Peer pressure, supervisory objectives, mindset, and the capacity to evaluate personal risk in relation to garbage risks are all examples of contextual variables.

BMW, in its operational sense, refers to garbage produced at various points in the healthcare process, including diagnostics, therapy, vaccination, manufacturing, biological testing, and similar activities (Baverja et al., 2000). Even though HCWs can be either contagious or noninfectious, not all HCWs pose a health risk to patients. The Central Pollution Control Board places the percentages at 85% for non-infected garbage, 10% for contagious waste, and 5% for dangerous waste (CPCB, 2000). The goals of healthcare are twofold: to alleviate existing health issues and to forestall the onset of new ones. Waste that could be hazardous to workers, the general public, and the ecosystem is often created as a result. Unless otherwise determined by a medical professional, trash left behind by medical facilities or by healthcare employees during their jobs should be treated as potentially contagious. There has been a significant increase in the production of biological waste and pollution of air, water, and land resources as a result of fast development, ever-increasing population growth, and the resulting increase in the need for healthcare services. Workers involved in the gathering, transfer, treatment, and removal of BMWs are at risk for a wide range of health problems and safety risks due to the nature of the wastes themselves and the methods used to dispose of them. If not handled correctly, the garbage produced by hospitals and other healthcare facilities poses serious threats to human health and the natural ecosystem. In particular, some biological pollutants have been shown to be highly contagious and poisonous (WHO, 2004). Unregulated garbage disposal has led to an increase in health risks, as well as the pollution of surface and subterranean water, which poses a significant environmental danger to human health.

The majority of hospitals and other medical facilities lack an effective method of dealing with biological debris. Most healthcare facilities have a problem with the commingling of general refuse and contagious waste. Since cause separation of garbage is crucial in hospital waste management, this issue requires priority consideration. The people who generate biological debris have a responsibility to guarantee that strict substance separation is carried out. Proper identification, storage, transit, and treatment of resources will be made possible. One crucial aspect of effective waste management is the source-separation of various types of biological refuse. Infectious waste (especially sharps), toxins, and other specific biological pollutants pose serious dangers to the health of healthcare employees, patients, refuse processors, garbage collectors, and the general public. Open dumping and unregulated burning are two examples of inefficient ways to dispose of BMW, both of which can lead to the spread of disease and the release of harmful gases due to their inefficient combustion. This is why it's important for strategies to handle biological debris to include measures to ensure workers' health and safety. It's crucial that all medical institutions, including hospitals, have access to a shredder or autoclave/shredding device that is capable of safely and effectively disposing of medical debris.

The proper categorization of biological waste provides a firm basis for assuring the effective safety of waste workers in full conformity with waste management laws, which is essential for meeting duty of care requirements. According to Muniafu and Otiato (2010), mishaps and other consequences for refuse workers remain an issue despite efforts to improve the situation through better classification of trash. Since biological waste is so difficult to categorize and manage, it is a major contributor to mishaps in the waste management industry, as discussed at length by Henry et al. (2006). Some of the most important factors that put waste handlers at risk include inadequate management, indiscriminate waste disposal at the source, poorly established waste handling policies, weak waste segregation practices, insufficient enforcement and awareness of effective waste management policies, and a lack of interest and commitment from the senior staff.

Henry et al. (2006) provide empirical proof that even mixing a tiny quantity of hazardous waste with nonhazardous waste results in waste that is 100% hazardous and presents significant risk to the worker. Problems arise when companies don't fully commit to and adhere to garbage classification and management best practices, which, as evidenced by both policy and practice, put workers at a higher risk of acquiring illnesses from improperly sorted trash. Waste that needs UN-approved packing is just one example; inadequate cleaning methods to prevent or minimize contact with germs and infectious fluids also contribute to the difficulty of refuse division. Similar proof was found in research on waste makeup conducted by Muniafu and Otiato (2010), who found that only 15–25% of medical refuse is classified as dangerous. When ordinary trash is mixed in with toxic materials, the result is refuse that is 100% dangerous (Ferri et al., 2015). Using quantifiable data, Muniafu and Otiato (2010) concluded that improper waste classification presents serious health dangers to waste workers as well as to other medical personnel, patients, the general public, and the environment. Several factors contribute to hospitals improperly discarding their waste: improper waste classification, noncompliance with local and national laws, and noncompliance with WHO recommendations on waste management. According to Muniafu and Otiato (2010), this is because of a lack of knowledge, a lack of law enforcement, a lack of process control, and a lack of continuous tracking of waste management techniques. The research by Muniafu and Otiato (2010) on the difficulties of classifying and managing BMW contains important voids and analytical defects, but the dialog correctly depicts issues that can be addressed to safeguard those who deal with BMW.

Policy, management dedication, effective planning, dependable execution and operations, performance metrics, and an audit review program are all components of the Workplace Safety and Health Management structure advocated for by Ferri et al. (2015). Management of BMW, inadequate attention to waste segregation from the direct sources of waste, insufficient knowledge of waste segregation methods, and a lack of efficient training on waste classification techniques for nurses and other sources of BMW are all identified as additional key challenges by Muniafu and Otiato (2010). Waste disposal and division locations often lack distinct signs and banners on the dumping of particular kinds of refuse in English and other comprehensible languages, as noted by Allison and Von Blottnitz (2010); and Ferri et al. (2015). This makes it easier to dump trash without regard for its contents. According to Guerrero et al. (2013), hospitals are a major contributor to waste categorization and management issues because they lack the resources to properly sort biological waste. The proper categorization of waste, including waste sharps, human anatomy waste, animal waste, microbiology, and biotechnology waste, chemical waste, combustion waste, various kinds of solid refuse, and abandoned medications and lethal drugs, has been found to have serious defects. Careless disposal of clinical waste within the non-infectious waste, insufficient training provided to the safety of waste managers, and the lack of a proper tracking and controls system all put employees at risk of exposure to waste dangers, as noted by Egondi et al. (2015). From a managerial standpoint, Allison and Von Blottnitz (2010) observed that waste segmentation not only lowers hazards for workers but also lowers the expense of removal, as some of the non-hazardous materials that are separated can be reused or repurposed. There is a legal obligation for businesses to deal with the management of waste, and yet 90% of Nairobi County's biological waste managers have badly handled audit reports with records of work-related mishaps, as reported by Ahmed et al. (2015). Among a sample of 100 businesses that produce biological debris, Ahmed et al. (2015) found that 73% failed to maintain records to avoid liability for workers exposed to the risk of mishaps and other dangers on the job. According to the findings, syringes and cutting cuts are the most common ways that pathogens like hepatitis B and HIV are spread, followed by contact with liquid debris, polymers, and crematorium ash.

A lack of information on how to implement preventative measures for human health and pollution with unprocessed bodily residues are two of the new accident-related problems and job-related mishaps noticed by Parizeau (2015). The Tabuk Regional Healthcare System in Saudi Arabia served as a case study for Ahmed et al.'s (2015) research, Hospital Employees Exposure Hazards, and Knowledge on Inadequate Medical Garbage Management. Sharp and syringe injury injuries were found to be the most significant exposure risk factors for healthcare workers (Patan and Mathur, 2015).

Poor waste management and control increase the risk and probability of nosocomial illnesses due to the consumption of repurposed medications, and the absorption of air and particulate particles from biological refuse. Siddharudha and Sowmyashree's (2015) research Work Exposure to Illness on Healthcare Waste Handlers of a Tertiary Care Hospital in South India found that 41.8% of the healthcare waste handlers had been exposed to healthcare refuse and needle puncture injuries. Pharmaceutical wastes, heavy metal wastes, compressed receptacles, and chemical wastes including laboratory chemicals, outdated disinfectants, and organic chemical wastes were named by Siddharudha and Sowmyashree (2015) as the primary causes of mishaps involving biological waste workers. According to research by Ahmed et al. (2015), poorly maintained incinerators are one of the biggest issues in healthcare institutions. Maintainable incinerators are essential for eliminating or greatly reducing the exposure of workers to harmful substances like heavy metals and dioxin. It was suggested by Siddharudha and Sowmyashree (2015) that when dealing with biological waste or other waste-related crises, strict adherence to rules for managing specialized equipment—including emergency measures in case of accidents—was necessary (Garg and Sarkar, 2013). Researchers urged healthcare providers and other institutions that produce biological debris to implement training and education initiatives. They also came to the conclusion that mishaps at work were greatly aided by the hazardous gathering and dumping of various kinds of biological waste.

5.3 HANDLERS OF BIOLOGICAL WASTE: THEIR UNDERSTANDING OF AND COMMITMENT TO WORKPLACE SAFETY AND HEALTH

According to Othigo (2014), the understanding and mindset of waste workers decide the breadth and degree of conformance in the BMW gathering, transit, and dumping laws and standards, making them a distinguishing element of successful mishap management. Ministry of Health, National Guidelines on Safe Disposal of Pharmaceutical Waste in different countries, which incorporates hazard protection elements, is consistent with such mandates. These guidelines require the use of protective gear such as face masks, mandatory overalls, leg protection, disposable gloves, mandatory industrial aprons, and eye protectors. After analyzing the most common factors that lead to accidents in the workplace, Siddharudha and Sowmyashree (2015) recommended designing methods to address issues like inadequate warning signals, lack of immunization, ineffective management, improper waste packaging, and inadequate waste transportation. In order to reduce the number of mishaps, penalties should be imposed on those who move potentially hazardous machinery. Companies that do not provide sufficient caution signs to users of machines or those operating in hazardous settings with dangerous machinery will be punished under the rules in an effort to reduce the number of mishaps that occur. Othigo (2014) found that one of the main reasons for workplace mishaps was the lack of initiatives to reduce tension in the workplace. No danger communication strategies were found in the research to meet the issues that biological waste workers encounter on the job.

According to Othigo (2014), proper guidance on how to manage biological debris, as well as information on group protective measures and knowledge of general danger avoidance policies, could be disseminated to employees through effective communication. The impact of a system strategy and a guilt-free setting on corporate success was investigated by Pietra et al. (2005). It was found that improving corporate performance was much more feasible through a systemic strategy and a blame-free setting. Successful methods to achieve the aim of patient safety within healthcare companies include the use of technology, the availability of information, communication, patient cooperation, and multi-professional coordination. Parizeau (2015) recommends first aid measures, such as rinsing eyes and treating wounds, as part of broader strategies. These include fostering an organizational culture and structure that supports waste handlers, retaining items involved in accidents for documentation, and implementing medical surveillance. Safe use of cytotoxic drugs among pharmacists in healthcare facilities, use of safe product handling procedures, implementation of emergency response procedures, clarification using product data sheets that show detailed hazards handling procedures, and education of cytotoxic drug handlers are all recommendations from a study concluded by Singh and Gupta (2009).

5.4 USE OF PERSONAL PROTECTIVE EQUIPMENT

Wilkins (2016) suggests that using safety equipment to prevent direct pollution due to equipment failure is necessary to reduce the frequency of mishaps in the workplace. According to Singh and Gupta (2009), laws are a resource that can help safeguard employees by laying out standards and principles. This is in line with Petersen's Accident/Incident theory, which maintains that workers should be shielded from potentially harmful materials so that they can enjoy a long and healthy career. The research recommends that the Kenya National Bio-safety Authority create a strategy for the use of PPE in BMW sites and make it mandatory for all employees to comply with it. A biomedical or healthcare center has an obligation to inform its staff of the necessity of wearing PPE and supply them with the equipment. Wilkins (2016) investigated the role of PPE in the humanitarian administration of Ebola: between individual patient care and global biosecurity in Ghana and found that such equipment plays a crucial role in balancing between the two, thereby sparing lives. Based on the results, it was suggested that the government conduct yearly health and safety inspections of the BMW plant to ensure that all employees are meeting health and safety requirements. The risk of exposure to harmful substances, such as insecticides, can be mitigated through the use of personal protective equipment, as argued by Singh and Gupta (2009) in a comprehensive evaluation of workplace dangers and security among biological waste workers.

5.5 CONCLUSION

The following is what the authors of the present research have determined. Awareness of waste classification and management challenges, ineffective segregation of waste at source, inappropriate waste management practices, use of indiscriminate waste disposal, ineffective enforcement of waste handling policies, lack of interest by the management, and lack of commitment were identified as potential challenges with significant impact on the occupational health and safety of biomedical waste (BMW) handlers. The healthcare facilities in this research report a high rate of workplace accidents. It was pointed out that it would be possible to drastically cut down on, if not eradicate, these mishaps. Physical injuries such as pricks, cuts, and scrapes were among the most common, followed by those caused by moving large items, working in high temps, pollution, an inefficient accident management program, breathing toxic fumes, falling, and ripping muscles or tendons. There were no mechanisms in place for reporting incidents, and in some instances, there was no record-keeping of workplace mishaps. Third, all buildings were found to be listed with the Employment Safety section as Workplace, meeting all legal and regulation standards.

5.6 RECOMMENDATIONS

The following are some recommendations made by this research. The study's main results on the nature of mishaps, the mindset of biological waste workers, and the extent to which they adhere to applicable laws and guidelines for the disposal of such waste are important criteria to set recommendations. The waste management authority must take preventative steps and establish proper mechanisms to educate workers on BMW management handling and management best practices to mitigate the negative effects of accidents, improve workers' perceptions of the importance of safe waste handling practices, and ensure that the laws governing such practices are enforced.

Waste management firms should give this issue their full attention and implement a stringent rule of control for their staff to help prevent it. Protecting workers' health from hazardous waste is a moral and ethical obligation of the respective governments.

Health and safety guidelines for the use of PPE in biological waste sites should be formulated, implemented, and enforced. A biomedical or healthcare center has an obligation to inform its staff of the necessity of wearing PPE and supply them with the equipment.

Every year, the government must conduct a thorough health and safety inspection of the BMW facility to ensure that the health and safety regulations are being adhered to by the facility's employees. If a BMW center is found to be violating health and safety regulations, the concerned Authority may revoke its license. According to the policy structure established by the

World Health Organization (WHO), all workers in any particular company have the right to a clean and secure working environment.

The National Bio-safety Commission is responsible for monitoring the industry to guarantee compliance with the HWM Regulations amended from time to time in various countries. It would be in line with the National Bio-safety Authority's proposed policy of fining businesses for not providing employees with protective gear. It is also responsible for overseeing the training programs provided by bio-medical solid refuse sites to ensure that all personal sanitation safety requirements are met.

Among other statutory laws and requirements, facilities that handle biomedical materials must provide a workplace free of seriously recognized hazards, as outlined in the occupational safety and health policy guidelines for the health sector.

KEYWORDS

- **biological waste handlers**
- **biomedical waste**
- **disposal methods**
- **occupational health risks**
- **personal protective equipment**
- **toxins**

REFERENCES

Ahmed, N. B., Zeyad, T., Thakir, A., & Mohammed, E. (2015). Analysis of hospital staff exposure risks and awareness about poor medical waste management: A case study of the Tabuk Regional Healthcare System, Saudi Arabia. *ADR Journals, 47*(2), 1–10.

Allison, K., & Von Blottnitz, H. (2010). *Solid Waste Management in Nairobi: A Situation Analysis*. Report for City Council of Nairobi, UNEP.

Andrews, R. N., Lord, W. B., O'Toole, L. J., & Requena, L. F. (1993). *Guidelines for Improving Wastewater and Solid Waste Management* (WASH Technical Report No. 88). Office of Health, Bureau for Research and Development, U.S. Agency for International Development.

Baveja, G., Muralidhar, S., & Aggarwal, P. (2000). Hospital waste management: An overview. *Hospital Today, 5*(9), 485–486.

Boss, J. C., Poyyamoli, G., & Gautam, R. (2014). Biomedical waste management practices in Puducherry region, India: A case study of three selected health care facilities. *International Journal of Advanced Research, 2*(8), 780–799.

Central Pollution Control Board, India (CPCB). (2000). *Manual on Hospital Waste Management*. CPCB.

Dohare, R., Prakash, S., Sahu, R. K., & Srivastava, A. (2013). A study of hospital waste management status in health facilities of an urban area. *International Journal of Pharma and Bio Sciences, 4*(1), 1107–1112.

Egondi, T., Muindi, K., Kyobutungi, C., Gatari, M., & Rocklöv, J. (2016). Measuring exposure levels of inhalable airborne particles (PM2.5) in two socially deprived areas of Nairobi, Kenya. *Environmental Research, 148*, 500–506.

Environmental Protection Agency (EPA). (2013). *Hazard Standards for Lead in Paint, Dust, and Soil (TSCA Section 403)*. Available online at: www.epa.gov/lead/hazard-standards-lead-paint-dust-and-soil-tsca-section-403 (accessed on 18 September 2024).

Ferri, G. L., Chaves, G. L. D., & Ribeiro, G. M. (2015). Analysis and location of centers for storage and sorting of municipal solid waste for the logistics network reversal: A case study in the municipality of São Mateus, ES. *Production (São Paulo), 25*(1), 27–42.

Fisher, A. A., Laing, J. E., & Strocker, J. E. (1998). *Handbook for Family Planning Operation Research Design in Sampling*. Population Council.

Garg, D. S., & Sarkar, B. K. (2013). A study of waste management status in health facilities of an urban area. *International Journal of Pharma and Bio Sciences, 4*(1), 1107–1112.

Guerrero, L. A., Maas, G., & Hogland, W. (2013). Solid waste management challenges for cities in developing countries. *Waste Management, 33*(1), 220–232.

Harhay, M. O., Halpern, S. D., Harhay, J. S., & Olliaro, P. L. (2009). Hazardous waste management: A neglected and growing public health problem worldwide. *Tropical Medicine & International Health, 14*(11), 1414–1417.

Hassan, M. A., & Misra, R. P. (2011). Knowledge, attitude, and practices about biomedical waste management among healthcare personnel: A cross-sectional study. *Indian Journal of Community Medicine, 36*(2), 143–145.

Heldal, K. K., Halstensen, A. S., Thorn, J., Eduard, W., & Halstensen, T. S. (2003). Airway inflammation in waste handlers exposed to bioaerosols assessed by induced sputum. *European Respiratory Journal, 21*(4), 641–645.

Henry, R. K., Yongsheng, Z., & Jun, D. (2006). Municipal solid waste management challenges in developing countries: Kenyan case study. *Waste Management, 26*(1), 92–100.

Hildebrandt, V. H., Bongers, P. M., Kemper, H. C., & Dul, J. (2002). The influence of climatic factors on non-specific back and neck-shoulder disease. *Ergonomics, 45*(7), 456–475.

Kanki, M. T., Tsukamoto, T., & Shibata, T. (2002). Klebsiella pneumoniae produces no histamine: Raoultella planticola and Raoultella ornithinolytica strains are histamine producers. *Applied and Environmental Microbiology, 68*(7), 3462–3466.

Kelkar, R. (1998). A practical approach to hospital waste management in India. *Industrial Safety Chronicle*, Oct–Dec, 67–70.

Kumar, G. V., Duvvuri, K., & Bakki, V. K. (2007). A critical analysis of healthcare waste management in developed and developing countries: Case studies from India and England. *International Conference on Sustainable Solid Waste Management*. Chennai, India.

Kunitoshi, S. (1990). Improvement of solid waste management in developing countries. *Japan International Cooperation Agency (JICA)*.

Lehtinen, L. (2002). Occupational health and safety. *African Newsletter on Occupational Health and Safety, 12*, 32–34.

Mazrui, N. (2010). Medical waste management in Kenya: Opportunities for improvement. *Tropical Resources, 29*.

Michel, D., Marjan, F., Juliette, C., Richard, G., Lisette, S., & Michel, W. (2009). Safety and risk management in hospitals. IQ Scientific Institute for Quality of Healthcare, Radboud University Nijmegen Medical Centre. Retrieved from: http://www.health.org.uk/sites/default/files/SafetyAndRiskManagementInHospitals.pdf (accessed on 18 September 2024).

Ministry of Health, Kenya. (2001). *Kenya National Guidelines on Safe Disposal of Pharmaceutical Waste*.

Morais, V. P., Daporta, M. T., Bao, A. F., & Campello, M. G. (2009). Enteric fever-like syndrome caused by Raoultella ornithinolytica. *Journal of Clinical Microbiology, 47*(3), 868–869.

Mugenda, O. M., & Mugenda, A. G. (2003). *Research Methods: Quantitative and Qualitative Approaches*. African Centre for Technology Studies.

Muniafu, M., & Otiato, E. (2010). Solid waste management in Nairobi, Kenya: A case study of emerging economies. *The Journal of Language, Technology & Entrepreneurship in Africa, 2*(1), 342–351.

Muthee, M. M., Onguso, M. J., Kiiyukia, C., Makhonge, P., & Nyakundi, W. O. (2014). Biosafety and biomedical hazards among the support staff of Kenyatta National Hospital, Mbagathi District Hospital, and Kiambu District Hospital in Kenya. *Journal of Agriculture Science and Technology, 16*(1).

National Environment Management Authority (NEMA). (2005). *State of Environment Report (SoE)*. Nairobi.

National Institute for Occupational Safety and Health (NIOSH). (2004). *Preventing Occupational Exposures to Antineoplastic and Other Hazardous Drugs in Health Care Settings* (pp. 10–19).

Occupational Safety and Health Administration (OSHA). (2003). *Personal Protective Equipment*. Retrieved from: http://www.osha.gov/Publications/osha3151.pdf (accessed on 18 September 2024).

Olima, W. H. A. (2001). Community participation in urban development planning in Kenya. In G. Kroes & J. Y. Kokor (Eds.), *Community Development in Sub-Saharan Africa* (Spring Research Series Number 31). Dortmund, Germany.

Othigo, E. O. (2014). Evaluation of biomedical waste management: A case study of Mater Hospital in Nairobi County, Nairobi University of Nairobi.

Parizeau, K. (2015). Urban political ecologies of informal recyclers' health in Buenos Aires, Argentina. *Health & Place, 33*, 67–74.

Patan, S., & Mathur, P. (2015). Assessment of biomedical waste management in government hospitals of Ajmer city. *International Journal of Research in Pharmacy and Science, 5*(1), 6–11.

Pescod, M. B., & Saw, C. B. (1998). Hospital waste management in four major cities: A synthesis report. *Urban Waste Enterprise Programme Working Document No. 8*.

Pietra, L., Calligaris, L., Molendini, L., Quattrin, R., & Brusaferro, S. (2005). Medical errors and clinical risk management: State of the art. *Acta Otorhinolaryngologica Italica, 25*(9), 339–346.

Practice Greenhealth. (2013). *Waste*. Retrieved from: https://practicegreenhealth.org/topics/waste (accessed on 18 September 2024).

Reinach, S., & Viale, A. (2006). Application of a human error framework to conduct train accident/incident investigations. *Accident Analysis and Prevention, 38*, 396–406.

Rushbrook, P., Giroult, E., & Pruss, A. (1999). *Safe Management of Wastes from Healthcare Activities*. World Health Organization.

Sargent, V. E., & Gallo, F. (2009). Use of personal protective equipment for respiratory protection. *ILAR Journal, 44*(1), 52–56.

Sharma, S. (2010). Awareness about biomedical waste management among healthcare personnel of some important medical centers in Agra. *International Journal of Environmental Science and Development, 1*(3), 251–255.

Siddharudha, S., & Sowmyashree, H. (2015). Occupational exposure to infection: A study on healthcare waste handlers of a tertiary care hospital in South India. *Journal of the Association of Physicians of India, 63*. Retrieved from: http://www.japi.org/november_2015/03_oa_occupational_exposure.pdf.

Singh, B., & Gupta, K. M. (2009). Pattern of use of personal protective equipment and measures during application of pesticides by agricultural workers in a rural area of Ahmednagar district, India. *Indian Journal of Occupational and Environmental Medicine, 13*(3), 127–130.

United Nations Development Programme (UNDP). (2009). *Needs Assessment for Hospitals in African Countries in Relation to Infectious Waste Treatment: Final Report*.

Wickens, C. D., Hollands, J. G., Banbury, S., & Parasuraman, R. (2013). *Engineering Psychology and Human Performance* (4th ed.). Pearson.

Wilkins, P. P. (2016). Personal protective equipment in the humanitarian governance of Ebola: Between individual patient care and global biosecurity. *Third World Quarterly, 37*(2).

World Health Organization (WHO). (2011). *Waste from Health-Care Activities* (Fact sheet No. 253, November 2011).

World Health Organization. (2004). *Safe Health-Care Waste Management Policy Paper*. World Health Organization.

CHAPTER 6

Impact of Biomedical Waste on Human Health and Environment: Management Strategies

JAVID MANZOOR,[1] UROOJ NISSAR,[2] HEENA QADIR,[3] and KHURSHEED AHMAD WANI[4*]

[1]*Department of Environmental Science, JJT University, Churela, Jhunjhunu, Rajasthan, India*

[2]*Government Medical College, Baramulla, Jammu and Kashmir, India*

[3]*Department of Sociology, Government Degree College Awantipora, Jammu and Kashmir, India*

[4]*Department of Environmental Science, Government Degree College Thindim Kreeri, Jammu and Kashmir, India*

*Corresponding author.

ABSTRACT

Healthcare is critical to saving lives, but the biomedical waste (BMW) it generates is a major issue for both humans and the environment. BMW handling, treatment, and disposal are critical components of infection control procedures in healthcare settings. The environment, healthcare professionals, and the general public are all directly impacted by improper waste management at healthcare institutions. Healthcare facilities around the world produce a sizable amount of potentially harmful and infectious waste each day. The environment and human health are at risk from the careless disposal of BMWs, which requires special handling before final disposal. Everyone has a responsibility to manage waste management for their health and the environment. According to their significant quality, the numerous BMW

Biomedical Waste Management: Recycling and Remediation Strategies. Younis Ahmad Hajam, Javid Ahmad Malik, Anu Pandita, and Deepu Pandita (Eds.)

elimination procedures are anticipated, reduced, reused, enhanced, treated, and finally organized. Definitions, examples, and problems pertaining to BMW are covered in this chapter. It explores legal frameworks, management techniques, and disposal procedures for BMW. Additionally, it concludes by urging healthcare professionals to be more aware of the issue and states that additional research is required in light of BMW's rising generation.

6.1 INTRODUCTION

One of the toughest issues facing society today is solid waste management, particularly in urban areas. Due to the diverse nature of BMW as shown in Figure 6.1, India is still falling behind in the correct management of BMW and its risks, aside from tight norms and regulations and adequate oversight (Javeed and Kallihal). Between 2017 and 2019, BMW production in India climbed from 559 to 613 tons per day (CSE India, 2021). The rapid depletion of natural resources and the accompanying threat to India's sustainable and ecological development are direct results of the country's exploding population and strong economic expansion (CPCB, 2000, 2003). The fast increase in the human population has led to a drastic deterioration of the natural environment. Pollution of the environment and natural resources may have negative effects on people, animals, and vegetation (Sadeghi, 2002). Due to rising patient demand, several different types of hospitals and clinics are now in operation. It's undeniable that people are always trying to better people's health conditions. Providing more options for public and private healthcare services is one method to improve quality. As a vital part of their mission to serve the community, hospitals must maintain a spotless facility for their staff and patients. However, the costs incurred by healthcare facilities are not necessarily proportional to the advantages gained.

The usage of more single-use items in healthcare settings, as well as the wide range of activities that take place inside such settings, may result in the production of a wide variety of biomedical wastes (BMWs) as healthcare technology advances. The creation of healthcare service centers may result in the generation of relatively considerable volumes of waste with a wide variety of compositions and features. Given the contagious nature of medical waste, it is among the most hazardous forms of pollution (Sadeghi, 2002). Though hospitals are intended to improve patients' health, there is still a problem with how they handle trash. Sharps, pathological wastes, and infectious wastes are all examples of BMW that provide a significant risk of damage or infection if handled improperly. PVC plastics, mercury-containing devices, and radioactive wastes are some of the other types of pollutants that accumulate in hospitals and clinics (Remy, 2001). Healthcare personnel, scavengers, and municipal employees are most at risk from issues like blood-borne diseases caused by the poor treatment of BMW. Furthermore, BMW has the potential to harm both wildlife and ecosystems. Medical waste that isn't properly disposed of poses numerous risks to public health. Infectious diseases like typhoid, cholera, and hepatitis can be spread through contact with contaminated utensils and instruments, as well as through airborne droplets from discarded syringes and needles (Henry and Heinke, 1996). According to the World Health Organization (WHO, 1999, 2001, 2004), around 10% of hospital wastes are infectious, while approximately 5% are non-infectious yet harmful. About 15% of US hospitals' trash must be handled as infectious waste. Depending on the overall garbage produced in India, this number might be anywhere from 15% to 35% (Glenn and Garwal, 1999). Although there is a very small quantity of infectious and hazardous waste, improper waste management, in which infectious waste is mixed in with normal rubbish, might result in the vast majority of waste being potentially contagious.

Hospitals that aren't able to manage their waste to even the most basic requirements are adding stress to already overburdened healthcare systems. Access to trustworthy information and analysis related to hospital waste is crucial for this management. For Indian hospital trash, however, there hasn't been enough thorough research or statistical analysis to make this determination. Careful waste management is an essential component of any effective infection prevention or hygiene program. Despite variations in how

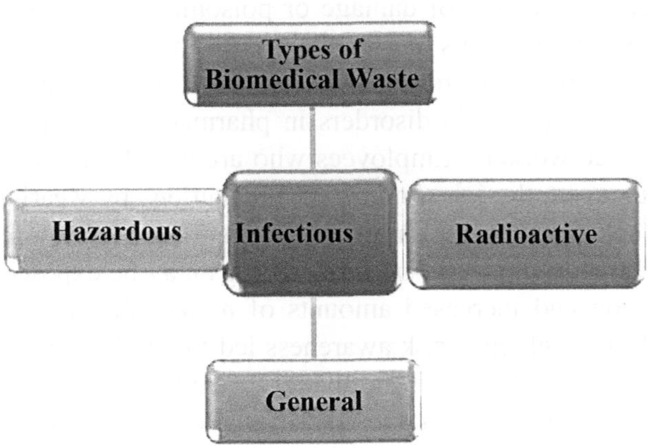

FIGURE 6.1 Major types of biomedical waste.

medical waste is handled now, most hospitals have similar challenges. Waste management technologies are gradually standardizing after the notification of the BMW (Handling and Management) Rules (1998). This is achieved through the use of appropriate methods of source separation, collection, proper packaging, storage, safe transportation, disinfection, treatment, and disposal to prevent environmental pollution.

6.2 THE INFLUENCE OF INFECTIOUS AGENTS ON HUMAN AND ENVIRONMENTAL HEALTH

Because of the potential for contamination, hospital waste has traditionally been seen as dangerous. Since infectious illnesses have been the leading cause of population morbidity and death for millennia, and since the vast majority of hospital patients suffer from infectious diseases, this threat was prioritized. People who work with the garbage and those in the region around the disposal site are at risk of contracting and spreading infectious illnesses if the waste is not properly segregated (Patil and Shekdar, 2001; Kelkar 1998). Disease transmission includes the spread of an infectious agent from its original point of introduction into a vulnerable host. Direct contact with an infected person (including their secretions, excretions, bodily fluids, or tissues), breathing contaminated air, ingesting contaminated food or drink, and acquiring an infection via vectors or other objects are the four most common ways diseases are spread. Epidemiological research shows that HIV is spread via contaminated medical waste and that the HBV and HCV viruses are spread through syringe and needle injuries. The danger of infection is highest for those working in healthcare, especially nurses. There have been several reports of damage or poisoning caused by the careless administration of chemicals and medicines in hospitals and other medical facilities. Exposure to chemicals including vapor, aerosols, and liquids may induce respiratory and skin disorders in pharmacists, anesthetists, nurses, and maintenance workers. Employees who are at risk of being exposed to these substances on the job should take precautions by wearing protective gear and by switching to less dangerous alternatives wherever feasible.

Those who work with anti-neoplastic drugs may be exposed to a higher rate of abortion and increased amounts of mutagenic substances in their urine. A lack of heightened risk awareness led to a reduction in the number of precautions taken by these individuals. Hand hygiene is an important preventative measure against the transmission of nosocomial infections, which may benefit both patients and healthcare personnel. Employees are required to wash their hands after removing gloves, after handling items that may be contaminated with blood, body fluids, excretions, or secretions, and in other situations as outlined by departmental policy. Using gloves, gowns, masks, and other forms of PPE, barrier protection is a method of infection control that aims to prevent the spread of disease. Protective gear reduces the likelihood that a person may be harmed by physical, chemical, or infectious hazards.

6.3 BIOMEDICAL WASTE (BMW; MANAGEMENT AND HANDLING) RULES

Anyone involved in the generation, collection, storage, transportation, treatment, disposal, or handling of BMW must adhere to the standards established in BMW. Biomedical facilities have a responsibility to ensure that their waste is properly managed to prevent harm to human health and the environment. This includes but is not limited to hospitals, nursing homes, clinics, dispensaries, veterinary institutions, animal houses, pathological labs, blood banks, and any other name under which these facilities may be known. Compared to when the BMW criteria were specified, it seems that proper BMW administration has improved and that the segregation of BMW has vastly improved (Agarwal and Singh, 2005).

6.3.1 CHARACTERISTICS OF BIOMEDICAL WASTE (BMW)

Infectious, radioactive, chemical, heavy metal and typical municipal garbage are all included in the broad category of hospital trash (DoE, 1998). Injury and infection risk are taken into account when classifying BMW. Sharps, pathological wastes, and infectious wastes all need special handling and disposal measures. Aside from the obvious, radioactive trash, mercury-containing devices, and (PVC) plastics are also common in healthcare facilities' waste streams. Some of the most ecologically dangerous waste products in healthcare are those listed below (Remy, 2001).

Large volumes of trash with a wide variety of compositions and features are produced by the healthcare industry. These waste products pose a greater risk of infection and harm than those generated by any other industry. Many hospitals present waste management processes that include inappropriate segregation and mixing of potentially contagious, office, general, food, construction debris, and hazardous chemical items during transit and disposal or treatment. When hospitals don't have proper segregation methods and infrastructure in place, the wastes they release into the community are at risk of spreading disease and posing other dangers (chemical). Employees whose jobs include dealing with garbage are at the most danger. All three of these risks are referred to as secondary risks. It is essential that wastes be separated (ideally at the place of creation) before treatment and disposal, regardless of the ultimate waste management approach chosen. The first and foremost action that has to be performed to protect the health of healthcare professionals is to implement appropriate safety measures. Mercury and other toxic metals, chemical solvents, and carcinogenic preservatives are all found in the garbage. Carcinogens and polymers (such as PVC) generate dioxins and other pollutants when burned, endangering the health of both workers and members of the public. Mandatory segregation procedures help healthcare institutions because they reduce the amount of biological and chemical hazardous wastes and increase the amount of clean solid waste (90%) that can be readily, safely, and cheaply handled via recycling, composting, and landfilling. Much of this waste is organic and recyclable, whereas only a small percentage is infectious waste. Separation programs have been implemented at many Indian hospitals. To ensure segregation occurs and is maintained, personnel must be trained, use the appropriate containers, post appropriate signs, and wear appropriate safety gear. The tiny percentage of trash that requires special treatment (incineration, autoclave, microwave, sanitary disposal, and so forth,) may be managed with less effort if adequate segregation is accomplished by training, clear standards, and strict enforcement.

6.4 IMPACT OF INFECTIOUS AGENTS ON HUMAN HEALTH AND ENVIRONMENT

Given the contagious nature of medical waste, it is among the most hazardous forms of pollution (Sadeghi, 2002). There is a risk that pathogens will be present in hospital garbage. Some of the pathogens are particularly worrisome because of the fact that they are difficult to cure and are themselves highly harmful. Polluting the environment, creating an unpleasant odor, and encouraging the growth and multiplication of insects, rodents, and worms are all side effects of improper waste management, not to mention the increased risk of disease transmission from injuries caused by syringes and needles contaminated with human blood (Henry and Heinke, 1996).

Healthcare personnel, scavengers, and municipal employees are most in danger from issues like blood-borne diseases caused by the improper disposal of BMW. HIV and hepatitis B and C infections are of particular concern since there is considerable evidence that they may be transmitted via hospital waste (WHO, 1999). Pollution of soil and groundwater from the transport of mixed hospital waste to trash disposal sites poses serious health risks to living organisms (Shirazinejad, 1996). The appropriate collection and disposal of trash is crucial because it may mitigate both direct and indirect threats to human health and ecological stability (Centers for Disease Control and Prevention (CDC), 2001).

6.5 BMW STRATEGIES

In addition, it includes instructions for sorting biological waste, labeling containers, transporting trash, and treating it using methods including incineration, autoclave, and microwave. Managing biological waste at a facility safely and scientifically calls for meticulous attention to detail throughout the whole process, from collection to mutilation to disinfection to storage to final disposal (Acharya and Singh Meeta, 2000). The most effective method for distinguishing between various sorts of garbage is to place BMW in distinct, color-coded plastic bags or containers (Patil and Shekdar, 2001). Biological waste is kept in utility rooms by 60% of surgical, medical, and laboratory departments and by 40% of critical care units. When biological waste is generated in a hospital department like labor and delivery, operating room, or dialysis unit, it is not kept in that specific location but instead is sent to the hospital's central waste storage facility. When looking at five hospitals and 10 basic healthcare settings in Egypt, the average time biological wastes were kept in storage was 4.6 to 8.1 days. This was arrived at after doing research into the matter. After conducting an inquiry, we came to this realization as a result of the findings.

As part of the EC-LIFE program-funded study, Muehlich et al. (2003) compared the ways in which five European hospitals dealt with their garbage. Different approaches to classifying infectious hospital waste were found in a comparison of the rules guiding current waste management. In this context, we analyzed the disparities in infrastructure and explored the economic and environmental repercussions of improper garbage sorting and disposal. The quantity of rubbish produced daily in the Khuzestan area of southern Iran has increased due to the region's growing population, as reported by Mohammad Karamouz et al. (2007). It was recommended that the framework for a master plan be developed to manage hospital solid wastes in the Iranian province of Khuzestan. This plan would take into account many criteria for assessing the contamination of hospital solid waste loads. Additionally, the effectiveness of the management strategies is examined and analyzed. The analytical hierarchy method (AHP) ranks hospitals and determines the relative contribution of each to the total pollution burden caused by hospital solid waste by using a weighted combination of numerous criteria. The study found that pollution from hospital solid waste accounts for over 43% of the entire pollution load in the province, with the majority of this pollution coming from hospitals in and around Ahvaz, the provincial capital. Results show that better management practices are more significant than new construction. Using the proposed method, a complete plan is created for handling waste from hospitals and other healthcare institutions. Needlestick transmission of infectious illnesses such as HIV and hepatitis is demonstrated by the work of Tamplin et al. (2005). Direct exposure and environmental contamination are both clear concerns. Document analysis and summarization methods were utilized to investigate the nexus between syringe reuse and clinical waste disposal and identify challenges and potential solutions for the secure destruction and disposal of spent injectable materials. The authors argue that comprehensive strategies for syringe disposal and clinical waste management are necessary. It is necessary to conduct pilot projects in certain nations to test this framework. The methods and equipment under consideration have the potential to be implemented in remote parts of poor nations.

The waste management technique was conducted in Japan by Miyazaki and Une (2005). Infectious waste management received its first regulation in 1992, and in 2004, the Ministry of the Environment issued updated guidelines for managing infectious waste. There are three classifications for infectious waste: waste type, waste origin, and infectious illness kind (Manzoor and Sharma, 2019). The conclusion was as follows, A medical facility may hire a specialized waste management firm to collect and sort its infectious trash before sending it to be incinerated. However, the expense of disposal is rising. As a result, hospitals and other medical centers should do all they can to cut down on the amount of infectious waste they produce. The goal of the new rule is to safeguard garbage employees from being sick while also encouraging the minimization of infectious waste. Blenkharn (2006) conducted research on the methods that 26 hospitals in the UK use to dispose of the vast volumes of clinical waste that they generate. It was common practice to leave garbage cans unsecured and positioned in open areas where anybody might get them. In other instances, the garbage cans and storage areas in the clinical setting were in a state of disrepair. Similarly, Patil and Shekdar (2001) state that the management of waste in the healthcare industry is not just a matter of technical concern but is also strongly influenced by economic considerations. The simple act of passing legislation will not, on its own, result in an increase in efficiency. It is possible that identifying long-term solutions will be facilitated by including local entities in waste management and ensuring that these bodies adhere to the principles of good management. In order for medical waste to be repurposed, recycled, or disposed of properly, it must first be sterilized and mangled. Under no circumstances is it acceptable to reuse plastic disposables such as needles, syringes, intravenous sets, or other similar items made of plastic. An adequate strategy must be developed in light of the current circumstances in order to minimize waste. In the end, sufficient financial provision must be made. Researchers at Government General Hospital (GH) in Puducherry, led by Chandira Bose and colleagues, 2009, analyzed the kind and volume of BMW production there. Disposing of non-segregated BMW in Puducherry is very unhealthy for the local populace and scavengers. These days, a growing number of foreign patients are choosing to go to India for cutting-edge medical care because of the much lower costs associated with such procedures there. As the market for "medical tourism" grows (Connell, 2006), facilities must ensure that their BMW is well-managed to protect locals and wildlife (Mudur, 2004). The correct management of BMW and the segregation of BMW seem to have improved once the recommendations were stated.

6.6 QUANTITY OF BIOMEDICAL WASTE (BMW) GENERATION

Consistent economic expansion in India, a prospective economic global leader, is threatening the country's sustainable and ecological development since it is causing a large decrease in the country's natural resource stock. It is estimated that urban areas in India produce 0.115 million tons of trash daily, or 42 million tons yearly (CPCB, 2003). The average daily garbage produced by an Indian city resident is between 0.2 and 0.6 kilograms, which is far less than the waste produced in affluent nations. However, economic development and rapid urbanization have led to changes in consumer habits that have contributed to a yearly increase in per capita trash creation of roughly 1.3%. The Energy Resources Institute (TERI) predicts that by 2047, annual garbage output will have increased by more than five times the current level, reaching over 260 million tons (CPCB, 2003). Around 15 Iranian hospitals were studied by Mehrdad et al. (2004). The findings revealed a daily waste creation rate of 4.45 kilograms per bed, with a total of 1830 kilograms of garbage produced (71.44% household waste, 712 kilograms of infectious waste, and 19.6 kilograms of sharps). The separation of trash into its many components is imperfect. Only two hospitals (13.3% of the total) transport garbage using containers without covers. There are incinerators at nine hospitals (60%), although only six of those hospitals are fully functional. The researchers came to the conclusion that the hospitals in their sample weren't offering enough training on how to handle hospital trash and the risks that come with it. The available training options are either inadequate or inappropriate. An Indian government hospital with 360 beds has varying amounts and types of garbage, as described by Jayanthi and colleagues (2002). The average daily garbage produced by the hospital is 285 kg. The remaining 41% is comprised of biological waste. An equation was created by Mohammad Reza Sabor et al. in 2007 to predict the amount of (infectious) hospital waste produced in any given year. The model predicts a total of 6,98,937 metric tons of infectious hospital waste will be produced in 2008 (in the near future) and 3,494,387 metric tons will be produced in 2028. The quantity of infectious wastes created may be decreased by 15.1% if only the true infectious wastes are collected individually. The physical investigation of the hospital garbage revealed that 67.3% of the waste was infectious, 8.8% medicinal, 1.8% biological, and 22.1% municipal waste. Hospital employees must be trained in an acceptable collecting strategy, and the necessary facilities must be prepared before the process can begin. Patil and Shekdar (2001) examined several aspects of waste management in the healthcare industry, including the amount and composition of wastes generated by different facilities, and their treatment and disposal (HCUs). Around 27 hospitals and 16 primary care clinics were given a questionnaire on healthcare waste management. Hospital waste production was reported to be 1.13 ± 0.96 kg/bed/day on average. Healthcare risk waste creation in primary care settings averaged 0.080.08 kg/visitor/day. Annually, the Kingdom of Saudi Arabia generates an average of 25,207 metric tons of healthcare risk waste. In order to develop guidelines for the disposal of medical waste, a new initiative is being launched.

Mohee (2005) argued that around 10% of the solid wastes generated at these locations are infectious, pathological, or chemical in origin, according to a waste audit. Findings from the waste quantification project also showed that daily water consumption was 0.654 m^3 per patient and wastewater production was 500 m^3d^{-1} at SSRN Hospital. Analyzes of the wastewater showed that it included excessive levels of COD, BOD5, TSS, and coliforms. Chitnis et al. (2006) claimed that a considerable quantity of BMW is produced by pathology, microbiology, blood bank, and other diagnostic labs (BMW). Having a thorough BMW audit will allow for more informed decision-making. Needles are autoclaved and sent to the needle pit after being gathered in puncture-proof containers. Tissues preserved with formalin must not be incinerated due to the risk of harmful gas leakage, necessitating revisions to the regulations governing BMW. In addition to putting waste reduction, reuse, and recycling into correct contexts, it is essential to segregate trash where it is generated. Dental solid wastes have a relatively low percentage of hazardous materials, however improper disposal of these materials still poses risks of cross-contamination and environmental harm. Research from a Turkish university hospital's dentistry school looked at the make-up of eight of the clinic's solid wastes. As could be predicted, waste composition differs across facilities. Nonetheless, the collected data suggests that rubber gloves account for about half of the total solid waste in the vast majority of hospitals and medical centers. About 30% of all solid trash is made up of paper. The entire amount of garbage generated by medical facilities is usually proportional to the total number of patient operations. The fact that just a small percentage of trash is dangerous shows that the hazardous waste collection regulations are being followed at the Hacettepe University School of Dentistry. Byeong-Kyu Lee et al. (2006) analyzed medical waste from three of Massachusetts's five typical urban hospitals to determine its generation volume, sources, content, treatment, and disposal techniques. Moreover, the research evaluated the RMW production rates and totals between the hospital and the medical school. One hospital's RMWs annual operating treatment and disposal expenses were examined using a variety of treatment and disposal scenarios. The research concludes that the most cost-effective method for lowering the treatment and disposal costs of RMWs is to combine on-site combustion with microwave technology.

6.7 BIOMEDICAL WASTE (BMW) TREATMENT FACILITIES

Rules 14 (as modified in June 2000) established a central treatment facility for all types of medical waste, making it the responsibility of local governments to collect and dispose of both untreated and processed trash. CBMWTF owners are service providers that arrange for the collecting of BMWs from healthcare facilities for ultimate disposal at their facility. "Autoclave, shredder, compactor, and incinerator for anatomical waste; secured landfill facility; laboratory; trucks for transportation of wastes; these are the components of a typical BMW treatment and disposal facility" (CBMWTFs). So far, the Board has compiled an inventory of 317 public hospitals and

1,835 non-public facilities. The Board has mandated that public and private hospitals work together with the Indian Medical Association to find a central location for biological waste disposal and set up shared facilities within a certain period of time. PCB and the Coimbatore chapter of the Institute of Management Accountants (IMA), together with the municipal government of Coimbatore, have been making progress toward their goals. The order authorizing this was issued on May 21, 2003. Using private haulers to carry medical waste to treatment facilities was the most common method of treating and disposing of medical waste (61.5% of the time). Pouring into municipal sewage systems (46.6%), dumping in landfills (41.6%), and autoclaving (32%) followed. The use of (ETD), hydropulping, microwaving, and grinding prior to disposal in municipal sewage are some of the other procedures utilized by hospitals. In response to stricter rules on air pollution emissions, most hospitals in Oregon and Washington have shut down their incinerators. However, despite the lack of hospital-specific state laws, incinerators continued to be used by Idaho hospitals. The various BMWM approaches are shown in Figure 6.2.

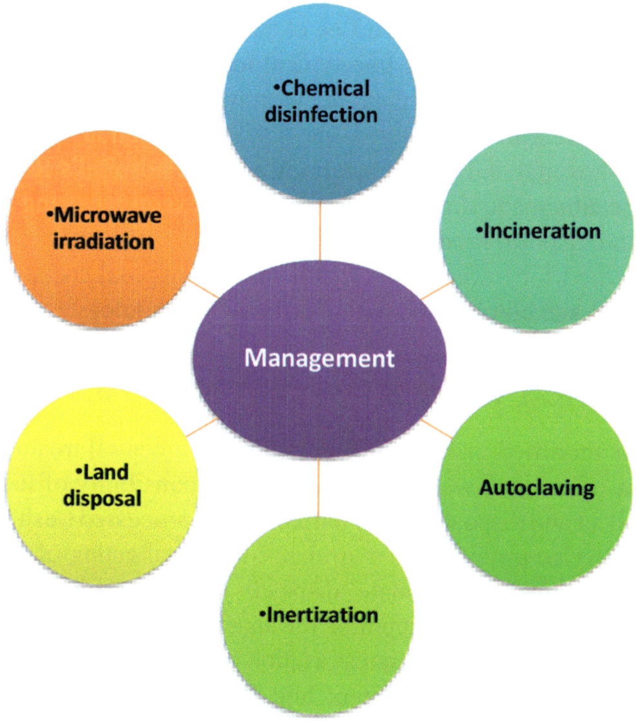

FIGURE 6.2 Biomedical waste management approaches.

6.7.1 INCINERATION

The 10th International Congress focused on the sources, characteristics, and health effects of fine and ultrafine particles produced by combustion, as well as mercury and dioxin emissions, as well as the creation and implementation of novel analytical and diagnostic techniques. In addition, mercury and dioxin emissions were discussed (2008). This was the conclusion reached by the majority of researchers. Although oxidative stress caused by ultrafine particles is thought to be the origin of the reported health effects, many crucial biological and chemical aspects, as well as probable catalytic cycles, remain unclear. Important secondary findings included the following: (i) Combustion emissions are responsible for 70% of airborne fine particles in urban areas, and 50% of these are related to primary emissions from combustion sources; (ii) Combustion generates one or two types of nanoparticles with mean sizes of 10 nm and 1 nm, respectively, in addition to soot; and (iii) Surface area, sulfate concentration, total carbon, and organic carbon, the most often used metrics for describing particle toxicity, do not adequately explain reported health outcomes. When metal-containing nanoparticles, organic carbon molecules, and PFRs are all present in the same environment, oxidative stress may be generated in the organisms that are subjected to it. Lee and Huffman (1996) analyzed the rules for getting rid of medical waste. Hospitals need to reevaluate their waste collection and disposal policies in light of rising public awareness of AIDS and other infectious illnesses, as well as right-to-know laws. Satnam and Vinit (2007) medical waste incineration's harmful emissions were analyzed; they included organic emissions, as well as inorganic emissions and ashes containing toxic metals. Efforts to curb or remove emissions that have been made by different researchers are also mentioned. Large quantities of energy are generated during the burning (incineration) of garbage; this energy must be used wisely. Garbage-to-energy systems may partially replace traditional energy production facilities powered by fossil fuel and, as a result, help to address global environmental issues, in addition to the fundamental objective of incineration, which is processing the stipulated quantity of waste. With the use of this computational tool, a mathematical model was developed by combining standard auxiliary procedures with fundamental thermodynamic models of heat engines. The creation of this product was influenced by both business needs and existing practices. The most significant challenge of efficiently operating incineration facilities is the cost-effective usage of generated energy. To investigate the existing state of medical waste disposal in several Alexandria hospitals and to accurately evaluate hazardous waste management, Hosny and El-Zarka

(2007) conducted a survey. With over 3,911 medical establishments serving the residents of Alexandria, a significant quantity of medical waste is produced every day, totaling almost 208 tons per month.

Kanemitsu et al. (2005) conducted preliminary tests to validate the effectiveness of muffled furnace incineration against *bacillus* spores. Stainless steel Petri plates containing 106 spores of *Bacillus* stearothermophilus were heated in furnaces to kill any potentially harmful bacteria. Incineration times and temperatures ranged from 15 minutes at 300°C to 30 minutes at the same temperature, 500°C to 15 minutes at the same temperature, and 3 minutes at 1100°C. We determined that all *B. stearothermophilus* spores were eliminated throughout the board. In terms of the effect that they have, sterilization and incineration seem to have a comparable degree of similarity based on the acceptable sterilization assurance level of 10–6. Using emission requirements that were established by Portuguese law, Alvim-Ferraz and Afonsot (2005) determined the quantity of air pollutants that were produced as a result of the burning of healthcare wastes. For the sake of the study, the investigators made use of a single controlled-air incinerator that was devoid of the standard air pollution precautions. The most important takeaways from this are as follows: I emissions from incineration exceed legal limits, which poses a threat to the safeguarding of human health; and (ii) emissions from incineration exceed legal limits when there is not enough equipment to regulate air pollution, which is a problem. Both of these findings are problematic. It was only possible to meet the legal limit for pollutant concentrations for NO_x, while all other concentrations were greater than the maximum that was permitted: dioxins, 93–710 times higher than the legal limit for dioxins; furans, 93–710 times higher than the legal limit for furans; and NO_x. If everyone stuck to a tight segregation policy, the amount of waste that needed to be burnt might be reduced in half, which would save a significant amount of time and money. Reduced garbage burning results in decreased levels of pollutants, with reductions of arsenic (As), cadmium (Cd), chromium (Cr), manganese (Mn), and nickel (Ni) by 90%, 92%, 84%, 77%, and 92%, respectively; hg and pb are nearly eradicated; SO_2 and No_x are decreased by 93%; and CO and HCl are reduced by 98%.

6.7.2 PLASMA TECHNIQUES

Mosse and Savchin (2006) review the findings from international studies to determine the makeup and toxicity level of medical and biological wastes. Studies have demonstrated that these wastes pose serious ecological risks, highlighting the necessity for a standardized approach to dealing with them.

For the purpose of plasma-thermal treatment of medico-biological waste, they designed and tested a plasma chamber incinerator. They built a thermal calculation model of the plasma chamber incinerator to improve the plant's efficiency and lessen its impact on the environment. According to Amouroux et al. (2005), the European Union is aggressively regulating air and water pollution as well as waste treatment. New depollution methods using plasma or laser technology are required for compliance with these rules. Time is a key issue in pollution monitoring, but this difficulty may be mitigated by using time-resolution laser-induced breakdown spectroscopy, a technology that allows for high-sensitivity, real-time measurement of any element in the periodic table with no need for a sample. By employing a mass spectrometer or an optical emission spectrometer in combination with plasma-enhanced desorption from fly ashes or contaminated soils, it is possible to rapidly and accurately monitor volatile organic compounds (VOCs) and organochlorine species. Several plasma-based methods for treating VOCs were presented, including methods for removing or trapping PAH molecules.

6.7.3 NON-INCINERATION TECHNOLOGIES

EPA Part V (2002) provides advice on alternatives to incineration for dealing with medical waste (Procedures for Microbiological Testing) This protocol was developed to guarantee that sterilization and disinfection methods other than burning (such as the Autoclave and the microwave) are effective for dealing with biological waste. This is confirmed both during initial equipment commissioning at a new location and during periodic testing of operational equipment. Before the properly treated BMW may be disposed of, it must first undergo rigorous verification processes and be subjected to careful examination of test findings.

6.7.4 AUTOCLAVE

Pimshtein and Zhukova (2003) results from stress state calculations for both the monolithic and multilayer forms of autoclaves are used to evaluate the feasibility of external heating. In the analysis, the effects of temperature, temperature gradient, internal pressure, and the closeness of fit between the layers of the multilayer wall on the material's mechanical characteristics are considered. On the basis of the calculations, we propose the following settings for the autoclaves' heaters:

1. The heating rate should not exceed 30°C h^{-1}.

2. Monolithic autoclaves may be heated from the outside. Here, the allowed temperature gradient (DT) is calculated using the pressure within and the surface temperature inside.
3. It is acceptable to operate the rolled-sheet autoclave 1-r with external heating, an average initial interlayer clearance of 0.03 millimeters, an inner-surface wall temperature of 338°C, and a temperature gradient of no more than 32°C while maintaining an internal working pressure of 70 megapascals.
4. Rolled-sheet autoclaves 2-r and 3-r with an average starting clearance of 0.03 mm are only allowed to be heated outside while they are being brought up to operational temperatures. For an internal pressure of more than 90 MPa, external heating is not permitted.

Williams and Ewel (1984) recovered six times as much Mn from a Typic Dystrandept in Costa Rica during heat sterilizing, reaching amounts that are often considered harmful. After delays of 1, 8, 15, and 28 days, seeds from eight species, comprising six successional taxa and two cultivars, were planted in sterilized soil and unsterile soil, respectively. There were no significant changes in germination or mortality between the two soils, suggesting that steam-sterilized soil may be used safely in seed traps. Two natural taxa grew much faster in sterilized soil, but six species fared better in unsterile soil. Lee et al. developed an operational model of a high-pressure polyethylene plant (2000).

The consequences of varying the flow of the initiator are detailed, and a realistic yet complete model is presented that can forecast the temperature at each reaction zone. The stability study helped them determine the optimal working parameters for LDPE autoclaves, allowing them to minimize the risk of breakdown events (runaway reactions) and achieve the highest possible rates of polymer conversion. In his 2006 article, Sadykov outlines the findings of research on the autoclave technique used to treat zinc deposits in Kazakhstan. In order to create the appropriate sulfates and elementary sulfur, the autoclave leaching procedure depends on an interaction between the constituents of a crushed concentrate, sulfuric acid, and oxygen. Excellent agreement between the results for zinc, copper, iron, lead, and sulfur in the input and output materials of autoclave leaching shows the reliability of the sampling and analytical techniques utilized.

6.7.5 MICROWAVE

Microorganisms on surfaces are destroyed by two different types of UV sources produced by microwave radiation, according to Barkhudarov et al.

(2008). In both instances, microwaves from a power source akin to a home microwave oven are used to power gas discharge electrode-less lamps (Ar/Hg) that emit ultraviolet (UV) light. One of these sources uses a coaxial line with a truncated outer electrode for UV lamp excitation, making it possible to excite gases and gaseous mixtures across a wide range of pressures with just a little microwave power input. The second one makes use of microwave-enclosed UV lamps. Thin films and drops of water were treated with ultraviolet light from both sources to kill vegetative Escherichia coli bacteria. Lamps that emit ozone and UV light have been proven to be more effective in killing germs in films on surfaces than in droplets. The UV sources indicated may be used to quickly kill and cost-effectively treat food and other surfaces, as well as to eradicate airborne viruses and bacteria. Moreover, the lights may be used to quickly kill bacteria and other microbes in water and other liquids.

The waste extraction test (WET) was implemented in California by Angela and Ramon (2002) to supplement the (TCLP). Shaking a sample in citrate buffer and measuring the resulting solution for metal concentration is what the WET technique is all about. There must be a 1:10 ratio of waste to liquid, and the extraction time must be 48 hours. The WET procedure is effective in determining leaching and mobility; however, it is a time-consuming process. Since the relative extraction efficiency achieved by the WET technique was a desirable target, a microwave extraction process was designed to simulate it. ICPMS or ICP-AES was used to measure the amounts of lead, arsenic, and copper, with a stringent quality assurance system in place. The novel approach yielded results statistically on par with those of the WET procedure. This microwave extraction method was quick and easy to use, cutting down on sample preparation by about 28%. Production of trash, use of resources, and utilization of labor and chemicals are all drastically cut down on. Therefore, the microwave extraction procedure is proposed as a quick and cost-effective monitoring tool for waste samples, either alone or in combination with the usual WET approach.

6.8 HOSPITAL LIQUID WASTE TREATMENT

Two hospitals' wastewater was collected from Rio Grande do Sul, RS, Brazil. Samples of the surface water were taken from bodies of water that were exposed to this effluent discharge. Disk-diffusion testing was used to evaluate antibiotic susceptibility profiles, and genotyping was accomplished by amplification of Enterobacterial Repetitive Intergenic Consensus sequences (ERIC-PCR). Strains isolated from hospital wastewater had greater levels of antibiotic resistance, and the multi-resistant phenotype was unique to these bacteria. Genetic similarity between wastewater and surface water *P. aeruginosa* bacteria was not detected using ERIC-PCR profiles. Contrarily, they demonstrated that genetically unique populations had developed in these various habitats, suggesting that another pollution source may be contributing to the existence of resistant strains in these water sources.

Saurabh and Ram (2006) biological waste management techniques at Balrampur Hospital, a top hospital in the Indian city of Lucknow. According to the findings, both contagious and non-infectious trash are routinely deposited together inside hospital grounds, leading to a mingling of the two types of waste before being sent to municipal disposal sites. It is collected every several days by the Lucknow Municipal Corporation. There is nowhere for infectious trash to be treated at the hospital. Without adequate disinfection, germs in laboratory waste get up in the Gomti River after being flushed down the drain. Rag pickers go through garbage cans and dumps at hospitals and other public facilities, separating any plastics. Due to the open nature of the trash disposal, ragpickers are often injured by sharps, needles, and other materials used in the injection process, putting their health at risk. Healthcare plastic waste was studied by Byeong-Kyu Lee (2002), who determined how much of it could be recycled. Information on waste generation and recycling rates was collected from three Massachusetts veterinary clinics and five urban medical facilities. Plastic trash from cafeterias, operating rooms, labs, emergency rooms, ambulance services and facilities, and animal clinics was analyzed for its potential for recycling in urban healthcare settings. The majority of plastic trash in hospitals comes from the hospital's facilities, labs, operating rooms, and cafeterias. Plastic waste from hospital cafeterias was shown to have a much higher potential for recycling than that from any other hospital department. The low risk of infection or contamination and the ease of sourcing plastic parts were major factors in this shift. Improved recycling of plastics from hospitals was proposed as a possible solution. According to the results of this research, hospitals may benefit by sorting their plastic wastes into several categories based on the potential for infection and/or the kind of plastic used in their production.

The filters consistently removed at least 97.60% of the iron from the solution. There was a noticeable decrease in efficiency while trying to remove manganese from the solution in comparison to iron. The filters were able to remove anything from 22.22% to 100% of the manganese. The removal rates of iron and manganese did not change regardless of the filter or solution used. For the duration of the 7-day trial, iron precipitate could be seen; nevertheless, the effectiveness of the element removal was not impacted by the presence of the armor. All the water samples had a pH level below 7.7. This means that the wetland ecology shouldn't suffer any ill impacts from water with a little higher pH. According to Lim et al. (2004), MSWIF production has been on the rise in recent years, and the high concentration of leachable heavy metals in MSWIF makes its disposal difficult and expensive. In a leaching test, natural clay was shown to be effective in transforming MSWIF into a non-hazardous substance. Different leaching behavior from pure MSWIF was observed in all MSWIF-clay combinations. In contrast, the stabilized MSWIF-clay mixes were not sensitive to the oxidizing and alkaline conditions.

Sethi and Di Molfetta (2007) aerobic and anaerobic exothermic processes within the trash itself have been documented to be the primary cause of heat production inside MSW landfills. Temperatures within a sanitary landfill vary from mesophylic to thermophilic as a consequence of the formation and transfer of heat. Liner systems may be badly harmed by the high temperatures at the landfill's base. It is possible that the temperature of the surrounding geological strata and the underlying aquifer will rise as a result of the increase in convective and conductive heat transmission. In this study, the team examines the potential causes of a temperature anomaly in the aquifer under a Northern Italian municipal dump. The similarity between heat and mass transmission in porous medium has been used in the heat transport model. The thermal anomaly was shown to originate in the landfill and spread to the aquifer by convective and conductive heat transfer, as demonstrated by the model.

Biosolids, on the other hand, are poised to play a significant role in the renewable energy sector and the creation of "carbon credits" as part of the growingly popular low-carbon economy agenda. Currently, a significant amount of biosolids end up in landfills. The nutrients and energy found in biosolids may be put to good use. New methods have been developed to better handle biosolids and create valuable products from them. Consequently, mono-incineration and co-combustion will continue to be viable choices for biosolids treatment. When the leftover ash is utilized to create building supplies, these solutions provide a full suite for biosolids management.

Mufide Banar et al. (2006) conduct a thorough evaluation of the characterization and identification of significant pollutant parameters. When looking at each parameter and each sample point independently, however, winter season analysis findings were typically lower. Correlation analysis shows that the solids and dissolved oxygen concentrations in leachate have the greatest influence on all the other variables. In addition, there is a strong relationship between chloride and heavy metal content, and changes in sodium and potassium are proportionate to those in the same parameters (suspended particles, fixed solids, dissolved oxygen).

6.9 INTEGRATED BIOMEDICAL WASTE (BMW) MANAGEMENT PLAN

Hospitals and nursing home administrators aren't the only ones concerned about biomedical waste (BMW) management. Authorities in the environmental and law enforcement sectors, as well as the media and the general public, have also taken notice. Currently, BMW accounts for 2%–4% of all municipal waste. Some of this waste may pose risks to both human health and ecosystems. The type and volume of biological waste generated varies widely, not only across countries but even within regions. The waste stream produced by healthcare institutions is highly diverse due to the broad range of services they provide, with hospitals being the primary source. Healthcare workers (HCWs) and those handling biomedical waste are frequently at risk of needle-stick injuries, which expose them to infectious diseases such as HIV/AIDS, Hepatitis B, and Hepatitis C due to the sharp and infectious components of the waste. Every healthcare facility carries significant risks related to BMW. In response to growing global concerns about the disposal of biomedical waste, the Government of India enacted the "Biomedical Waste (Management and Handling) Rules (1998)" under the provisions of the Environment (Protection) Act (1986). However, despite these regulations, managing biological waste remains a complex issue. There is a need for further improvement, both in waste management practices and in changing the mindset of healthcare professionals.

6.10 NEED FOR BIOMEDICAL WASTE (BMW) MANAGEMENT

The management of public services, such as garbage collection, has become more important as the global population rises. Because of its contagious and potentially dangerous nature, BMW takes on a unique role within the realm of municipal solid trash. It has the potential to do damage to people or transmit illness. Although infectious trash accounts for just around 20% of total hospital waste, it has the potential to contaminate the whole hospital's waste stream if it is not properly separated or transferred. Hospital and other healthcare facility trash varies from typical municipal garbage. Depending on the specifics, just one of these systems may be necessary for collecting, transporting, and finally disposing of municipal trash. BMW, on the other hand, contains things like organs, human and animal tissue, radioactive waste, gauze, cotton, plastics, plaster casts, infectious liquid waste, blood, and laboratory wastes, therefore additional systems may be necessary to deal with it. There is a threat to public and environmental health from the medical waste produced by hospitals and other healthcare institutions. Diseases including TB, hepatitis, enteric fever, HIV infection, and AIDS may be transmitted via improper treatment and disposal. The Indian government and its health care administrators have neglected to give BMW disposal the attention it deserves, despite it being a key area of infrastructural growth. The developed world has mostly affected prior efforts, although there have been some successes. Since population density, socioeconomic features, and climatic circumstances varied considerably, the results were not adequate. Policies and priorities at the federal level are also distinct. The strategy used for dealing with BMW was inappropriate for the conditions in India. For this reason, hospitals need a waste management system that is distinct from the one used for municipal garbage. The approach used in this study is to first characterize the nature and scope of the issues plaguing the research domain, and then to examine those issues in order to provide workable answers. As a first step, we need to learn more about the complex nature of the issues surrounding the handling of medical waste. BMW management consists of the following steps, as determined by a review of the relevant literature: waste generation; waste classification; waste storage and collection; waste transportation; waste treatment; and waste disposal. The thorough research exposed the current situation for BMW in the metropolis, as well as the significant issues connected to it. Data for this came from in-depth sample surveys, in addition to primary and secondary sources. The framework is then used to conduct a thorough examination of the BMW-related difficulties and concerns revealed via the use of waste management-related research and methodology.

6.11 SOCIOLOGICAL CONCERNS OF BIOMEDICAL WASTE (BMW)

Concern about the societal effects of BMW is urgent. When these BMWs are discarded without proper planning, they constitute a risk to human health and safety, the environment, the possibility of genetic diseases, workplace dangers, and the welfare of animals. The effects of BMW on society are a mirror image of the social hierarchy that produced them. Since the most at-risk populations are those who did not contribute to the garbage's production, international waste trade exacerbates environmental injustice, social hierarchy, and exploitation along lines of class and race. Waste management is seen as a colonial problem. Underprivileged communities are disproportionately hit hard by environmental hazards including improper trash disposal and its repercussions. Most incinerators, dumps, landfills, and burn facilities are situated in close proximity to economically disadvantaged neighborhoods, communities of color, and marginalized groups, posing a serious hazard to their health via their emissions.

Waste piles, compactors, and excavators, the smell of decaying trash, falling property prices, and negative health consequences all add up to a devastating impact for those living near landfills. Recent research found that 34% of residents living near a landfill had trouble breathing due to the site's generated odors and dust. Cancer, congenital impairments, low birth weight, genetic mutations, asthma, wounds, stomachache, recurrent flu, cholera, malaria, cough, skin irritation, and tuberculosis, were all diseases that were more common among this population than among others.

Despite the importance of recycling, the hazardous job of collecting trash for this reason presents significant risks to the health of the workers and is not covered by any legal or regulatory framework. Potentially infectious microbes in healthcare waste may infect hospital patients, healthcare professionals, sanitation employees, visitors accompanying patients, and the general public. Infections from BMW are especially dangerous for low-income scavengers and trash handlers who work at disposal sites. They are in imminent danger of being poked with a needle or being exposed to harmful or infectious substances. The release of BMWs into the environment has negative consequences for human health, the value of surrounding property, and the feasibility of future expenditures in things like public works projects and educational opportunities. When property prices are low, governments are unable to raise enough money to adequately fund infrastructure projects like new roads, new schools, and new municipal buildings. According to data from the Ministry of Environment and Forests, only 5–10% of institutions in India have followed the guidelines for the disposal of BMW.

It is crucial that both hospitals and city planners use effective strategies for managing their facilities. All the way from the physicians and nurses to the ward boys and the cleaning staff must be included in the waste management process. A breakdown in duty or carelessness at any point may have far-reaching consequences.

6.12 CONCLUSIONS

In spite of India's stringent laws and regulations governing the disposal of BMW safely, immediate action is required to boost system capacity and enhance financing and commitment. A sudden increase in biological waste generation due to a pandemic ailment should include optional strategies for its treatment. Creative measures made by the state pollution control board and steps done by the local municipal system make it possible to exercise greater control over BMW management. The primary goal of BMW management is to develop a system that considers geographically specific weather, physical, economic, and social factors. The goal is to create a healthier atmosphere via long-term methods. Waste producers should always be responsible for segregation, which should happen as near to the point of generation as feasible and be maintained in storage places and throughout transit. Assigning different colors to different types of medical trash is the most efficient approach to keeping track of what goes where. When it comes time to dispose of trash, wastes from general health care should be mixed in with the rest of the household garbage. Regardless of whether or not the sharps are infected, they should all be gathered together in the same location. It is essential that storage containers have coverings and are unable to be punctured (often constructed of metal or high-density plastic). They should be hard and impermeable in order to properly contain not only the sharps but also any remaining liquids from syringes. This ensures that the patient is not exposed to any potential danger. In order to reduce the likelihood of misuse, packaging should be tamper-proof (meaning it should be challenging to open or break) and needles and syringes should be made ineffective. Cytotoxic waste should be kept in a safe facility that is designated for it, distinct from other garbage from health care facilities. The proper way to store radioactive waste is in sealed containers that are shielded by lead and kept in a secure location. Within the hospital or other facility, BMWs should be transported using wheeled trolleys, containers, or carts that are not used for any other purpose and meet the following specifications: they should be simple to load and unload; there should be no sharp edges that could damage waste bags or containers during loading and unloading; and they should be simple to clean. A suitable disinfectant needs to be used on a regular basis for the purpose of cleaning and sanitizing the vehicles. At the conclusion of the transportation process, each waste bag seal should still be in place and unbroken.

KEYWORDS

- **anatomical waste**
- **autoclave**
- **biological waste disposal**
- **biomedical waste**
- **healthcare facilities**
- **medical waste**

REFERENCES

Acharya, D. B., & Singh, M. (2000). *The Book of Hospital Waste Management*. Minerva Press.

Agarwal, A. G., & Singh, R. (2005). *Understanding and Simplifying Biomedical Waste Management*. Toxics Link.

Alvim-Ferraz, M. C. M., & Afonso, S. A. V. (2005). Incineration of healthcare wastes: Management of atmospheric emissions through waste segregation. *Waste Management, 25*, 638–648.

Amouroux, J., Morvan, D., Cavadias, S., Ph, A., Gonnord, M. F., Coulibaly, K., Vincent, A., Morel, S., Daou, F., Ognier, S., Rousseau, P., & Martin, L. (2005). Pollution control and depollution processes by plasma techniques. *Technical Physics, 50*(5), 603–612.

Askarian, M., Vakili, M., & Kabir, G. (2004). Results of a hospital waste survey in private hospitals in Fars province, Iran. *Waste Management, 24*, 347–352.

Banar, M., Zkan, O., Mine, K. U., & Uog, L. U. (2006). Characterization of the leachate in an urban landfill by physicochemical analysis and solid-phase microextraction. *Environmental Monitoring and Assessment, 121*, 439–459.

Blenkharn, J. I. (2006). Standards of clinical waste management in UK hospitals. *Journal of Hospital Infection, 62*, 300–303.

Bopp, F., Einsfeld, A., Ferreira, G., & Corção, G. (2011). Antibiotic-resistant *Pseudomonas aeruginosa* from hospital wastewater and superficial water: Are they genetically related? *Journal of Environmental Management, 92*(1), 250–255.

Bose, J. C., Poyya, U. M., Roy, G., Prasad, G., & Devi, K. V. (2009). Biomedical waste generation in Puducherry government general hospital and its management implications. *Journal of Environmental Health, 12*(3), 200–205.

Centers for Disease Control and Prevention. (2001). *Healthcare Infection Control Practices Advisory Committee (HICPAC)*. Draft guideline for environmental infection control in healthcare facilities (pp. 96–101).

Central Pollution Control Board (CPCB). (2000). *Guidelines for Design and Construction of Biomedical Waste Incinerator*. Ministry of Environment and Forests, Parivesh Bhawan, Delhi, India.

Central Pollution Control Board (CPCB). (2003). *Guidelines for Design and Construction of Biomedical Waste Incinerator*. Ministry of Environment and Forests, Parivesh Bhawan, Delhi, India.

Centre for Science and Environment (CSE). (2021). 46% increase in COVID-19 biomedical waste generation in India in April-May 2021, says CSE's new statistical analysis. Retrieved from: https://www.cseindia.org/46-per-cent-increase-in-covid-19-biomedical-waste-generation-in-india-in-april-may-2021-10849 (accessed on 18 September 2024).

Chitnis, V., Chitnis, S., Patil, S., & Chitnis, D. (2006). Solar disinfection of infectious biomedical waste: A new approach for developing countries. *Lancet, 362*, 1285–1286.

Connell, J. (2006). Medical tourism: Sea, sun, sand, and surgery. *Tourism Management, 27*(6), 1093–1100.

Environmental Protection Agency (EPA). (2002). *Non-Incineration Technologies for Treatment of Biomedical Waste (Procedures for Microbiological Testing)*. Legislative Authority: Sections 19 and 27; Part XVII, Section 197; Ministry of the Environment Section 27, and Part II, Section 9.

Gayathri, Patil, V., & Pokhrel, K. (2005). Biomedical solid waste management in an Indian hospital: A case study. *Waste Management, 25*, 592–599.

Ghaly, A. E., Kamal, M. A., Mahmoud, N. S., & Cote, R. (2007). Treatment of landfill leachate using limestone/sandstone filters under aerobic batch conditions. *American Journal of Environmental Sciences, 3*(2), 43–53.

Glenn, M. C., & Garwal, R. (1999). Clinical waste in developing countries: An analysis with a case study of India and a critique of the Basle-TWG guidelines.

González, M. A., & Barnes, R. (2002). Comparison of microwave-assisted extraction and waste extraction test (WET) preparation for inductively coupled plasma spectroscopic analyses of waste samples. *Analytical and Bioanalytical Chemistry, 374*, 255–261.

Gupta, S., & Boojh, R. (2006). Report: Biomedical waste management practices at Balrampur Hospital, Lucknow, India. *Waste Management Research, 24*, 584–591.

Henry, G., & Heinke, G. W. (1996). *Environmental Science and Engineering* (2nd ed.). Prentice-Hall.

Hosny, G., & El-Zarka, E. M. (2007). A comparative study on the medical waste disposal in some hospitals in Alexandria. *Waste Management, 27*(6), 746–759.

Javeed, A. G., & Kallihal, K. G. (2021). Awareness, attitude, and practices of biomedical waste management amongst public healthcare staff in Karnataka, India. *Journal of Human & Applied Social Science, 3*(1), 49–63.

Jayanthi, S., Padmasri, M., & Arunkumar, S. (2002). Biomedical waste management practices in a hospital and strategies for safe disposal as per biomedical waste (management and handling) rules 1998. In *Proceedings of ENVIRONICS 2002, National Conference on Pollution Control and Industrial Ecology*, Thiagaraja Engineering College, Madurai.

Kanemitsu, K., Indena, K., Kunishima, H., Ueno, K., Hatta, M., Gunjib, Y., Watanabe, I., & Kaku, M. (2005). Does incineration turn infectious waste aseptic? *Journal of Hospital Infection, 60*, 304–306.

Karamouz, M., Zahraie, B., Kerachian, R., Jaafarzadeh, N., & Mahjouri, N. (2007). Developing a master plan for hospital solid waste management: A case study. *Waste Management, 27*, 626–638.

Kelkar, R. (1998). A practical approach to hospital waste management in India. *Industrial Safety Chronicle*, Oct–Dec, 67–70.

Klangsin, P., & Harding, A. K. (1998). Medical waste treatment and disposal methods used by hospitals in Oregon, Washington, and Idaho. *Journal of Air and Waste Management, 48*(6), 516–526.

Lee, B. K., Ellenbecker, M., & Moure-Eraso, R. (2006). Analyses of the recycling potential of medical plastic wastes. *Waste Management, 22*, 461–470.

Lee, C. C., & Huffman, G. L. (1996). Review: Medical waste management incineration. *Journal of Hazardous Materials, 48*, 1–30.

Lim, T. T., Tay, J. H., Tan, L. C., & Choa, V. C. I. (2004). Changes in mobility and speciation of heavy metals in clay-amended incinerator fly ash. *Environmental Geology, 47*, 1–10.

Manzoor, J., & Sharma, M. (2019). Impact of biomedical waste on the environment and human health. *Environmental Claims Journal, 31*(4), 311–334.

Miyazaki, M., & Une, H. (2005). Infectious waste management in Japan: A revised regulation and a management process in medical institutions. *Waste Management, 25*, 616–621.

Mohandasundaram, V., Bunch, J. M., Suresh, V. M., & Kumaran, T. V. (2003). The environmental implications and economic issues in biomedical waste management in urban Coimbatore, Tamil Nadu, India. In *Proceedings of the Third International Conference on Environment and Health* (pp. 270–275). Chennai, India.

Mohee, R. (2005). Medical waste characterization in healthcare institutions in Mauritius. *Waste Management, 25*, 575–581.

Mosse, A. L., & Savchin, V. V. (2006). Analysis and simulation of the process of medicobiological waste treatment in a plasma chamber incinerator. *Journal of Engineering Physics and Thermophysics, 79*(4).

Mudur, G. (2004). Hospitals in India woo foreign patients. *British Medical Journal, 328*, 7452.

Muehlich, M., Scherrer, M., & Daschner, F. D. (2003). Comparison of infectious waste management in European hospitals. *Journal of Hospital Infection, 55*, 260–268.

Nicholson, W. (2003). Using thermal inactivation kinetics to calculate the probability of extreme spore longevity: Implications for paleomicrobiology and bioterrorism. *Journal of Applied Microbiology, 94*, 246–254.

Ozbeka, F., & Dilek, S. (2004). A study of the dental solid waste produced in a school of dentistry in Turkey. *Waste Management, 24*, 339–345.

Patil, A. D., & Shekdar, A. V. (2001). Health-care waste management in India. *Journal of Environmental Management, 63*, 211–220.

Pavlas, M., & Tous, M. (2008). Efficient waste-to-energy system as a contribution to clean technologies. *Clean Technology and Environmental Policy, 10*(3), 231–233.

Pimshtein, P. G., & Zhukova, V. N. (2003). Stress state and allowable temperature gradients in high-pressure autoclaves. *Chemical and Petroleum Engineering, 39*, 9–10.

Remy, L. (2001). *Managing Hospital Waste is a Big Nasty Deal*. Great Western Pacific Coastal Publication.

Sabour, M. R., Mohamedifard, A., & Kamalan, H. R. (2007). A mathematical model to predict the composition and generation of hospital wastes in Iran. *Waste Management, 27*, 584–587.

Sadeghi, A. (2002). Evaluation, collection, transportation, and disposal of Mashhad's hospital wastes. *Journal of Environmental Management, 13*, 223–246.

Sethi, R., & Molfetta, A. D. (2007). Heat transport modeling in an aquifer downgradient municipal solid waste landfill in Italy. *American Journal of Environmental Sciences, 3*(3), 106–110.

Singh, S., & Prakash, V. (2007). Toxic environmental releases from medical waste incineration: A review. *Environmental Monitoring and Assessment, 132*, 67–81.

Tamplin, S. A., Davidson, D., Powis, B., & Leary, Z. (2005). Issues and options for the safe destruction and disposal of used injection materials. *Waste Management, 25*, 655–665.

Wang, H., Brown, L. S., Magesan, N., Slade, A., Quintern, M., Clinton, W. P., & Payn, T. (2008). Technological options for the management of biosolids. *Environmental Science Pollution and Research, 15*, 308–317.

Williams-Linera, G., & Ewel, J. J. (1984). Effect of autoclave sterilization of a tropical andept on seed germination and seedling growth. *Plant and Soil, 82*, 263–268.

World Health Organization (WHO). (1999). In Pruss, A., Giroult, E., & Rushbrook, P. (Eds.), *Safe Management of Wastes from Health-Care Activities*. World Health Organization.

World Health Organization (WHO). (2001). *Review of Health Impacts from Microbial Hazards in Health Care Wastes (Draft)*. World Health Organization.

World Health Organization (WHO). (2004). *Preparation of National Health-Care Waste Management Plans in Sub-Saharan Countries: Guidance Manual*. World Health Organization and the Secretariat of the Basel Convention.

CHAPTER 7

Bioremediation of Pharmaceutical Compounds: An Overview

DEBRAJ BISWAL*

Department of Zoology, Government General Degree College, Mangalkote, Burdwan, West Bengal, India

*Corresponding author.

ABSTRACT

Advancements in medical sciences have deeply impacted human and animal lives by offering relief to several ailments and rare diseases that were previously unknown. This has been possible because of the tremendous growth of the pharmaceutical industry in the last few years. The effluents from this industry being dumped into the environment have also surged concomitantly. Additionally, several drug metabolites excreted from the human/animal body also end up in domestic wastewater. Earlier studies in this field show the wastewater treatment plants not to be well equipped for the removal of these pharmaceutical compounds. Humans, animals, and other living organisms are thereby exposed to these potentially toxic substances whose effects on living systems have been poorly documented. Unregulated release of these substances has contaminated the environment with high proportions being reported from different aquatic bodies, sediments, and even drinking water. The commonly reported pharmaceuticals include anti-inflammatory drugs, analgesics, anti-diabetics, antibiotics, and beta-blockers. The main problem arises from their poor degradability, prolonged persistence, and varied physico-chemical properties facilitating their entry into living systems. Deleterious effects of these compounds on aquatic organisms (e.g., fishes, crustaceans, copepods, algae) and mammals (e.g., rats) have been observed. They have also been reported to trigger antibiotic resistance in bacteria

Biomedical Waste Management: Recycling and Remediation Strategies. Younis Ahmad Hajam, Javid Ahmad Malik, Anu Pandita, and Deepu Pandita (Eds.)

which may seriously affect the health of humans and animals in the long run. However, the extent of damage they can cause has not been elaborated yet. As such, strict regulations on their disposal are still pending. Moreover, current techniques may not be sufficient enough to identify all the pharmaceutical products contaminating the environment making the situation far more complicated. Currently, they are being regarded as "emerging contaminants of environmental concern." This has drawn research interests to devise mechanisms of filtering out detoxifying or removing these pharmaceutical compounds from industrial and domestic wastewater by affordable and eco-friendly means. Application of certain algae (e.g., *Chlorella vulgaris*), fungi (e.g., *Trametes versicolor*, *Ganoderma lucidum*), and bacteria (e.g., *Acidobacteria, Firmicutes*) have been found to be beneficial in this regard. These bioremediation agents commonly employ mechanisms like biomineralization, bioprecipitation, biodegradation (BD), biosorption, photodegradation, volatilization, and bioaccumulation for detoxification or elimination of the pharmaceuticals and their derivatives. The oxidases, peroxidases, and hydrolytic enzymes play remarkable roles in the process. Though they have some restrictions, they are preferred over chemical processes which often generate toxic by-products. Additionally, they are cheaper than the advanced membrane technologies. Thus, bioremediation can be considered a good mitigation strategy for these polluting agents. However, this field needs rigorous research for further development.

7.1 INTRODUCTION

The development of the health sector has helped in increasing the life expectancy of humans. It has also increased livestock productivity by promoting their growth, and reproduction, and boosting their immunity against various diseases. Consequently, the amounts of pharmaceutical compounds and their metabolites being dumped into the sewage system have multiplied tremendously in the last few years (Arnold et al., 2014). Growing evidence shows the accumulation of pharmaceutical products and their metabolites from industrial and domestic sources in the aquatic and terrestrial ecosystems. Their deposition rates have been reported to be higher than their natural decomposition capabilities exposing humans and wildlife to the toxic effects of these compounds (Akerman-Sanchez and Rojas-Jimenez, 2021). Lack of elaborate research in this field and insufficient evidence of the negative impacts of these substances on human and veterinary health have failed the endeavor to frame regulations for their restricted disposal to the environment (Silva et al., 2015). Failure to detect their presence in soil, water, and other environmental samples adds to the problem (Oulton et al., 2010). As such, they are polluting the environment continuously and have been included under the category of emerging pollutants of growing concern (Akerman-Sanchez and Rojas-Jimenez, 2021). The primary sources include domestic wastewater, hospital wastewater, and wastewater from livestock and pharmaceutical industries (Arnold et al., 2014; Ji et al., 2012). Lipid-regulators, antiepileptics, β-blockers, antibiotics, tranquilizers, X-ray contrast media, and anti-phlogistics have been reported to be the most common pharmaceutical pollutants (Galichet et al., 2004). The development of analytical techniques to identify these compounds even when present in low concentrations has proved helpful in this regard (Oulton et al., 2010).

Pharmaceutical compounds threaten environmental health because they are bioactive substances and mostly hydrophilic to enable their easy uptake by living systems. Additionally, they are designed in such a way that they are not readily degraded before they have exerted their desirable effects on the human or animal body (Silva et al., 2015). Moreover, these substances retain their biochemical and physiological activities even outside the body of humans and animals enabling their impacts on any biological system they come in contact with (Jjemba, 2006). Moreover, the metabolites of these substances can find their way outside the animal body through their feces and urine. Unfortunately, these degradation products of the primary pharmaceutical substances can be more poisonous and biologically active compared to their parent compounds raising further concerns (Akerman-Sanchez and Rojas-Jimenez, 2021). Kümmerer (2009) reported almost 90% of the antibiotics consumed by humans and animals are excreted as a mixture of metabolites. Subsequently, they have been reported from various aquatic bodies, wastewater samples, and sediments (Cui et al., 2020; Yuan et al., 2019). According to Cui et al. (2020), the sludge recovered from wastewater treatment plants is a rich source of breakdown products of antibiotics. This is alarming because of their toxic effects on aquatic organisms such as fishes, micro-crustaceans, and plankton (Baumann et al., 2015). Studies performed by Han et al. (2016) have also reported deleterious effects of similar substances in mammals.

The problems arise because conventional wastewater plants are not designed to detoxify these compounds. Additionally, excessive accumulation of these substances in the environment, at rates higher than their natural degradation processes, poses a threat to the biological systems exposed to them (Paíga et al., 2013). The wastewater treatment plants find it difficult to degrade them because of their high solubility, low hydrophobicity, and acidity (Verlicchi et al., 2012). Though the plants are not designed for removing them, some pharmaceuticals are eliminated to the extent determined by their structure and physicochemical properties (Saussereau et al., 2013). Though advanced membrane technologies and modern oxidative processes can remove these micro-pollutants they have not been popular because of the expenses involved (Joss et al., 2008). Moreover, they require huge energy inputs and use chemicals that generate harmful by-products frequently (Wu et al., 2012). In this regard, biological degradation of these compounds during the tertiary treatment stages of wastewater treatment has been found to be effective, eco-friendly, and affordable (Sahu et al., 2014). This has fueled research projects in this field to identify and use suitable biological agents such as algae, fungi, and bacteria for their bioremediation (Silva et al., 2019).

The current chapter documents the importance of considering pharmaceutical compounds and their derivatives as emerging pollutants. It tries to emphasize the role of various biological agents such as algae, fungi, and bacteria in the detoxification of harmful wastes containing such substances. It highlights their affordability, eco-friendliness, and sustainability over the conventional physical and chemical methods of treatment of wastewater containing pharmaceuticals. It also attempts to discuss the advantages and limitations of bioremediation. Last, but not least, it tries to explore the recent developments in this field and identify the existing research gaps that can be utilized for further improvement.

7.2 SOURCES OF PHARMACEUTICAL POLLUTANTS

The main sources of pharmaceutical compounds are wastewater originating from domestic, agricultural, industrial, hospital, and veterinary sources (Figure 7.1). These contaminants may also make their way to the soil and sediments thereby polluting them (Xiong et al., 2018). Not only the parent compounds but also their metabolites are contaminants. For instance, the antibiotics after being consumed are metabolized within the animal body and the end products are released into the environment through their urine and feces (Zhang et al., 2014). The metabolites, however, depend on the shape, size, physicochemical characteristics, solubility, and electronic nature of the pharmaceuticals (Jjemba, 2006). The routes of elimination depend on the routes of drug administration and their metabolic pathways within the biological systems (Jjemba, 2006). In many cases, improper disposal of antibiotics, incomplete metabolism of antibiotics within the animal body, or discarding expired antibiotics have been observed to contaminate the environment (Grabert et al., 2018).

There are some differences regarding the disposal of pharmaceuticals in rural and urban areas (García et al., 2020). In rural areas, the excretory products of medicated cattle in animal farms are the major routes of entry into the environment. Additionally, the manures used as fertilizers or for soil amendments in organic farming cause contamination of the soil with pharmaceuticals, antibiotics, or their derivatives. Along with this, the runoffs from agricultural farms, fish farming, and biosolids present in sewage treatment plants (STPs) are also important sources of release of pharmaceuticals to the environment (Bird et al., 2019; Sapkota et al., 2008). On the contrary, wastewater effluents from hospitals and direct disposal of effluents from STPs to aquatic bodies have been reported as the major routes of entry of pharmaceuticals into the environment in urban areas (Hocquet et al., 2016). Wastewater from livestock, lack of proper disinfection of septic tanks, and agricultural runoffs have been reported as important sources of contamination of aquatic bodies with antibiotics (Bird et al., 2019; Figure 7.1).

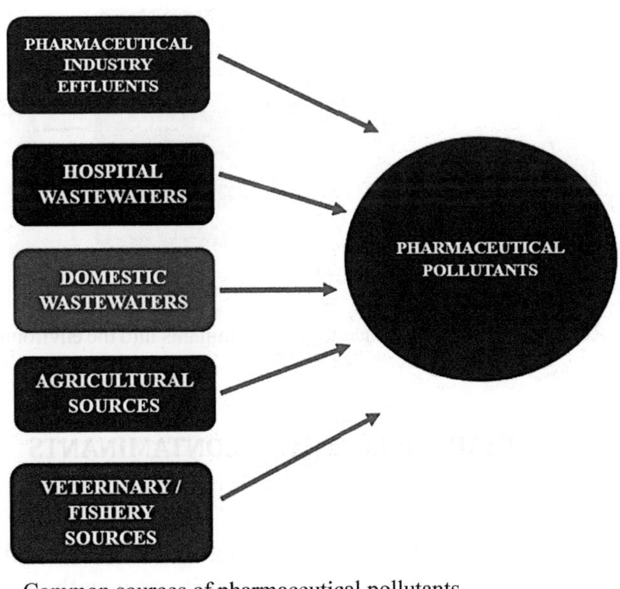

FIGURE 7.1 Common sources of pharmaceutical pollutants.

Direct disposal of unused or expired pharmaceutical products is also a source of environmental contamination (Kümmerer et al., 2016). These compounds are present in high proportions in domestic and hospital wastewater which are directed to the wastewater treatment plants. After the treatment process is over, the water is discharged to surface waters (Gogoi et al., 2018). Since wastewater treatment plants are not capable of removing or detoxifying pharmaceutical compounds, they become one of the primary sources of environmental contamination (Akerman-Sanchez and Rojas-Jimenez, 2021; Figure 7.2). Administration of veterinary medicine leads to the presence of their metabolites in the excretory products of livestock. As such, fertilizers manufactured from these excretory products become a source of contamination of soil and groundwater with pharmaceutical products (Biel-Maeso et al., 2018; Wohde et al., 2016). Eventually, these contaminants find their way to drinking water through the polluted surface and ground waters (Figure 7.2).

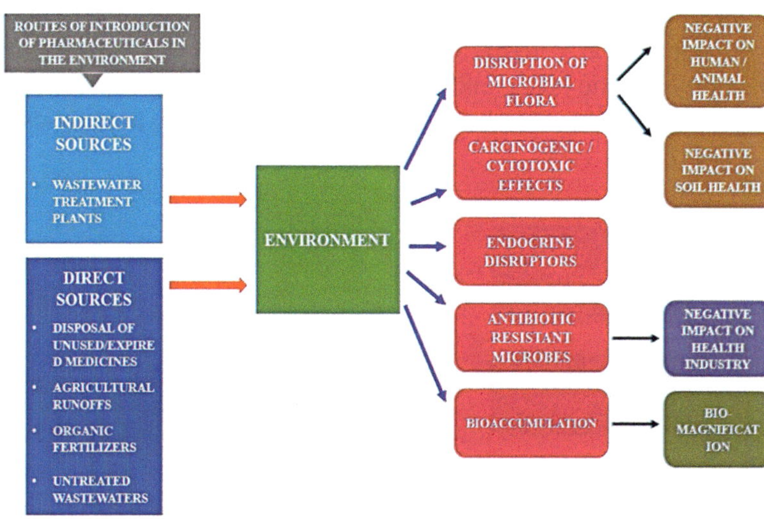

FIGURE 7.2 Routes of entry of pharmaceutical contaminants into the environment and their toxic effects.

7.3 COMMON PHARMACEUTICAL CONTAMINANTS IN THE ENVIRONMENT

Emerging contaminants refer to those natural or artificial chemical compounds whose presence in the environment is not usually monitored but are capable of exerting adverse effects on the environment and human health (Ramírez-Durán et al., 2017). They include several chemical substances including pharmaceuticals, nanomaterials, and even plastics. Apart from the pharmaceutical compounds related to the human health sector, antibiotics, antifungals, antimicrobials, hormones, and growth promoters utilized by the veterinary industry are also included under the pharmaceutical contaminants. Personal hygiene products like soaps, sanitizers, and antimicrobials also fall under this category (Chaturvedi et al., 2021). The pharmaceutical compounds or their derivatives or even the personal care and hygiene products can interact with each other and generate complex mixtures that are even more toxic than their parent compounds. Since this field has been poorly studied not much is known about these interactions and the mixtures obtained. Consequently, their impact on human health remains to be documented (Fuhrman et al., 2015).

Some of the most commonly reported pharmaceutical compounds from aquatic systems include antidepressants (e.g., benzodiazepines), antiepileptics (e.g., carbamazepine), β-blockers (e.g., propranolol, atenolol), analgesics and anti-inflammatory drugs (e.g., ibuprofen, diclofenac, paracetamol), antibiotics (e.g., macrolides, penicillins, sulfonamides, fluoroquinolones, chloramphenicol, quinolones) and lipid-lowering agents (e.g., fibrates; Bush, 1997; Figure 7.3). Natural (e.g., testosterone) and synthetic hormones (e.g., trenbolone, 17α-ethynylestradiol) have also been reported from aquatic ecosystems (Table 7.1). Alarmingly, they have been reported to interfere with the normal physiochemical mechanisms of steroid production and hormonal actions in aquatic animals (Ramírez-Durán et al., 2017; Table 7.2).

Verlicchi et al. (2012) reported antibiotics, anti-inflammatory drugs, antihypertensive drugs, beta-blockers, lipid regulators, and diuretics to be the dominant pharmaceuticals in the water entering the wastewater treatment plants because of their high consumption levels. Among these, drugs such as carbamazepine, naxopren, ibuprofen, acetaminophen, and tramadol top the list of drugs recovered from untreated wastewater (Akerman-Sanchez and Rojas-Jimenez, 2021; Ramírez-Morales et al., 2020). The results are worrisome because conventional wastewater treatment plants are not equipped with mechanisms to eliminate pharmaceutical compounds. In some cases, the concentration of these compounds is higher in the effluent water from wastewater treatment plants compared to the influent water further aggravating the conditions (Sgroi et al., 2017). According to Akerman-Sanchez and Rojas-Jimenez (2021), the list of pharmaceuticals or their derivatives reported from various environmental samples is incomplete because several

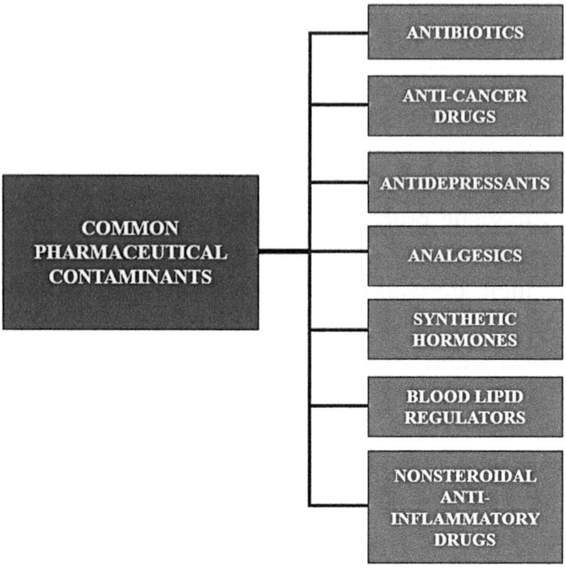

FIGURE 7.3 Major groups of pharmaceutical contaminants.

TABLE 7.1 Some Pharmaceutical Compounds Recovered from Various Samples

Sample Recovered from	Pharmaceutical Compound	Concentration	References
Drinking water	Acetylsalicylic acid	0.3–1.4 ngL^{-1}	Rabiet et al. (2006)
	Carbamazepine	13.9–43.2 ngL^{-1}	Rabiet et al. (2006)
	Diclofenac	1.0–1.6 μgL^{-1}	Radjenović et al. (2009)
	Naproxen	359–2208 ngL^{-1}	Rosal et al. (2010)
	Lincomycin	2.0–4.4 ngL^{-1}	Wang et al. (2011)
	Sulfamethoxazole	12.7 ngL^{-1}	Padhye et al. (2014)
River water	Ibuprofen	0.3–4.5 ngL^{-1}	Rabiet et al. (2006)
	Paracetamol	10.6–72.3 ngL^{-1}	Rabiet et al. (2006)
	Cytarabine	13 ngL^{-1}	Martín et al. (2011)
	Gemcitabine	2.4 ngL^{-1}	Martín et al. (2011)
Estuarine water	Bezafibrate	4–6 ngL^{-1}	Mijangos et al. (2018)
	Ciprofloxacin	3–540 ngL^{-1}	Mijangos et al. (2018)
	Eprosartan	8–183 ngL^{-1}	Mijangos et al. (2018)
	Norfloxacin	4–62 ngL^{-1}	Mijangos et al. (2018)
Seawater	Indomethacin	3 ngL^{-1}	Gros et al. (2012)
	Trimethoprim	1 ngL^{-1}	Gros et al. (2012)
	Phenazone	2 ngL^{-1}	Gros et al. (2012)
	Atorvastatin	1 ngL^{-1}	Gros et al. (2012)

TABLE 7.1 *(Continued)*

Sample Recovered from	Pharmaceutical Compound	Concentration	References
Groundwater	Sulfamethoxazole	3 ngL^{-1}	Vulliet and Cren-Olivé (2011)
	Fenofibric acid	0.4 ngL^{-1}	Vulliet and Cren-Olivé (2011)
	Enrofloxacin	65.2–264 ngL^{-1}	López-Serna et al. (2013)
	Doxycycline	27.6–188 ngL^{-1}	López-Serna et al. (2013)
	Pravastatin	12.2 ngL^{-1}	López-Serna et al. (2013)
	Ibuprofen	92 ngL^{-1}	Petrović et al. (2014)
	Phenazone	23.4 ngL^{-1}	Petrović et al. (2014)
Domestic wastewater	Acetaminophen	50.2 to 58.7 μgL^{-1}	Hom-Diaz et al. (2017)
	Ibuprofen	39.0–52.8 μgL^{-1}	Hom-Diaz et al. (2017)
	Codeine	33 ± 1 ngL^{-1}	Hom-Diaz et al. (2017)
	Atenolol	7.8 μgL^{-1}	Hom-Diaz et al. (2017)
	Lorazepam	3.7 μgL^{-1}	Hom-Diaz et al. (2017)
	Hydrochlorothiazide	228 ngL^{-1}	Hom-Diaz et al. (2017)
Urban wastewater	Propranolol	0.06 μgL^{-1}	Cruz-Morató et al. (2013)
	Citalopram	0.04 μgL^{-1}	Cruz-Morató et al. (2013)
	Acetaminophen	1.56 μgL^{-1}	Cruz-Morató et al. (2013)
	2-hydroxycarbamazepine	163.8 μgL^{-1}	Cruz-Morató et al. (2013)
Hospital effluent	Tamoxifen	26.3–133.4 ngL^{-1}	Ferrando-Climent et al. (2013)
	Azathioprine	14.5–187.9 ngL^{-1}	Ferrando-Climent et al. (2013)
	Ifosfamide	6–86,200 ngL^{-1}	Gómez-Canela et al. (2014)
	Acetaminophen	109.3–114.4 μgL^{-1}	Cruz-Morató et al. (2014)
	Ofloxacin	3.34–31.9 μgL^{-1}	Cruz-Morató et al. (2014)
	Metformin	0–1159 ngL^{-1}	Papageorgiou et al. (2019)
Veterinary hospital effluent	Antibiotics	34.4 μgL^{-1}	Badia-Fabregat et al. (2016)
	Analgesics and anti-inflammatories	1.16 μgL^{-1}	Badia-Fabregat et al. (2016)
	Psychiatric drugs	2.26 μgL^{-1}	Badia-Fabregat et al. (2016)
Effluents from wastewater treatment plant	Meloxicam	735 ngL^{-1}	Gros et al. (2012)
	Estrone	11.8 ± 1.2 ngL^{-1}	Zhou et al. (2014)
	Progesterone	2.4 ± 0.8 ngL^{-1}	Zhou et al. (2014)
	Salicylic acid	23–419 ngL^{-1}	Biel-Maeso et al. (2018)
	Gemfibrozil	518–3720 ngL^{-1}	Biel-Maeso et al. (2018)
	Amitriptiline	4–59 ngL^{-1}	Biel-Maeso et al. (2018)
Soil and sediment	Fenofibrate	13.20–17.23 ng g^{-1}	Vazquez-Roig et al. (2010)
	Ciprofloxacin	42 ng g^{-1}	Wu et al. (2014)
	Ofloxacin	0.09–2.99 ng g^{-1}	Osorio et al. (2016)
	Citalopram	0.23–7.79 ng g^{-1}	Osorio et al. (2016)
	Sulfamethoxypyridazine	0.11–0.37 ng g^{-1}	Biel-Maeso et al. (2017)

TABLE 7.2 Toxic Effects of Some Pharmaceutical Compounds in Aquatic Organisms

Pharmaceutical Compounds	Toxic Concentrations Reported	Species Studied	Toxic Effects	References
Ibuprofen	0.0001–25 mgL^{-1}	Zebra fish (*Danio rerio*)	Oxidative stress	Bartoskova et al. (2013)
	5.95–7.44 mgL^{-1}	Green neon shrimp (*Neocaridina denticulata*)	Increased mortality	Sung et al. (2014)
	2.9 mgL^{-1}	*Daphnia magna*	DNA damage	Gómez-Oliván et al. (2014)
	500 μgL^{-1}	Zebra fish (*Danio rerio*)	Delay in hatching of eggs; neurotoxic effects on embryo	Xia et al. (2017)
	0.1–10 μgL^{-1}	American catfish (*Rhamdia quelen*)	Nephrotoxic and immunosuppressive effects	Mathias et al. (2018)
Naproxen	269.15 mgL^{-1}	*Cyprinus carpio*	Increased mortality	Gheorghe et al. (2016)
	98.3 mgL^{-1}	Zebra fish (*Danio rerio*)	Malformation of embryo	Li et al. (2016)
	5 mgL^{-1}	*Oryzias latipes*	Increased mortality of juveniles	Kwak et al. (2018)
	74.13 mgL^{-1}	*Moina macrocopa*	Loss of mobility	Kwak et al. (2018)
	85.34 mgL^{-1}	*Daphnia magna*	Loss of mobility	Kwak et al. (2018)
Amitriptyline	3000 μgL^{-1}	Zebra fish (*Danio rerio*)	Increased mortality of embryo	Sehonova et al. (2019)
	3000 μgL^{-1}	*Xenopus tropicalis*	Cardiotoxic effects in embryo	Sehonova et al. (2019)
Simvastatin	1000 ngL^{-1}	Zebra fish (*Danio rerio*)	Reduced survival rate; decreased levels of triglycerides; decreased body weight	Barros et al. (2018)
	8 and 40 ngL^{-1}	Zebra fish (*Danio rerio*)	Developmental abnormality of fish embryo	Barros et al. (2018)
5-Fluorouracil	50–500 μgL^{-1}	Fathead minnow (*Pimephales promelas*)	Teratogenic effects with dwarfed embryos	DeYoung et al. (1996)
	50–500 μgL^{-1}	African clawed frog (*Xenopus laevis*)	Teratogenic effects with malformations of embryos	DeYoung et al. (1996)

TABLE 7.2 *(Continued)*

Pharmaceutical Compounds	Toxic Concentrations Reported	Species Studied	Toxic Effects	References
Tamoxifen	10–8000 mg kg^{-1}	Japanese medaka (*Oryzias latipes*)	Alterations in vitellogenin levels in males and females	Chikae et al. (2004)
	0.019 μgL^{-1}	Sea urchin (*Strongylocentrotus purpuratus*)	Teratogenic effects with reduced embryonic development	Roepke et al. (2005)
	3.71 μgL^{-1}	African clawed frog (*Xenopus laevis*)	Altered gene expression in the liver; fluctuations of steroid levels in plasma	Urbatzka et al. (2007)
	1 mg kg^{-1}	Black goby (*Gobius niger*)	Anti-estrogenic effects were noticed in the livers of males.	Maradonna et al. (2009)
	50–100 mg kg^{-1}	Tench (*Tinca tinca*)	Increased mortality; masculinization of fishes	Hulak et al. (2010)
	10 μgL^{-1}	Mangrove killifish (*Krytolebias marmoratus*)	Age-specific hindrance of DNA repair in certain tissues	Rhee et al. (2011)
	0.12–5.26 μgL^{-1}	*Daphnia pulex*	Reduced size; poor surviv-ability rates; impaired reproduction	Borgatta et al. (2016)
Cyclophosphamide	200 mg kg^{-1}	Walking catfish (*Clarias batrachus*)	Immunosuppressive effects	Kumari and Sahoo (2005)
	1–1000 μgL^{-1}	Snail (*Lymnea stagnalis*)	No adverse effects on the immune system	Boisseaux et al. (2017)
Diclofenac	500 μgL^{-1}	Zebra fish (*Danio rerio*)	Reduced swimming capabilities	Xia et al. (2017)
Acetaminophen	5.33-7.84 mgL^{-1}	Green neon shrimp (*Neocaridina denticulata*)	Increased mortality	Sung et al. (2014)
Ketoprofen	4.03 mgL^{-1}	*Scenedesmus obliquus*	Growth inhibition	Wang et al. (2020)
Aspirin	103.05 mgL^{-1}	*Scenedesmus obliquus*	Growth inhibition	Wang et al. (2020)
Fluoxetine	44.99 ± 1.76 $\mu g\ L^{-1}$	*Scendesmus acutus*	Inhibition of growth	Johnson et al. (2007)
Sertraline	763.66 ± 25.42 μgL^{-1}	*Chlorella vulgaris*	Inhibition of growth	Johnson et al. (2007)
Fluvoxamine	4002.88 ± 142.52 μgL^{-1}	*Pseudokirchneriella subcapitata*	Inhibition of growth	Johnson et al., 2007

TABLE 7.2 *(Continued)*

Pharmaceutical Compounds	Toxic Concentrations Reported	Species Studied	Toxic Effects	References
Paroxetine	10 μgL^{-1}	Zebra fish (*Danio rerio*)	Abnormal larval development; reduced egg hatching	Nowakowska et al. (2020)
Sertraline	10 μgL^{-1}	Zebra fish (*Danio rerio*)	Reduced multiplication of hepatocytes in larvae	Nowakowska et al. (2020)
Atorvastatin	1.2 μgL^{-1}	Mussels (*Mytilus edulis*)	Depletion of energy reserves due to increased basal metabolic rate; Depletion of stored lipid, protein, and carbohydrates	Falfushynska et al. (2019)

others may have escaped detection. This may be due to the absence of analytical techniques to detect them or the failure of the existing techniques to identify them when present in low concentrations (Akerman-Sanchez and Rojas-Jimenez, 2021). The development of sensitive techniques to detect them in the environment is a challenge because of the diverse physico-chemical nature of these substances. As such, most of them can pass through wastewater treatment plants unaltered and end up contaminating surface water, groundwater, and drinking water (Mahmood et al., 2019; Figure 7.4).

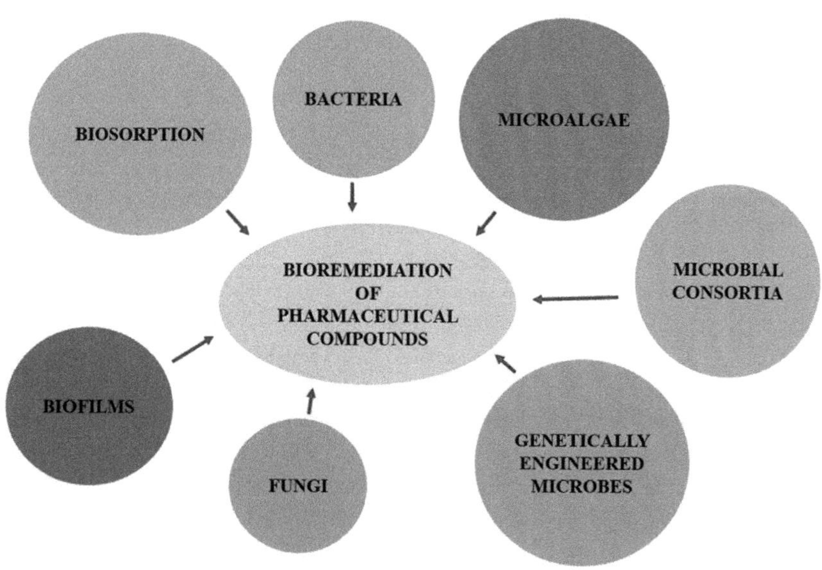

FIGURE 7.4 Commonly used agents for bioremediation of pharmaceutical compounds.

7.4 NEGATIVE IMPACTS OF PHARMACEUTICAL CONTAMINANTS

Pharmaceutical compounds are complex chemicals with a molecular mass lower than 500 Daltons and are highly variable in terms of structure, function, and molecular weight. Some are moderately hydrophilic while others are lipophilic in nature. They are polar molecules with numerous ionizable groups. Recently, several pharmaceutical compounds have been reported from various environmental samples raising concerns about the safety of human and animal health (Tables 7.1 and 7.2). The problem arises because of their ability to persist for long. Pharmaceutical compounds like sulfamethoxazole, naproxen, and erythromycin have been reported to last for over a year in the environment while others like clofibric acid can persist for several years and can undergo bioaccumulation (Rivera-Utrilla et al., 2013). The pharmaceutical compounds and their derivatives occur at concentrations ranging from ng per liter to 100 μg per liter (approximately) in the environment (Ramírez-Durán et al., 2017; Table 7.1). They are mostly present as complicated solutions making it very difficult to discern the effect of these compounds individually on living systems. This makes their toxic effects on living organisms under real conditions largely unknown.

The presence of these compounds in the environment increases their probability of coming in contact with living organisms. The situation is exacerbated by the ability of pharmaceuticals to retain their pharmacological potential even outside the animal body. This allows them to generate adverse impacts on animals when they come in contact with them (Tran et al., 2019). Shockingly, the pharmaceuticals have been reported to be harmful at very low concentrations (lower than 10^3 ng/L; Fuhrman et al., 2015; Kong et al., 2015; Table 7.2). The effects generated depend on the species being exposed as well as the nature of the pharmaceutical compounds and/or their mixtures. The interactions are, however, influenced by several factors that hinder the understanding of their effects on living organisms (Niemuth et al., 2015). The available reports show that these contaminants can disrupt the bacterial flora within living organisms or impede their biochemical and metabolic pathways (Caracciolo et al., 2015). The presence of steroidal estrogens (e.g., 17α-ethynylestradiol, 17β-estradiol) in water has been reported to induce feminization in fishes (e.g., *Pimephales promelas*) and obstruct their hormonal pathways (Caracciolo et al., 2015; Kramer et al., 1998). Several pharmaceutical compounds such as doxylamine, carbinoxamine, and ranitidine can produce the antecedents of N, N-nitrosodimethylamine which are known for their carcinogenic properties. They can have serious effects on human health (Kramer et al., 1998). Furthermore, pharmaceuticals can hinder microbial growth, destroying the activities of the soil microbial flora and impacting the denitrification rate of bacteria (Xiong et al., 2018). These may affect the overall microbial community of various ecosystems and cause alterations in their composition (Xiong et al., 2018). Table 7.2 enlists the toxic effects of some pharmaceutical compounds on aquatic creatures.

The presence of antibiotics in the environment can also promote the development of antibiotic resistance in bacteria with serious outcomes on human and veterinary health in the long run (Dijkshoorn et al., 2007). For instance, the development of antibiotic resistance in *Acinetobacter baumannii* due to continuous exposure to antibiotics in hospitals has made it difficult to control this opportunistic pathogen in humans (Dijkshoorn et al., 2007). Existing reports show the inhibitory effects of antibiotics on bacteria, algae, fungi, mollusks, and arthropods in aquatic and terrestrial ecosystems (Fonseca et al., 2018; Figure 7.4). The effect of antibiotics on bacteria has attracted special research attention because of their ability to survive adverse situations by coping with environmental alterations quickly. This is facilitated by their rapid generation time and capability to obtain novel traits through horizontal gene transfer (Tran et al., 2018). Additionally, they can also pass on the antibiotic resistance genes to their offspring by vertical transfer during their multiplication (Salyers et al., 2004). This can be exemplified by the occurrence of 38 varieties of tetracycline resistance genes in bacteria (Grabert et al., 2018). It is, therefore, evident that regular antibiotic consumption can augment the presence of antibiotic resistance in the environment (Gillings and Stokes, 2012). Not only antibiotics but also the free DNAs carrying antibiotic resistance genes have been reported to persist in the environment for a long time. Cummings et al. (2010) found huge amounts of antibiotic resistance genes in the urban wetlands due to poor treatment or management of wastewater. Antibiotic resistance genes and bacteria have been reported from coastal waters as well (Belding and Boopathy, 2018).

Conventional STPs that utilize activated sludge biological treatment promote the development of antibiotic-resistant genes and antibiotic-resistant bacteria. Actually, the antibiotics in the activated sludge exert selective pressures on the population of microorganisms allowing only the resistant strains to survive (Bird et al., 2019). This is possible because the antibiotics are present in non-lethal doses permitting the bacterial population to undergo physiological or molecular adjustments as a result of which the bacteria gradually adapt to the antibiotics. This is strongly evidenced by the presence of multidrug-resistant bacteria from polluted environments (Bhullar et al., 2012). It is not an encouraging find because of the tendencies of bacteria to evolve resistance to almost all antibiotics in the future which may seriously impact the health of animals and the environment.

Drugs applied in the chemotherapeutic treatment of cancers can also be released to the environment by similar pathways and affect living systems. Some of these drugs can exert cytotoxic effects by damaging the structure and function of DNA in living cells (Azuma, 2018; Table 7.2). Others can affect biological systems by disrupting the endocrine pathways (Azuma, 2018). Eventually, the growth, survival, and reproduction of non-target aquatic organisms can be severely affected (Akerman-Sanchez and Rojas-Jimenez, 2021; Table 7.2). The lipophilic nature and long persistence of several pharmaceutical compounds make them suitable candidates for accumulation in organisms belonging to different trophic levels and undergoing biomagnification (Xiong et al., 2018). This represents a serious condition because the ecotoxicological effects do not remain restricted to an individual organism, but entire populations are affected (Nassour et al., 2020).

7.5 BIOREMEDIATION OF PHARMACEUTICAL COMPOUNDS

Bioremediation can be described as a biological treatment that utilizes microbial enzymes to detoxify xenobiotics and other difficult pollutants to shorten their longevity in the environment or bring about their complete mineralization (Boopathy, 2011). It is to be noted that the mineralization reaction results in the formation of water and carbon dioxide as the primary end products (Boopathy, 2011). However, the process is influenced by environmental factors as well as several physical and chemical factors (Gillespie and Philip, 2013). Some of these include structural features, physicochemical properties, and concentration of the contaminants together with their nature of toxicity and degradation conditions (Misal et al., 2011). The reactions are also governed by the microbial species being used and the duration they are allowed to act upon the contaminants (Misal et al., 2011).

Using natural microbial flora for the bioremediation of complex organic compounds has been a common practice for a long. They can utilize the pollutants as their source of nutrition and the breakdown products in turn can nourish other organisms in the food chain (Caracciolo et al., 2015). Sometimes, it has been observed that dumping certain organic compounds in the environment can alter the natural microbial flora and disrupt the interactions among them. This can be injurious to the environment in the future. It is, therefore, necessary to have a thorough understanding of the pathways of metabolic breakdown of pollutants while preserving the microbial community (Lahti and Oikari, 2011). Often, the microbial community as a whole is more beneficial in degrading pollutants because of the co-metabolic reactions (Caracciolo et al., 2015). The problem arises when the rate of generation and accumulation of the pollutants exceed the microbial breakdown rate. Thus, bioremediation heavily relies on the natural detoxification mechanisms of the microbes, but care should be taken to maintain their community structure to derive the maximum benefits (Megharaj et al., 2011). This requires careful choosing of the microorganisms and understanding the interactions among them (Paul et al., 2005). This area has great prospects in the future because of the biodiversity of microorganisms in nature and the variations in their catalytic mechanisms to metabolize organic compounds (Megharaj et al., 2011). Moreover, they can be adapted to various environmental fluctuations allowing them to function even under several constraints (Megharaj et al., 2011). Table 7.3 represents various mechanisms employed by microbes to detoxify pharmaceutical pollutants.

Several authors find bioremediation to be more effective than conventional methods because it helps in the complete degradation or transformation of the contaminants instead of simply moving them from one medium to another (Mashi, 2013). Additionally, it is eco-safe and affordable compared to other methods (Andrade et al., 2010). However, bioremediation is not a foolproof method because it is effective on compounds that are biodegradable. Moreover, it often results in partial detoxification of pollutants (Andrade et al., 2010). Above all, not all microbes are equally efficient in bioremediation. This necessitates the search for suitable microbes that can bring about the degradation of pharmaceutical compounds by reactions that can withstand harsh reaction conditions (Andrade et al., 2010). Several genetically engineered microbes have also been used for this purpose (Gaylarde et al., 2005). Fungi, bacteria, and microalgae are the most commonly used agents for bioremediation (Figure 7.4).

7.5.1 FUNGI

Fungi have long been used for bioremediation processes because they are rich in both intracellular and extracellular enzymes which being non-specific in nature can exert their effects on a broad range of compounds. As such,

TABLE 7.3 Common Techniques and Corresponding Microbes Used to Detoxify Pharmaceutical Contaminants

Reaction/Technique Applied	Pharmaceutical Compounds Detoxified	Biological Agent(s)	Bacteria/Fungi/ Microalgae	References
Biodegradation	Paracetamol	*Stenotrophomonas* sp., *Pseudomonas* sp.	Bacteria	Zhang et al. (2013)
	17α-ethinylestradiol	*Trametes versicolor*	Fungi	Blánquez and Guieysse (2008)
	Sulfamethoxazole	*Chlorella vulgaris*	Algae	Bodin et al. (2016)
	Ibuprofen	*Lessonia nigrescens*	Algae	Eriksen (2008)
Biotransformation	Moxifloxacin	*Labrys portucalensis* F11	Bacteria	Carvalho et al. (2016)
	17α-ethynylestradiol	*Desmodesmus subspicatus*	Algae	Maes et al. (2014)
	Estradiol valerate, ethinylestradiol	*Chlorella vulgaris*	Algae	Lai et al. (2002)
	Ethinylestradiol	*Chlorella ellipsoidea*, *Scenedesmus vacuolatus*	Algae	Greca et al. (2008)
Bioaccumulation	Carbamazepine	*Pseudokirchneriella subcapitata*, *Scenedesmus obliquus*	Algae	Joss et al. (2006)
Biosorption	Penicillin G	*Rhizopus arrhizus*	Fungi	Navarro et al. (2014)
	Sulfapyridine	*Trametes versicolor*	Fungi	Rodriguez-Rodriguez et al. (2012)
	Ketoprofen	*Pleurotus ostreatus*	Fungi	Palli et al. (2017)
	Carbamazepine	*Macrocystis integrifolia*	Algae	Eriksen (2008)
	Tetracycline	*Chlorella vulgaris*	Algae	De Godos et al. (2012)
Biostimulation	Sulfamethazine	*Geobacillus* sp. strain S-07	Bacteria	Pan et al. (2017)
	Ciprofloxacin, norfloxacin	*Labrys portucalensis* F11	Bacteria	Carvalho et al. (2005)

they can bring about the oxidative breakdown of several pharmaceutical compounds as well as their metabolites (Durairaj et al., 2015). They are highly profitable in the sense that their enzymes can remain functional in a wide range of pH and humidity allowing them to withstand sudden alterations in the reaction conditions (Anastasi et al., 2013). These properties enable them to degrade highly complicated organic compounds (Anastasi et al., 2013).

The enzymatic system of white-rot fungi (WRF) finds wide application in the biodegradation (BD) of pharmaceutical pollutants because they are non-stereo-selective, non-specific, and are influenced by the concentration of free radicals (Pointing, 2001). Though they are naturally lignin-degrading enzymes, they can be used to break down toxic and recalcitrant pollutants. This is brought about by a system of extracellular enzymes consisting of mainly peroxidases (e.g., lignin peroxidase, manganese peroxidase) and laccases (Wesenberg et al., 2003). WRF secretes peroxidases during secondary metabolism and has higher redox potential compared to laccases (Gasser et al., 2012). They are, however, susceptible to deactivation by hydrogen peroxide unlike laccases (Gasser et al., 2012). WRF produces laccases during primary metabolism which are primarily glycosylated multi-copper oxidoreductases (Tortella et al., 2005). Though they are more stable compared to peroxidases, they are affected by the presence of chloride ions (Kim and Nicell, 2006). Additionally, the WRF hyphal structure helps to purify water through biosorption. In this process, the drugs or their derivatives are attached to the hyphal surface or can be taken up by the cells thereby helping the water to get rid of these pollutants (Lu et al., 2016).

Lignin peroxidases act by oxidizing different aromatic compounds followed by the formation of cationic aryl radicals which are eventually degraded by several enzymatic pathways (Sayadi and Ellouz, 1995). Manganese peroxidase, on the other hand, can oxidize various phenolic substrates by bringing about the oxidation of Mn^{2+} to Mn^{3+} (Sayadi and Ellouz, 1995). The main reason for encountering fungal laccases so commonly in literature is because of their nature to use only oxygen as the terminal electron acceptor generating water as the sole derivative (Hofmann and Schlosser, 2016). Some of the most widely used WRF for bioremediation purposes include *Trametes versicolor*, *Phlebia tremellosa*, *Bjerkandera adusta,* and *Phanerochaete chrysosporium* (Rodarte-Morales et al., 2011). Among these, *Trametes versicolor* is highly beneficial because it possesses the cytochrome P450 complex together with laccase and lignin peroxidase enzymes. This makes *Trametes versicolor* highly versatile and effective in the breakdown of pharmaceutical contaminants (Asif et al., 2017). Their efficiency has been proved by their ability to degrade wastewater-derived drugs such as ibuprofen, carbamazepine, and naproxen (Marco-Urrea et al., 2009; Rodríguez-Rodríguez et al., 2010). The efficiency of *Phanerochaete chrysosporium* and *Bjerkandera adusta* in the degradation of fluoxetine and citalopram present in wastewater has been reported by Rodarte-Morales et al. (2011).

Several workers have suggested the use of isolated fungal enzymes instead of the fungi themselves in BD reactions (Jebapriya and Gnanadoss, 2013). This helps to overcome the lag phase of fungal growth and quickens the process by reducing the treatment time (Jebapriya and Gnanadoss, 2013). Here, it needs to be mentioned that long growth cycles of fungi and their ability to produce spores are regarded as their shortcomings (Spina et al., 2012). The isolated enzymes also serve to lower the amount of sludge production and facilitate proper regulation of the entire process (Jebapriya and Gnanadoss, 2013). Marco-Urrea et al. (2009) experimentally demonstrated the degradation of carbamazepine and clofibric acid by *Trametes versicolor* were brought about by the intracellular cytochrome P450 enzyme system. However, Nguyen et al. (2014) found that presence of inhibitors of the intracellular enzyme system can negatively impact the bioremediation properties of fungi. The extracellular enzymes of fungi come to the rescue (Nguyen et al., 2014). This is exemplified by the inhibitory effects of piperonyl butoxide on cytochrome P450 suppressing the degradation of sulfathiazole by *Trametes versicolor* (Zhang and Geißen, 2012). Here lies the importance of the existence of intracellular and extracellular enzymatic pathways in fungi which work independently.

Studies by Jelic et al. (2012) revealed that reactions involving elimination of carbamazepine by *Trametes versicolor* were initially catalyzed by manganese peroxidase and laccase. The importance of laccase and manganese peroxidase enzymes in BD of 17α-ethinylestradiol by another strain of WRF, *Pleurotus* sp. P1 have also been reported (Santos et al., 2012). The necessity of lignin peroxidase enzymes in these reactions by an unidentified basidiomycetes WRF strain BNI have been highlighted by Santos et al. (2012). Studies by Rodarte-Morales et al. (2011) have emphasized the role of manganese peroxidase enzymes in the detoxification of pharmaceutical compounds such as naproxen, citalopram, sulfamethoxazole, carbamazepine, and diclofenac by different WRF strains like *Phanerochaete chrysosporium* and *Bjerkandera adusta* (Rodarte-Morales et al., 2011).

Apart from inhibitors, researchers have also focused on facilitators or mediators such as 1-hydroxybenzotriazole which can improve the enzymatic activities of laccases from *Trametes versicolor* (Nguyen et al., 2014).

Therefore, removal of inhibitors from the degradation system coupled with addition of mediators can facilitate faster and more efficient degradation of pharmaceutical contaminants. According to Fabbrini et al. (2002), the mediators or facilitators act by getting oxidized to their free radical forms by laccases which can help in oxidizing the pollutants in a more non-specific manner thereby augmenting the BD rates of the enzymes. This is exemplified by the studies of Becker et al. (2016) using immobilized laccase derived from *Trametes versicolor* to degrade several antibiotics both in the presence and absence of the mediator, syringaldehyde. The results show that the degradation activities of laccases were significantly improved in the presence of the mediator. Attempts have also been made to create artificial laccase mediators such as 1-hydroxybenzotriazole to facilitate the enzymatic degradation of pharmaceutical compounds and their derivatives (Cañas and Camarero, 2010). However, natural mediators such as acetosyringone and syringaldehyde have been found to be more acceptable because they are affordable and eco-friendly compared to the synthetic mediators (Cañas and Camarero, 2010).

According to many scientists, the morphological and biochemical properties of fungi make them excellent candidates for bioremediation of pharmaceutical pollutants. However, using suitable combinations of fungi belonging to different species can be more advantageous in this regard because the pharmaceutical derivatives often react amongst themselves to form complexes and mixtures of diverse nature (Lu et al., 2016). The enzymatic machinery of a single fungal species may not be enough to combat these complicated cocktail of contaminants (Rahman et al., 2014). Bioprospecting of novel fungal species or strains and their suitable applications could help increase the functional efficiency of wastewater treatment plants (Cruz del Álamo et al., 2020). For instance, an assortment of fungi consisting of *Trichoderma longibrachiatum*, *Rhizopus microsporus*, *Aspergillus niger* (AN), *Trametes polyzona,* and *Mucor circinelloides* have been reported to efficiently detoxify pharmaceutical compounds such as diclofenac, ibuprofen, and carbamazepine along with their derivatives (Kasonga et al., 2020). Additionally, their mechanism of introduction and culture within the wastewater treatment plants also need proper research because dispersed growth of fungal mycelia can hinder the smooth functioning of the plants (Mir-Tutusaus et al., 2018). Accordingly, Akerman-Sanchez and Rojas-Jimenez (2021) have proposed the restriction of fungi within small detachable chambers within the wastewater treatment plants to enhance their functionality. Earlier reports on bioremediation efficiency of immobilized

T. versicolor and *P. chrysosporium* have been provided by Dalecka et al. (2020); and Li et al. (2015), respectively.

According to Saeed and Iqbal (2013), the immobilization matrix should ideally be porous with open ends to enable it to come in contact with the aqueous medium containing pollutants. It should also be affordable, easy to handle, highly stable and have the ability to withstand wear and tear due to repeated usage. Silica-alginate-fungus biocomposites and gel matrices are good representatives of such matrices (Carabajal et al., 2016). Cobas et al. (2013) studied the efficacy of *Trichoderma longibrachiatum* immobilized with nylon sponge in reactive bio-barriers in removal of polycyclic aromatic compounds and found the results quite satisfactory. Lu et al. (2016) opined that the saprophytic nature of the fungi helps them to utilize the pharmaceutical compounds as their source of nutrition eventually degrading them. However, maintenance of appropriate physico-chemical characteristics of their growth medium is essential for their proper growth and functions compared to the bacterial population present in the medium (Lu et al., 2016). This mainly includes temperature, pH, and C: N ratio (Carbon:Nitrogen ratio) of the medium (Lu et al., 2016). Nevertheless, identification of appropriate fungal consortia may go a long way in developing bioremediation strategies for wastewaters contaminated with pharmaceuticals in real life situations (Cruz del Álamo et al., 2020).

Mycoremediation has been reported to be severely impacted by the presence of bacteria in the reaction medium because they often compete for the same substrate and have destructive influences on the fungal mycelia reducing their growth and disrupting their enzymatic activities (Asif et al., 2017). As a remedy, several simple mechanisms need to be followed while controlling the reaction conditions of the biological treatment system. For instance, regulating the pH can favor the growth of fungi over bacteria since both are pH-sensitive with fungi preferring acidic conditions compared to bacteria (Asif et al., 2017). Treatment of wastewater with suitable disinfectants such that they kill or inactivate the bacteria selectively can also be a sustainable mitigation strategy. The problem can also be overcome by treating the wastewater with appropriate coagulation and flocculation agents before fungal BD. This mechanism has been reported to decrease the bacterial count in the medium (Asif et al., 2017). Application of micro-screens to wash out the bacteria from wastewater can also help to mitigate the process. Moreover, immobilization of fungi on suitable carriers and periodically replacing them can assure their functionality over bacteria (Asif et al., 2017). One of the greatest advantages of using fungi-mediated bioremediation is its non-dependency on sterile conditions (Cruz-Morató et al., 2014). For instance, *Trametes versicolor* has been reported to retain their ability of bioremediation both under sterile and non-sterile conditions (Cruz-Morató et al., 2013).

7.5.2 BACTERIA

The ease of using bacteria for bioremediation of wastewaters lies in the fact that they are one of the most abundant organisms found in many different ecosystems including the living systems and possess unraveled potentialities for BD of recalcitrant pollutants (Kallmeyer et al., 2012). Members of the bacterial phyla Bacteroidetes, Actinobacteria, and Proteobacteria are reportedly very efficient in removal of pharmaceutical compounds from wastewaters (Amorim et al., 2018). Furthermore, the bacterial groups Cyanobacteria, Firmicutes, Chloroflexi, Epsilonproteobacteria, Chlamydiae, Tenericutes, Verrucomicrobia, Acidobacteria, and Gemmatimonadetes are also very beneficial as BD agents of pharmaceuticals (Amorim et al., 2018). Muter et al. (2017) reported that biostimulation increases the BD efficiency of bacteria belonging to phyla Firmicutes, Chloroflexi, Bacteroidetes, Nitrospirae, Verrucomicrobia, Proteobacteria, Actinobacteria, Gemmatimonadetes, Chlorobi, and Acidobacteria. At times, biostimulation could enhance the BD efficiency of pharmaceuticals like venlafaxine, bisoprolol, salbutamol, norfluoxetine, gemfibrozil, metoprolol, and 17β-estradiol by some bacteria by as much as 90% (Muter et al., 2017). Some bacteria can mediate biotransformation of pharmaceuticals instead of degrading them (Table 7.3). However, they have not yet been studied in detail (Verlicchi et al., 2012).

Bacterial bioremediation of pharmaceutical compounds is primarily based on biostimulation and bioaugmentation (BA; Alexandrino et al., 2017; Fernandes et al., 2020). Biostimulation involves addition of external nutrient sources to the contaminated site to promote the growth and multiplication of bacterial communities by overcoming nutrient shortage (Abed et al., 2014). BA, on the other hand, involves direct addition of the microbial inocula to the polluted site to increase the rate of BD of the contaminants (Abed et al., 2014; Yu et al., 2005). In many cases, a combination of both the techniques have been found to be advantageous because the direct addition of bacterial inocula to the polluted site may exhaust the nutrients in the medium. This may lead to competition among the bacteria for nutritional sources lowering the outcomes. External source of nutrients may help to tackle the problems to a large extent.

Bacterial strains employed for bioremediation are usually the ones collected from heavily polluted zones or extreme environments. Their ease of adaptation to hostile situations can be exploited to our advantage. For instance, the bacterial strain *Labrys portucalensis* F11 isolated from a highly polluted industrial site can be used for the degradation of antibiotics such as norfloxacin, ciprofloxacin, and ofloxacin (Carvalho et al., 2005). According to Amorim et al. (2014), the addition of sodium acetate as an external carbon source could enhance the BD process. Moreira et al. (2014) studied this further and concluded that BD rates of enantiomeric pharmaceutical compounds vary greatly. *Labrys portucalensis* F11 has been reported to dehalogenate diclofenac and moxifloxacin in the presence of the co-substrate sodium acetate (Carvalho et al., 2016; Moreira et al., 2018). Other bacterial strains like *Rhizobium* sp. C12 and *Brevibacterium* sp. D4 has been reported to remove carbamazepine and diclofenac, respectively in the presence of acetate (Bessa et al., 2017). The efficiency of the bacterial strains, *Geobacillus thermoleovorans* and *Geobacillus* sp. strain S-07 in the removal of sulfamethazine using glucose as the biostimulating agent has been reported by (Pan et al., 2017).

The bacterium, *Thermus* sp. strain C419 recovered from the sludge of a factory manufacturing antibiotics is reportedly thermophilic with abilities to degrade ciprofloxacin even at temperatures as high as 70°C (Pan et al., 2018). It was further reported that BD of ciprofloxacin could be stimulated by the addition of sodium acetate to the medium (Pan et al., 2018). Mulla et al. (2018) found that sulfamethoxazole could be degraded by *Gordonia* sp. SCD14, *Ochrobactrum* sp. SA1 and *Labrys* sp. SC11. Using a bacterial consortium composed of the strains *Stenotrophomonas* sp., *Pseudomonas* sp. f2, and *Pseudomonas* sp. fg-2. Zhang et al. (2013) performed experiments on paracetamol degradation and observed that the mixture was more effective than the individual strains. This was because of the inhibitory actions of high concentrations of paracetamol on *Stenotrophomonas* sp. and *Pseudomonas* sp. f2 while *Pseudomonas* sp. fg-2 could retain their bioremediation properties under such conditions. Based on these observations, Zhang et al. (2013) opined that bacterial mixtures when selected properly could enhance the BD of pharmaceutical contaminants by utilizing their synergistic interactions.

Bacterial consortia have been found to be more beneficial in removing toxic pollutants over single bacterial strains because of the synergistic interactions among them (Wang and Wang, 2016). However, in some cases the concentration of toxic pollutants may be such that the normal functioning of bacteria or the cooperative interactions among them may be inhibited

(Caracciolo et al., 2015). This necessitates the maintenance of reaction conditions within the wastewater treatment plants. Hay (2001) studied triclosan BD using a bacterial consortium recovered from activated sludge and reported that although the consortium could utilize triclosan as the only source of carbon, the bacteria obtained from it could not. They mainly belonged to the bacterial phylum Proteobacteria and consisted of *Agrobacterium*, *Alcaligenes*, *Sphingomonas*, *Pseudomonas*, and *Rhodanobacter*. Khunjar et al. (2011) reported that the removal of 17α-ethinylestradiol improved in the presence of a mixed bacterial culture composed of ammonia-oxidizing and heterotrophic bacteria. Alexandrino et al. (2017) obtained a microbial consortium from the rhizosphere of plants growing in constructed wetlands and analyzed the BD of different concentrations of ceftiofur and enrofloxacin. They observed that ceftiofur was removed completely irrespective of the presence of enrofloxacin while the latter never reached complete removal. It was further reported that bacteria belonging to the phyla Bacteroidetes and Proteobacteria could readily get acclimatized to different concentrations of antibiotics compared to others (Alexandrino et al., 2017).

Many workers like Topp et al. (2008); and Facey et al. (2018) isolated microbial consortia from forest and agricultural soils and found they were helpful in the BD and mineralization of pharmaceuticals like diclofenac and naproxen. Other workers enriched bacterial consortia with estuarine sediments or sludge and reported their usefulness in the removal of bezafibrate and paroxetine (Duarte et al., 2019; Fernandes et al., 2020). The most commonly encountered bacteria in these consortia included *Pseudomonas*, *Microbacterium*, *Acinetobacter*, and *Dyadobacter* (Fernandes et al., 2020). Despite the utility and efficiency of bacterial consortia in the removal of pharmaceutical compounds being encountered in several studies, there seems to be a lack of specific BD pathways (Fernandes et al., 2021). This is because of high variations in removal efficiencies observed across different studies based on the same pharmaceutical compound. The huge diversity of bacteria, their metabolic processes, and their proficiency in adaptation to harsh conditions make them efficient candidates for bioremediation. However, this field needs to be explored largely.

7.5.3 MICROALGAE

The concept of using microalgae for bioremediation of wastewater contaminated with pharmaceutical compounds is new and evolving because of the plethora of advantages it offers. They are highly adaptable to various growth conditions and can be heterotrophic, autotrophic, or mixotrophic. They can withstand severe environmental conditions such as extreme temperature and pH fluctuations. They can also grow in medium containing poor nutrient supplies. These are some advantages of using algae for bioremediation over strictly heterotrophic fungi (Subashchandrabose et al., 2013). Additionally, microalgae have the ability to acclimatize to alterations in light, temperature, salinity, and nutrient availability in the growth medium. This makes them resistant to harsh conditions often through genetic changes triggered by physiological adaptation or spontaneous mutations (Cho et al., 2016). This greatly enhances their ability to biodegrade (Cho et al., 2016). For instance, when *Chlorella vulgaris* is repeatedly exposed to increased salinity and sub-lethal doses of levofloxacin its ability to biodegrade levofloxacin is enhanced significantly (Xiong et al., 2017).

Cultivating microalgae has been reported to be cheap and simple since the rich nutrient contents of wastewaters themselves enable them to act as their growth medium (Silva et al., 2019). Moreover, they can utilize atmospheric carbon dioxide to produce their food by photosynthesis which also serves to clean the environment (Silva et al., 2019). Interestingly, adding bacteria to the medium containing microalgae has been found to boost BD activities due to the interactions among them. The bacteria generate carbon dioxide to be utilized by the microalgae which produces oxygen as a by-product of photosynthetic activities. The oxygen not only supports the bacteria but is also utilized for the oxidative degradation process reducing the dependency of the biological treatment process on the external supply of oxygen (Muñoz and Guieysse, 2006). Moreover, microalgae elevate the pH while carrying out their degradation activities which, in turn, checks the population of pathogens in the medium (Polprasert et al., 1983). Bioremediation of wastewater using microalgae generates several by-products that act as a resource for the production of biofuels (Pienkos and Darzins, 2009). This makes microalgal bioremediation of pharmaceutical compounds and their derivatives a promising alternative to conventional physical and chemical techniques.

The microalgae also utilize both intracellular and extracellular enzyme systems for BD of organic pollutants. The intracellular degradation pathway of pharmaceuticals depends primarily on cytochrome P450 and glutathione-S-transferase enzymes (Ding et al., 2017; Xiong et al., 2017). Extracellular degradation mechanisms depend on the secretion of extracellular polymeric substances (EPS) which forms a hydrated biofilm matrix around the cells maintaining the extracellular enzymes in close proximity and acting as an external digestive system (Flemming and Wingender, 2010). Some of the components of EPS include protein, lipids, polysaccharides, enzymes, and various substituents like acetyl groups (Flemming and Wingender, 2010).

Not much is known about the mechanism of action of microalgae during bioremediation. Apart from biotransformation and bioflocculation, BD, bioaccumulation, and bioadsorption have been widely studied in microalgae (Table 7.3).

7.5.3.1 BIOTRANSFORMATION

Biotransformation of target pollutants is one mechanism employed by microalgae in the bioremediation process. They have been reported to carry out biotransformation reactions by oxidation or reduction, hydroxylation, and degradation of sidechains of target compounds (Peng et al., 2014). For instance, *Chlorella pyrenoidosa* brings about the biotransformation of progestogens by these mechanisms (Peng et al., 2014). *Chlorella vulgaris*, on the other hand, has been reported to remove tetracycline from wastewater by biosorption and photodegradation methods (De Godos et al., 2012). Similarly, *Scenedesmus obliquus* and *Chlamydomonas mexicana* have been reported to detoxify carbamazepine by simultaneously assisting BD, adsorption, and bioaccumulation (Xiong et al., 2016). Degradation of natural and artificial steroid estrogens by *Chlorella vulgaris* has revealed that estradiol valerate (an artificial steroid estrogen) can undergo biotransformation to estradiol (another artificial steroid estrogen) and then to estrone (a natural steroid estrogen; Lai et al., 2002). On the other hand, ethinylestradiol (an artificial steroid estrogen) and hydroxyestrone, estriol, and estrone (natural steroid estrogens) can resist biotransformation (Lai et al., 2002).

7.5.3.2 BIOFLOCCULATION

The microalgae have also been reported to induce bioflocculation which results in the formation of extracellular biopolymers which assist in the flocculation of small particles naturally (Akhtar et al., 2004). Interestingly, some species of microalgae have been found to flocculate more easily than others. According to Taylor et al. (2012), they can induce other microalgal species to flocculate when intermingled with them. Despite being the eco-friendliest method, it is not foolproof because of the small size and high growth rates of microalgal cells (Silva et al., 2019). Recent developments involve the addition of suitable bacterial and fungal species to the microalgal culture medium for facilitating the process of flocculation (Muñoz and Guieysse, 2006). Though bacteria have been reported to enhance microalgal precipitation, additional nutrients are needed to be added to the medium to support bacterial growth and proliferation (Muñoz and Guieysse, 2006). Adding wastewater can support both the algae and bacteria because they offer several carbon resources for their nutrition (Su et al., 2011). However, some species of microalgae are dominant which may be a hurdle for the bacterial population to flourish. Additionally, the variability in the composition of the wastewater can hinder the growth of both organisms (Muñoz and Guieysse, 2006).

The fungi, on the other hand, do not require additional carbon sources when added (Zhang and Hu, 2012). Moreover, the fungal hyphae can trap the microalgal cells allowing simple harvest mechanisms using sedimentation and sieve filtration techniques (Zhou et al., 2012). Although adding fungi can combat the cost issues by not having to add extra carbon sources and reducing the dependency on high-energy inputs, the selection of proper strains is important (Gultom and Hu, 2013). This is because the rate of co-pelletization depends on the interactions between the microalgal and fungal strains along with the physicochemical features of the growth medium (Gultom and Hu, 2013). Moreover, the microalgae and fungi have been found to compete with each other for carbon resources in the medium under heterotrophic conditions. Under autotrophic conditions, the same situation arises due to a shortage of organic nitrogen. According to Wittenberg (1991), this competition prevents either of their population from reaching the maximal limit even though the conditions are favorable for both. This has created doubts regarding their co-culturing because the effect of fungi as well its extent on microalgal growth and multiplication still needs to be ascertained (Milledge and Heaven, 2013). Moreover, it is unclear whether bacteria-aided bioflocculation is superior to fungal-aided microalgal sedimentation (Milledge and Heaven, 2013). This leaves room for further research.

7.5.3.3 BIODEGRADATION (BD)

It has been observed that BD of pharmaceutical contaminants is a more efficient way of their removal compared to other methods. In this process, the complicated pollutants are broken down into simpler forms through catalytic reactions. For instance, *Chlorella pyrenoidosa* and *Scenedesmus obliquus* have been reported to cause 95% BD of progesterone present in their surrounding aquatic medium (Peng et al., 2014). Several other studies reported high rates of microalgae-mediated breakdown of pharmaceuticals such as carbamazepine, ibuprofen, and caffeine in wastewater from different sources (Ding et al., 2017; Hom-Diaz et al., 2017; Matamoros et al., 2016; Table 7.3).

The BD process carried out by microalgae involves their enzyme systems which involve two groups, phase I and phase II, that work in coordination with each other. The phase I enzyme is essentially the cytochrome P450 system that catalyzes oxidation, reduction, or hydrolytic reactions and produces derivatives that are more hydrophilic compared to their parent compounds. The basic mechanism involves adding hydroxyl groups to the parent molecule or exposing the already existing ones. Additionally, microalgae can catalyze several other reactions such as bromination, carboxylation, decarboxylation, demethylation, dehydroxylation, glycosylation, hydrogenation, and hydroxylation while degrading the pharmaceutical contaminants (Ding et al., 2017; Matamoros et al., 2016; Xiong et al., 2017). The phase II enzymes that aid in the conjugation reaction between glutathione and electrophilic compounds mainly include glutathione-S-transferases. The conjugation causes the epoxide ring to open and prevents destruction by oxidation. This is exemplified by Ding et al. (2017) through glycosylation experiments of ibuprofen involving *Navicula* sp. (a freshwater diatom). Several workers have reported the involvement of microalgal intracellular enzymes such as uroporphyrinogen III carboxylase, uroporphyrinogen III decarboxylase, glutamyl-tRNA reductase, acid phosphatase, alkaline phosphatase, violaxanthin de-epoxidase, mono(di)oxygenase, hydrolases, transferase, catalase, and laccases in microalgae-mediated BD of pharmaceutical compounds (Foflonker et al., 2016; Otto and Schlosser, 2014). The reactions involved are mainly biotransformations followed by detoxification. The biotransformation reactions are very complicated and have not yet been documented completely. This field needs further research.

The extrapolymeric substances (EPS) can not only aid in BD by acting as an external digestive system but also help to increase the bioavailability of recalcitrant pollutants by acting as emulsifiers and surface-acting agents (Xiao and Zheng, 2016). Moreover, the presence of various substitutes, proteins, and charged polysaccharide groups in the EPS enables the acquisition of nutrients from the surroundings and sorption of pharmaceuticals thereby purifying the environment (Flemming and Wingender, 2010; Table

7.3). These reactions mediated by different EPS and their interactions with microalgae can aid in the degradation of pharmaceutical compounds through extracellular mechanisms. Additionally, the derivatives obtained therefrom can also stimulate intracellular degradation pathways in microalgae (Xiao and Zheng, 2016).

7.5.3.4 BIOACCUMULATION

Microalgae have been reported to actively absorb organic contaminants from the medium along with their nutritional resources. These contaminants can be stored within their cells. This energy-driven process is termed as bioaccumulation. The phenomenon has been demonstrated experimentally by Maes et al. (2014) with the help of *Desmodesmus subspicatus* which can accumulate radiolabeled 17α-ethinylestradiol within their cells at rates as high as 23%. Xiong et al. (2016) went ahead to demonstrate the accumulation of carbamazepine by the microalgae, *Scenedesmus obliquus* and *Chlamydomonas mexicana.* Bai and Acharya (2017) reported that several micropollutants such as trimethoprim, sulfamethoxazole, and triclosan are removed by microalgae through bioaccumulation. The pollutants that accumulate within the microalgal cells generate reactive oxygen species (ROS) that tend to exert toxic stress, damage cellular elements, and even trigger cell death (Kumar et al., 2016). ROS, when present in excess, may also induce mutagenesis in algae (Kumar et al., 2016). However, the concentrations of the pharmaceutical compounds and their derivatives present in the external environment usually do not hinder the metabolic functioning of microalgae or their growth rates significantly (González-Pleiter et al., 2013). On the contrary, low concentrations of pharmaceutical pollutants in the surrounding medium can increase the tolerance of microalgae to these pollutants due to altered expression of stress-related genes and their respective signaling pathways induced by ROS (Kumar et al., 2016). Some micropollutants (e.g., amoxicillin (AMX), spiramycin) can also augment the expression of genes related to photosynthesis in microalgae (e.g., *Microcystis aeruginosa*; Liu et al., 2014). In many cases, the pollutant removal efficiency of microalgal species has been found to be enhanced if they have previously encountered them. As such, pre-adapted microalgae are more efficient in the removal of pharmaceutical pollutants (e.g., cefradine) compared to wild strains (Chen et al., 2015).

7.5.3.5 BIOADSORPTION

The cell wall of microalgae is abundant in polymers similar to pectins, extension, lignin, cellulose, hemicellulose, and arabinogalactan proteins. Moreover, their cell wall is negatively charged because of the presence of amine, carboxyl, phosphoryl, and similar functional groups. These structural features of the algal cell wall govern the interactions with the pharmaceutical compounds during bioadsorption. This, in turn, is influenced by the structure, functional sidechains, and hydrophobic nature of the pharmaceutical compounds. Bioadsorption being an extracellular process is therefore largely dependent on the structural features of the microalgal species as well as the pharmaceutical compounds. Pollutants with positively charged sidechains interact more readily with the cell wall of the microalgae because of electrostatic attractions. As such, the removal efficiency of microalgae varies for different pharmaceutical compounds (de Wilt et al., 2016). Interestingly, dead microalgal biomass can also adsorb pharmaceutical contaminants. For instance, the dead cell biomass of *Chlorella pyrenoidosa* and *Scenedesmus obliquus* have been reported to adsorb norgestrel and progesterone from the surrounding medium (Peng et al., 2014).

7.6 OTHER STRATEGIES FOR BIOREMEDIATION

7.6.1 MICROBIAL CONSORTIA

It has been observed that in many cases some pharmaceutical compounds are too complex to be removed by pure cultures of any one particular group of microbes. Instead, a mixed system of different microorganisms and/or an aquatic plant can be more efficient in the complete removal of such compounds. For instance, a mixed system of algae (e.g., *Spirulina platensis*, *Anabaena cylindrica*, *Scenedesmus quadricauda*, *Chlorococcus* sp., and *Chlorella* sp.) and duckweed (*Lemna* sp.) in an open pond system can remove natural and synthetic hormones very efficiently even when they are present in low concentrations (Shi et al., 2010). Further, the hormones present in wastewater can readily be sorbed on algae or duckweed followed by their microbial breakdown (Shi et al., 2010).

Advanced research has revealed that microalgae when mixed with bacteria and cyanobacteria become more efficient in the detoxification of inorganic and organic pollutants and the removal of contaminants from wastewater (Subashchandrabose et al., 2013; Xiong et al., 2017). These symbiotic relationships are based on the electron and energy flow promoted by microalgae.

For instance, a recently formulated consortium consisting of an alga and bacteria proved more effective than the individual strains in the removal of aromatic and aliphatic hydrocarbons present in crude oil (Wang et al., 2015). Such consortia can be developed through both natural and artificial means (Liu et al., 2017). In most of the cases, the microalgal cell surface provides refuge to the bacteria. The relationship between them is governed by the release of bacterial degradation products which enhance algal growth. For instance, the by-products of pollutant decomposition by bacteria are rich in vitamin B12 and indole acetic acid which help the microalgal population to flourish (Salim et al., 2014). On the other hand, the extracellular substances secreted by the algae are utilized by the bacteria as important sources of protein, lipids, and carbohydrates (Ramanan et al., 2016). These synergistic interactions have not only been observed in microalgae-bacteria but also in microalgae-microalgae consortia. The net outcome has been found to be beneficial in both cases (Wang et al., 2015).

Despite the multitude of advantages, there are several drawbacks to such consortia. The bacteria and algae can compete with each other for the same nutrient source, or the pH of the medium may be altered by the flourishing microalgae. In some instances, the bacteria may produce algicides or other harmful substances that may impede the microalgal electron transport system or transcription of genes related to photosynthesis (Liu et al., 2017). Wijffels et al. (2013), however, suggested that microalgal or microbial consortia can better withstand fluctuations in their surrounding medium because of the cooperative metabolic interactions among them.

7.6.2 BIOFILMS

Interestingly, microorganisms can assemble in sessile biofilms which offer additional protection from environmental factors, allow them to withstand different metabolic conditions, enhance communication amongst themselves, and facilitate the exchange of genetic materials. Additionally, biofilm also increases the ease of access to nutrients both by acquisition from the surroundings and by sharing among themselves. All these features make them superior to other solitary free-living planktons (Fux et al., 2005). The greatest advantage is that biofilm allows the co-existence of different microbial species, frequently from different kingdoms, with differences in their environmental and nutritional requirements together with their ability to function as electron donors (Baker et al., 2009). Often, they exhibit combinations of archaea, algae, fungi, and bacteria living together (Baker et al., 2009). Thus, the biofilm enables the component microbes to act like a multi-organ system because of their specialized functions (Fux et al., 2005).

The structural diversity of the biofilms reveals their variable origin to be directly correlated with the surrounding environment of the microorganisms. The participating microbes secrete extracellular polymeric substances (EPS) which gradually grow in size and develop water channels that aid in distributing nutrients and electron acceptors. This orientation, however, supports the better survival of cells in the outer layers compared to those occurring interiorly (Chen et al., 2013). The EPS can also provide surfactants and emulsifiers that can facilitate the dissolution of difficultly accessible hydrophobic or unmanageable pollutants (Wang et al., 2011). The occurrence of biofilms has been reported from various sites and geographical locations such as pipes that have been corroded, infections, tidal flats, and streambeds (Seo et al., 2009). They can be sessile or floating (e.g., activated sludge; Hall-Stoodley et al., 2004). The formation of biofilms by microorganisms has been found to be prevalent at sites of heavy pollution. This toughens them and allows their survival against several environmental odds (Gross et al., 2007).

Application of biofilms for bioremediation and biotransformation has been proved to be a successful strategy because of their ability to resist high concentrations of toxic chemicals which would otherwise kill solitary microorganisms despite their acclimation. Microbial aggregates can also thrive against several environmental stress factors such as temperature, pH, availability of water, and salinity (Hall-Stoodley et al., 2004). Onesios and Bouwer (2012) reported that biofilms are very efficient in removing several pharmaceutical compounds such as naproxen, gemfibrozil, chlorophene, valproic acid, ρ-chlorom-cresol, biphenylol, ibuprofen, and 5-fluorouracil. They further suggested that the application of biofilms for environmental cleaning can significantly reduce the negative impacts of pharmaceuticals on the environment. The activity of biofilms has, however, been found to vary with the altering climates. Sui et al. (2011) reported their activities to be reduced during the winter months as indicated by the rising levels of pharmaceutical compounds and their derivatives in wastewater effluents.

7.6.3 BIOSORBENTS

Another strategy for bioremediation is biosorption which is basically a type of adsorption. It uses biological materials, both alive and dead, to bind and concentrate contaminants present in a solution (Michalak et al., 2013). This method has been recommended by several workers because of its affordability. It also does not depend on the use of chemicals or nutrients as in conventional methods making it environment friendly (Michalak et al., 2013). Activated carbon is one of the most commonly used sorbents because of its porous structure and high surface area that can readily absorb different kinds of organic molecules. However, it is not cost-effective encouraging the researchers to look for natural and affordable sorbents (Babel and Kurniawan, 2003).

Interestingly, microorganisms such as microalgae and fungi have been found to be suitable and affordable alternatives that are abundant in nature (Tsezos and Bell, 1989). Their small size and high surface area make them very efficient adsorbents of various inorganic and organic contaminants from solutions (Tsezos and Bell, 1989). The process is facilitated by their complicated structure although the mechanisms have not been fully deciphered. It has been observed that the pollutants first interact with their cell wall followed by their deposition on the surface or getting trapped within the cell wall structure (Silva et al., 2019). For instance, the presence of different functional groups like sulfonate, carboxyl, amino, and hydroxyl groups in the fungal cell walls endows them with outstanding adsorption properties (Aksu, 2005). Similarly, the presence of sulfate, phosphate, amine, imidazole, and hydroxyl groups in algal cell walls makes them efficient adsorbents as well. The greater the variations in the cell wall structure of the microorganisms the greater their ability to remove pollutants from a solution because variations make them non-specific (Bhatti and Hamid, 2014). In many cases, the pollutants get adsorbed to the microbial surface by the formation of hydrogen bonds as observed by Navarro et al. (2014) during the removal of sulfacetamide and sulfamethoxazole by *Macrocystis integrifolia* and *Lessonia nigrescens*.

Biosorption is influenced by several factors such as temperature, pH, the concentration of the contaminants, time of interaction between the sorbates and sorbent, properties of the sorbates and sorbent, ionic strength of the solution and whether the biological agent being used is dead or alive (Park et al., 2010). The application of living organisms for biosorption may present certain restrictions because the metabolic processes of the living microbial cells may interfere with the physicochemical mechanisms of biosorption (Gadd, 2009). Additionally, the bioavailability of contaminants, their chemical speciation, transformation, and accumulation can also be affected by the metabolic processes within living cells (Gadd, 2009). In the case of organic pollutants, their diverse chemical structures, charge, molecular size, solubility, reactivity, and hydrophobicity dictate their extent of biosorption (Ikehata et al., 2006). They also lack specificity and have problems separating the biomass from the treated effluent (Liu and Liu, 2008). Further, the system is not robust enough and cannot be carried out for long periods (Eccles, 1999). Therefore, bioremediation cannot solely depend on biosorption. Combining it with other technologies can help in the complete removal of contaminants from wastewater (Bai and Acharya, 2016).

7.7 FACTORS INFLUENCING BIOREMEDIATION

The fate of the pharmaceutical compounds and their derivatives depends on their physicochemical properties. Their structural features also determine whether they will be degraded, transformed, or adsorbed by the sludge since these are the primary mechanisms of removal of pharmaceutical contaminants from polluted waters (Wang and Wang, 2016). The variety, nature, and abundance of the microbial population govern the extent of BD that such pollutants may undergo (Wang and Wang, 2016). Positive interactions among the microorganisms can increase the rate of BD or other removal processes of the pharmaceuticals. However, they are also susceptible to being affected by negative interactions among the component bacteria in a consortium or the inhibitory effects of any compound on a bacterial strain or recalcitrant complexes generated through reactions among the pharmaceutical derivatives (Zhao et al., 2015). Since the influent polluted water may contain several unidentified compounds other than the target pharmaceuticals, various antagonistic factors interplay within the reaction medium (Zhao et al., 2015). This explains the differences in observation of the removal efficiency rates with the same pharmaceutical compounds as targets.

The structure of the pharmaceutical compounds becomes important to understand their biodegradability because the presence of some functional groups or side chains renders them more biodegradable compared to others (Kümmerer and Al-Ahmad, 1997). Even if two or more pharmaceutical compounds belong to the same class, their structural differences can affect their removal efficiencies because of the reaction specificity of the enzymes involved (Onesios et al., 2009). The concentration of the pharmaceutical compounds is also an important factor for their biodegradability because high concentrations can inhibit the bacteria or any other member of the microbial consortia (Onesios et al., 2009). The toxic effects of a pharmaceutical compound depend on its concentration (Onesios et al., 2009). However, too low concentrations of the pharmaceutical compounds may fail to induce the enzymes necessary for microbial BD in cases where they are being used by the bacteria as the primary substrate (Ternes and Joss, 2015). The condition may be improved by adding an external source (e.g., a nutrient) such that they act as a co-substrate and help in the degradation of the target pharmaceuticals by stimulating the corresponding enzymes (Ternes and Joss, 2015). The time of incubation is another factor that dictates the biodegradability of the pharmaceutical compounds. It has been observed that the removal efficiency of the same pharmaceutical compound varies with the incubation time (Onesios et al., 2009).

Preadaptation of bacteria or microalgae to pharmaceutical compounds can improve the removal efficiencies of those pharmaceutical compounds (Aslam et al., 2017). This takes place because of the changes induced in the biochemical pathways of the bacteria or microalgae by the contaminant during their acclimatization period (Xiong et al., 2017). Eventually, the microbe can recognize the contaminant faster and bring about their quicker degradation (Aslam et al., 2017). The functional efficiency of microbes can also be affected by other pollutants present in the system. They include nanomaterials, pesticides, heavy metals, and so forth. Their effects on the microbial community as well as their interactions with pharmaceutical compounds should be considered seriously because they may inhibit the microbe directly or through the generation of toxic by-products by interacting with the target pharmaceutical compounds (Fernandes et al., 2021). According to Silva et al. (2019), the main problem of these microorganism-based bioremediation technologies is that most of them have been carried out under laboratory conditions with a few studies being conducted under field conditions. Therefore, more field-based studies should be performed to understand their practical applicability.

7.8 TECHNICAL ADVANCEMENTS FOR ENHANCING BIOREMEDIATION

Application of several physical and chemical methods to detoxify pharmaceutical compounds have been found to be beneficial but their high costs have prevented their widespread acceptance. In this context, bioremediation has been found to be an economic, eco-friendly, and sustainable alternative Shishir et al. (2019). Since bioremediation depends heavily on the enzymatic machinery and metabolic pathways of microbes, understanding their genetic structure and molecular mechanisms gives immense scope for enhancing their BD capabilities (Azubuike et al., 2016). The abundance and diversity of microorganisms and their presence in a wide range of environmental samples including the plant, animal, and human tissues make them potential candidates for BD. Interestingly, the differences in their origin endow them with different genetic, molecular, and metabolic properties that can be exploited for the BD of recalcitrant pollutants (Shekhar et al., 2020).

Despite these advantages, the main problem with microorganisms is that a number of them cannot be cultured which largely limits the knowledge of their structure, function, regulation, and adaptation (Wolf et al., 2020). For instance, Rossello-Mora and Amann (2001) reported that less than 1% of the bacteria present in soil can be cultured which leaves a huge proportion of bacteria unidentified. Gillbride et al. (2006) opined that the identification of bacteria based on their ability to be cultured limits the studies to the rapidly populating species. In these cases, the application of molecular technologies that do not depend on culturing the bacteria helps in elaborating the diversity and function of bacterial communities in polluted environments (Shakya et al., 2019).

Developments in biotechnology tools, molecular biology instruments, and bioinformatics have gone a long way to unravel the information about these microbes along with the ones that can be cultured (Yan et al., 2020). The information obtained with these technologies can be used for genetic engineering, recombinant, and hybridization strategies to obtain engineered microbes with enhanced capabilities for BD of pollutants (Puckett et al., 2020). Metagenomics, transcriptomics, proteomics, and several other advanced molecular techniques are faster, more sensitive, and have higher accuracy compared to conventional molecular techniques (Mishra et al., 2020). The information obtained therefrom can be utilized to develop engineered microbes that are much more efficient in BD than their naturally occurring variants (Mishra et al., 2020).

Currently, the omics-based approach has attracted the attention of researchers because of its potential to unravel the unexplored capabilities of bacteria that cannot be cultured. Though the knowledge of these viable but non-culturable (VNBC) bacteria is incomplete, several workers have suggested that they have important contributions to the BD of environmental contaminants (Bodor et al., 2020). The VNBC can be perceived by their viability since they have been reported to exhibit metabolic activities that largely depend on the regulation of their gene expression (Zhao et al., 2017). However, molecular methods that are not dependent on culturing the bacteria can be utilized to isolate and study their genetic materials (Bodor et al., 2020). Moreover, their genes can be amplified selectively giving elaborate details about their molecular functioning. Thus, these bacteria isolated from any source can be studied for the identification of beneficial genes to be used for recombination and hybridization experiments (Pinu et al., 2019).

7.9 CONCLUSION

Considering the current scenario of environmental crises and the need to detoxify or degrade pharmaceutical pollutants in the environment bioremediation seems to be the best alternative. It is affordable, eco-friendly, and sustainable. Current research has come up with several innovative bioremediation strategies with microalgae, fungi, and bacteria as the promising candidates. Their consortia have proved to be more effective compared to their individual application. However, their synergistic interactions have not yet been discerned completely. This field definitely requires further development. Additionally, the development of sensitive and budget-friendly techniques to identify the pharmaceutical compounds, their derivatives or the toxic products obtained from their interactions present in environmental samples at low concentrations should also be worked upon. Molecular tools and omics-based approaches can also strengthen bioremediation strategies. They are quite naïve and have not been explored thoroughly. However, care should be taken that bioremediation strategies should not disturb the natural microbial flora of different ecosystems. This can be achieved only through interdisciplinary research endeavors which is the current requirement in this field.

KEYWORDS

- **antibiotic resistance**
- **bioremediation**
- **microbial remediation**
- **pharmaceutical compounds**
- **phytoremediation**
- **pollutants**

REFERENCES

Abed, R. M. M., Al-Sabahi, J., Al-Maqrashi, F., Al-Habsi, A., & Al-Hinai, M. (2014). Characterization of hydrocarbon-degrading bacteria isolated from oil-contaminated sediments in the Sultanate of Oman and evaluation of bioaugmentation and bio stimulation approaches in microcosm experiments. *International Biodeterioration & Biodegradation, 89*, 58–66.

Akerman-Sanchez, G., & Rojas-Jimenez, K. (2021). Fungi for the bioremediation of pharmaceutical-derived pollutants: A bioengineering approach to water treatment. *Environmental Advances, 4*, 100071.

Akhtar, N., Iqbal, J., & Iqbal, M. (2004). Enhancement of lead (II) biosorption by microalgal biomass immobilized onto loofa (*Luffa cylindrica*) sponge. *Engineering in Life Sciences, 4*, 171–178.

Aksu, Z. (2005). Application of biosorption of the removal of organic pollutants: A review. *Process Biochemistry, 40*, 997–1026.

Alexandrino, D. A. M., Mucha, A. P., Almeida, C. M., Gao, W., Jia, Z., & Carvalho, M. F. (2017). Biodegradation of the veterinary antibiotics enrofloxacin and ceftiofur and associated microbial community dynamics. *Science of the Total Environment, 581–582*, 359–368.

Amorim, C. L., Alves, M., Castro, P. M. L., & Henriques, I. (2018). Bacterial community dynamics within an aerobic granular sludge reactor treating wastewater loaded with pharmaceuticals. *Ecotoxicology and Environmental Safety, 147*, 905–912.

Amorim, C. L., Moreira, I. S., Maia, A. S., Tiritan, M. E., & Castro, P. M. L. (2014). Biodegradation of ofloxacin, norfloxacin, and ciprofloxacin as single and mixed substrates by *Labrys portucalensis* F11. *Applied Microbiology and Biotechnology, 98*, 3181–3190.

Anastasi, A., Tigini, V., & Varese, G. C. (2013). The bioremediation potential of different ecophysiological groups of fungi. In E. M. Goltapeh, Y. R. Danesh, & A. Varma (Eds.), *Fungi as Bioremediators* (Vol. 32, pp. 29–49). Springer-Verlag.

Andrade, J. A., Augusto, F., & Jardim, I. C. S. F. (2010). Biorremediação de solos contaminados por petróleo e seus derivados. *Jornal Eclética Química, 35*.

Arnold, K. E., Brown, A. R., Ankley, G. T., & Sumpter, J. P. (2014). Medicating the environment: Assessing risks of pharmaceuticals to wildlife and ecosystems. *Philosophical Transactions of the Royal Society B: Biological Sciences, 369*.

Asif, M. B., Hai, F. I., Singh, L., Price, W. E., & Nghiem, L. D. (2017). Degradation of pharmaceuticals and personal care products by white-rot fungi – A critical review. *Current Pollution Reports, 3*, 88–103.

Aslam, A., Thomas-Hall, S. R., Mughal, T. A., & Schenk, P. M. (2017). Selection and adaptation of microalgae to growth in 100% unfiltered coal-fired flue gas. *Bioresource Technology, 233*, 271–283.

Azubuike, C. C., Chikere, C. B., & Okpokwasili, G. C. (2016). Bioremediation techniques—Classification based on site of application: Principle, advantages, limitations, and prospects. *World Journal of Microbiology and Biotechnology, 32*, 180.

Azuma, T. (2018). Distribution of anticancer drugs in river waters and sediments of the Yodo River Basin, Japan. *Applied Sciences, 8*(11).

Babel, S., & Kurniawan, T. A. (2003). Low-cost adsorbents for heavy metals uptake from contaminated water: A review. *Journal of Hazardous Materials, 97*, 219–243.

Badia-Fabregat, M., Lucas, D., Pereira, M. A., Alves, M. A., Pennanen, T., Fritze, H., Rodríguez-Mozaz, S., Barceló, D., Vicent, T., & Caminal, G. (2016). Continuous fungal treatment of non-sterile veterinary hospital effluent: Pharmaceuticals removal and microbial community assessment. *Applied Microbiology and Biotechnology, 100*, 2401–2415.

Bai, X., & Acharya, K. (2016). Removal of trimethoprim, sulfamethoxazole, and triclosan by the green alga *Nannochloris* sp. *Journal of Hazardous Materials, 315*, 70–75.

Bai, X., & Acharya, K. (2017). Algae-mediated removal of selected pharmaceutical and personal care products (PPCPs) from Lake Mead water. *Science of the Total Environment, 581–582*, 734–740.

Baker, B. J., Tyson, G. W., Goosherst, L., & Banfield, J. F. (2009). Insights into the diversity of eukaryotes in acid mine drainage biofilm communities. *Applied and Environmental Microbiology, 75*, 2192–2199.

Barros, S., Montes, R., Quintana, J. B., Rodil, R., André, A., Capitão, A., Soares, J., Santos, M. M., & Neuparth, T. (2018). Chronic environmentally relevant levels of simvastatin disrupt embryonic development, and biochemical and molecular responses in zebrafish (*Danio rerio*). *Aquatic Toxicology, 201*, 47–57.

Bartoskova, M., Dobsikova, R., Stancova, V., Zivna, D., Blahova, J., Marsalek, P., Zelníckova, L., Bartos, M., di Tocco, F. C., & Faggio, C. (2013). Evaluation of ibuprofen toxicity for zebrafish (*Danio rerio*) targeting selected biomarkers of oxidative stress. *Neuro Endocrinology Letters, 34*, 102–108.

Baumann, M., Weiss, K., Maletzki, D., Schüssler, W., Schudoma, D., Kopf, W., & Kühnen, U. (2015). Aquatic toxicity of the macrolide antibiotic clarithromycin and its metabolites. *Chemosphere, 120*, 192–198.

Becker, D., Varela Della Giustina, S., Rodriguez-Mozaz, S., Schoevaart, R., Barcelo, D., de Cazes, M., Belleville, M. P., Sanchez-Marcano, J., de Gunzburg, J., Couillerot, O., Völker, J., Oehlmann, J., & Wagner, M. (2016). Removal of antibiotics in wastewater by enzymatic treatment with fungal laccase—Degradation of compounds does not always eliminate toxicity. *Bioresource Technology, 219*, 500–509.

Belding, C., & Boopathy, R. (2018). Presence of antibiotic-resistant bacteria and antibiotic resistance genes in coastal recreational waters of southeast Louisiana, USA. *Journal of Water Supply: Research and Technology—Aqua, 67*, 800–809.

Bessa, V. S., Moreira, I. S., Tiritan, M. E., & Castro, P. M. L. (2017). Enrichment of bacterial strains for the biodegradation of diclofenac and carbamazepine from activated sludge. *International Biodeterioration & Biodegradation, 120*, 135–142.

Bhatti, H. N., & Hamid, S. (2014). Removal of uranium (VI) from aqueous solutions using *Eucalyptus citriodora* distillation sludge. *International Journal of Environmental Science and Technology, 11*, 813–822.

Bhullar, K., Waglechner, N., Pawlowski, A., Koteva, K., Banks, E. D., Johnston, M. D., Barton, H. A., & Wright, G. D. (2012). Antibiotic resistance is prevalent in an isolated cave microbiome. *PLoS One, 7*(4), e34953.

Biel-Maeso, M., Corada-Fernández, C., & Lara-Martín, P. A. (2017). Determining the distribution of pharmaceutically active compounds (PhACs) in soils and sediments by pressurized hot water extraction (PHWE). *Chemosphere, 185*, 1001–1010.

Biel-Maeso, M., Corada-Fernández, C., & Lara-Martín, P. A. (2018). Monitoring the occurrence of pharmaceuticals in soils irrigated with reclaimed wastewater. *Environmental Pollution, 235*, 312–321.

Bird, K., Boopathy, R., Nathaniel, R., & LaFleur, G. (2019). Water pollution and observation of acquired antibiotic resistance in Bayou Lafourche, a major drinking water source in southeast Louisiana, USA. *Environmental Science and Pollution Research, 26*, 34220–34232.

Blánquez, P., & Guieysse, B. (2008). Continuous biodegradation of 17β-estradiol and 17α-ethynylestradiol by *Trametes versicolor*. *Journal of Hazardous Materials, 150*, 459–462.

Bodin, H., Daneshvar, A., Gros, M., & Hultberg, M. (2016). Effects of biopellets composed of microalgae and fungi on pharmaceuticals present at environmentally relevant levels in the water. *Ecological Engineering, 91*, 169–172.

Bodor, A., Bounedjoum, N., Vincze, E. G., Kis, A. E., Laczi, K., Bende, G., Szilágyi, Á., Kovács, T., Perei, K., & Rákhely, G. (2020). Challenges of unculturable bacteria: Environmental perspectives. *Reviews in Environmental Science and Biotechnology, 19*, 1–22.

Boisseaux, P., Noury, P., Thomas, H., & Garric, J. (2017). Immune responses in the aquatic gastropod *Lymnaea stagnalis* under short-term exposure to pharmaceuticals of concern for immune systems: Diclofenac, cyclophosphamide, and cyclosporine A. *Ecotoxicology and Environmental Safety, 139*, 358–366.

Boopathy, R. (2011). Factors limiting bioremediation technologies. *Bioresource Technology, 74*, 63–67.

Borgatta, M., Waridel, P., Decosterd, L. A., Buclin, T., & Chèvre, N. (2016). Multigenerational effects of the anticancer drug tamoxifen and its metabolite 4-hydroxy-tamoxifen on *Daphnia pulex*. *Science of the Total Environment, 545–546*, 21–29.

Bush, K. (1997). Antimicrobial agents. *Current Opinion in Chemical Biology, 1*, 169–175.

Cañas, A. I., & Camarero, S. (2010). Laccases and their natural mediators: Biotechnological tools for sustainable eco-friendly processes. *Biotechnology Advances, 28*, 694–705.

Carabajal, M., Perullini, M., Jobbagy, M., Ullrich, R., Hofrichter, M., & Levin, L. (2016). Removal of phenol by immobilization of *Trametes versicolor* in silica-alginate-fungus bio composites and loofa sponge. *Clean – Soil, Air, Water, 44*(2), 180–188.

Caracciolo, A. B., Topp, E., & Grenni, P. (2015). Pharmaceuticals in the environment: Biodegradation and effects on natural microbial communities. A review. *Journal of Pharmaceutical and Biomedical Analysis, 106*, 25–36.

Carvalho, M. F., Ferreira Jorge, R., Pacheco, C. C., De Marco, P., & Castro, P. M. L. (2005). Isolation and properties of a pure bacterial strain capable of fluorobenzene degradation as sole carbon and energy source. *Environmental Microbiology, 7*, 294–298.

Carvalho, M. F., Maia, A. S., Tiritan, M. E., & Castro, P. M. L. (2016). Bacterial degradation of moxifloxacin in the presence of acetate as a bulk substrate. *Journal of Environmental Management, 168*, 219–228.

Chaturvedi, P., Shukla, P., Giri, B. S., Chowdhary, P., Chandra, R., Gupta, P., & Pandey, A. (2021). Prevalence and hazardous impact of pharmaceutical and personal care products and antibiotics in the environment: A review on emerging contaminants. *Environmental Research, 194*, 110664.

Chen, J., Zheng, F., & Guo, R. (2015). Algal feedback and removal efficiency in a sequencing batch reactor algae process (SBAR) to treat the antibiotic cefradine. *PLoS One, 10*(7), e0133273.

Chen, X., Suwarno, S. R., Chong, T. H., McDougald, D., Kjelleberg, S., Cohen, Y., Fane, A. G., & Rice, S. A. (2013). Dynamics of biofilm formation under different nutrient levels and the effect on biofouling of a reverse osmosis membrane system. *Biofouling, 29*, 319–330.

Chikae, M., Ikeda, R., Hasan, Q., Morita, Y., & Tamiya, E. (2004). Effects of tamoxifen, 17α-ethynylestradiol, flutamide, and methyltestosterone on plasma vitellogenin levels of male and female Japanese medaka (*Oryzias latipes*). *Environmental Toxicology and Pharmacology, 17*, 29–33.

Cho, K., Lee, C. H., & Ko, K. (2016). Use of phenol-induced oxidative stress acclimation to stimulate cell growth and biodiesel production by the oceanic microalga *Dunaliella salina. Algal Research, 17*, 61–66.

Cobas, M., Ferreira, L., Tavares, T., Sanroman, M. A., & Pazos, M. (2013). Development of permeable reactive biobarrier for the removal of PAHs by *Trichoderma longibrachiatum. Chemosphere, 91*(5), 711–716.

Cruz Del Álamo, A., Pariente, M. I., Martínez, F., & Molina, R. (2020). *Trametes versicolor* immobilized on rotating biological contactors as an alternative biological treatment for the removal of emerging concern micropollutants. *Water Research, 170*, 115313.

Cruz-Morató, C., Ferrando-Climent, L., Rodriguez-Mozaz, S., Barceló, D., Marco-Urrea, E., Vicent, T., & Sarrà, M. (2013). Degradation of pharmaceuticals in non-sterile urban wastewater by *Trametes versicolor* in a fluidized bed bioreactor. *Water Research, 47*, 5200–5210.

Cruz-Morató, C., Lucas, D., Llorca, M., Rodriguez-Mozaz, S., Gorga, M., Petrovic, M., Barceló, D., Vicent, T., Sarrà, M., & Marco-Urrea, E. (2014). Hospital wastewater treatment by fungal bioreactor: Removal efficiency for pharmaceuticals and endocrine disruptor compounds. *Science of the Total Environment, 493*, 365–376.

Cui, J., Fu, L., Tang, B., Bin, L., Li, P., Huang, S., & Fu, F. (2020). Occurrence, ecotoxicological risks of sulfonamides and their acetylated metabolites in the typical wastewater treatment plants and receiving rivers at the Pearl River Delta. *Science of the Total Environment, 709*, 136192.

Cummings, D. E., Archer, K. F., Arriola, D. J., Baker, P. A., Faucett, K. G., Laroya, J. B., Pfeil, K. L., Ryan, C. R., Ryan, K. R. U., & Zuill, D. R. (2010). Broad dissemination of plasmid-mediated quinolone resistance genes in sediments of two urban coastal wetlands. *Environmental Science & Technology, 45*, 447–454.

Dalecka, B., Juhna, T., & Rajarao, G. K. (2020). Constructive use of filamentous fungi to remove pharmaceutical substances from wastewater. *Journal of Water Process Engineering, 33*, 100992.

De Godos, I., Muñoz, R., & Guieysse, B. (2012). Tetracycline removal during wastewater treatment in high-rate algal ponds. *Journal of Hazardous Materials, 229–230*, 446–449.

de Wilt, A., Butkovskyi, A., Tuantet, K., Leal, L. H., Fernandes, T. V., Langenhoff, A., & Zeeman, G. (2016). Micropollutant removal in an algal treatment system fed with source-separated wastewater streams. *Journal of Hazardous Materials, 304*, 84–92.

DeYoung, D. J., Bantle, J. A., Hull, M. A., & Burks, S. L. (1996). Differences in sensitivity to developmental toxicants as seen in *Xenopus* and *Pimephales* embryos. *Bulletin of Environmental Contamination and Toxicology, 56*, 143–150.

Dijkshoorn, L., Nemec, A., & Seifert, H. (2007). An increasing threat in hospitals: Multidrug-resistant *Acinetobacter baumannii. Nature Reviews Microbiology, 5*, 939–951.

Ding, T., Yang, M., Zhang, J., Yang, B., Lin, K., Li, J., & Gan, J. (2017). Toxicity, degradation and metabolic fate of ibuprofen on freshwater diatom *Navicula* sp. *Journal of Hazardous Materials, 330*, 127–134.

Duarte, P., Almeida, C. M. R., Fernandes, J. P., Morais, D., Lino, M., Gomes, C. R., Carvalho, M. F., & Mucha, A. P. (2019). Bioremediation of bezafibrate and paroxetine by microorganisms from estuarine sediment and activated sludge of an associated wastewater treatment plant. *Science of the Total Environment, 655*, 796–806.

Durairaj, P., Malla, S., Nadarajan, S. P., Lee, P. G., Jung, E., Park, H. H., Kim, B. G., & Yun, H. (2015). Fungal cytochrome P450 monooxygenases of *Fusarium oxysporum* for the synthesis of ω-hydroxy fatty acids in engineered *Saccharomyces cerevisiae*. *Microbial Cell Factories, 14*, 45.

Eccles, H. (1999). Treatment of metal-contaminated wastes: Why select a biological process? *Trends in Biotechnology, 17*, 462–465.

Eriksen, N. T. (2008). The technology of microalgal culturing. *Biotechnology Letters, 30*, 1525–1536.

Fabbrini, M., Galli, C., & Gentili, P. (2002). Comparing the catalytic efficiency of some mediators of laccase. *Journal of Molecular Catalysis B: Enzymatic, 16*, 231–240.

Facey, S. J., Nebel, B. A., Kontny, L., Allgaier, M., & Hauer, B. (2018). Rapid and complete degradation of diclofenac by native soil microorganisms. *Environmental Technology & Innovation, 10*, 55–61.

Falfushynska, H., Sokolov, E. P., Haider, F., Oppermann, C., Kragl, U., Ruth, W., Stock, M., Glufke, S., Winkel, E. J., & Sokolova, I. M. (2019). Effects of a common pharmaceutical, atorvastatin, on energy metabolism and detoxification mechanisms of a marine bivalve *Mytilus edulis*. *Aquatic Toxicology, 208*, 47–61.

Fernandes, J. P., Almeida, C. M. R., Salgado, M. A., Carvalho, M. F., & Mucha, A. P. (2021). Pharmaceutical compounds in aquatic environments: Occurrence, fate, and bioremediation perspective. *Toxics, 9*(10), 257.

Fernandes, J. P., Duarte, P., Almeida, C. M. R., Carvalho, M. F., & Mucha, A. P. (2020). The potential of bacterial consortia obtained from different environments for bioremediation of paroxetine and bezafibrate. *Journal of Environmental Chemical Engineering, 8*, 103881.

Ferrando-Climent, L., Rodriguez-Mozaz, S., & Barceló, D. (2013). Development of a UPLC-MS/MS method for the determination of ten anticancer drugs in hospital and urban wastewaters, and its application for the screening of human metabolites assisted by information-dependent acquisition tool (IDA) in sewage samples. *Analytical and Bioanalytical Chemistry, 405*(18), 5937–5952.

Flemming, H. C., & Wingender, J. (2010). The biofilm matrix. *Nature Reviews Microbiology, 8*, 632–633.

Foflonker, F., Ananyev, G., Qiu, H., Morrison, A., Palenik, B., Dismukes, G. C., & Bhattacharya, D. (2016). The unexpected extremophile: Tolerance to fluctuating salinity in the green alga *Picochlorum*. *Algal Research, 16*, 465–472.

Fonseca, T. G., Auguste, M., Ribeiro, F., Cardoso, C., Mestre, N. C., Abessa, D. M. S., & Bebianno, M. J. (2018). Environmental relevant levels of the cytotoxic drug cyclophosphamide produce harmful effects in the polychaete *Nereis divers color*. *Science of the Total Environment, 636*, 798–809.

Fuhrman, V. F., Tal, A., & Arnon, S. (2015). Why endocrine disrupting chemicals (EDCs) challenge traditional risk assessments and how to respond. *Journal of Hazardous Materials, 286*, 589–611.

Fux, C. A., Costerton, J. W., Stewart, P. S., & Stoodley, P. (2005). Survival strategies of infectious biofilms. *Trends in Microbiology, 13*, 34–40.

Gadd, G. M. (2009). Biosorption: Critical review of scientific rationale, environmental importance, and significance for pollution treatment. *Journal of Chemical Technology & Biotechnology, 84*, 13–28.

Galichet, L. Y., Moffat, A. C., Osselton, M. D., & Widdop, B. (2004). *Clarke's Analysis of Drugs and Poisons* (3rd ed.). Pharmaceutical Press: London, UK.

García, J., García-Galán, M. J., Day, J. W., Boopathy, R., White, J. R., Wallace, S., & Hunter, R. G. (2020). A review of emerging organic contaminants (EOCs), antibiotic resistant bacteria (ARB), and antibiotic resistance genes (ARGs) in the environment: Increasing removal with wetlands and reducing environmental impacts. *Bioresource Technology, 307*, 123228.

Gasser, C. A., Hommes, G., Schaffer, A., & Corvini, P. F. (2012). Multi-catalysis reactions: New prospects and challenges of biotechnology to valorize lignin. *Applied Microbiology and Biotechnology, 95*, 1115–1134.

Gaylarde, C. C., Bellinaso, M. L., & Manfilo, G. P. (2005). Biorremediação – aspectos biológicos e técnicos da biorremediação de xenobióticos. *Biotecnologia, Ciência e Desenvolvimento, 34*, 36–43.

Gheorghe, S., Petre, J., Lucaciu, I., Stoica, C., & Nita-Lazar, M. (2016). Risk screening of pharmaceutical compounds in Romanian aquatic environment. *Environmental Monitoring and Assessment, 188*, 379.

Gillbride, K. A., Lee, D. Y., & Beaudette, L. A. (2006). Molecular techniques in wastewater: Understanding microbial communities, detecting pathogens, and real-time process control. *Journal of Microbiological Methods, 66*, 1–20.

Gillespie, I. M. M., & Philip, J. C. (2013). Bioremediation, an environmental remediation technology for the bioeconomy. *Trends in Biotechnology, 31*, 329–332.

Gillings, M. R., & Stokes, H. W. (2012). Are humans increasing bacterial evolvability? *Trends in Ecology & Evolution, 27*, 346–352.

Gogoi, A., Mazumder, P., Tyagi, V. K., Tushara Chaminda, G. G., An, A. K., & Kumar, M. (2018). Occurrence and fate of emerging contaminants in water environment: A review. *Groundwater for Sustainable Development, 6*, 169–180.

Gómez-Canela, C., Ventura, F., Caixach, J., & Lacorte, S. (2014). Occurrence of cytostatic compounds in hospital effluents and wastewaters, determined by liquid chromatography coupled to high-resolution mass spectrometry. *Analytical and Bioanalytical Chemistry, 406*(16), 3801–3814.

González-Pleiter, M., Gonzalo, S., Rodea-Palomares, I., Leganés, F., Rosal, R., Boltes, K., Marco, E., & Fernández-Piñas, F. (2013). Toxicity of five antibiotics and their mixtures towards photosynthetic aquatic organisms: Implications for environmental risk assessment. *Water Research, 47*(6), 2050–2064.

Grabert, R., Boopathy, R., Nathaniel, R., & LaFleur, G. (2018). Effect of tetracycline on ammonia and carbon removal by the facultative bacteria in the anaerobic digester of a sewage treatment plant. *Bioresource Technology, 267*, 265–270.

Greca, M. D., Pinto, G., Pistillo, P., Pollio, A., Previtera, L., & Temussi, F. (2008). Biotransformation of ethinylestradiol by microalgae. *Chemosphere, 70*, 2047–2053.

Gros, M., Rodríguez-Mozaz, S., & Barceló, D. (2012). Fast and comprehensive multi-residue analysis of a broad range of human and veterinary pharmaceuticals and some of their metabolites in surface and treated waters by ultra-high-performance liquid chromatography coupled to quadrupole-linear ion trap tandem. *Journal of Chromatography A, 1248*, 104–121.

Gross, R., Hauer, B., Otto, K., & Schmid, A. (2007). Microbial biofilms: New catalysts for maximizing productivity of long-term biotransformation's. *Biotechnology and Bioengineering, 98*, 1123–1134.

Gultom, S. O., & Hu, B. (2013). Review of microalgae harvesting via co-pelletization with filamentous fungus. *Energies, 6*, 5921–5939.

Hall-Stoodley, L., Costerton, J. W., & Stoodley, P. (2004). Bacterial biofilms: From the natural environment to infectious diseases. *Nature Reviews Microbiology, 2*, 95–108.

Han, H., Xiao, H., & Lu, Z. (2016). Short-term toxicity assessments of an antibiotic metabolite in Wistar rats and its metabonomics analysis by ultra-high performance liquid chromatography coupled to quadrupole time-of-flight mass spectrometry. *Toxicology and Applied Pharmacology, 293*, 1–9.

Hay, A. (2001). Growth of a bacterial consortium on triclosan. *FEMS Microbiology Ecology, 36*, 105–112.

Hocquet, D., Muller, A., & Bertrand, X. (2016). What happens in hospitals does not stay in hospitals: Antibiotic-resistant bacteria in hospital wastewater systems. *Journal of Hospital Infection, 93*, 395–402.

Hofmann, U., & Schlosser, D. (2016). Biochemical and physicochemical processes contributing to the removal of endocrine-disrupting chemicals and pharmaceuticals by the aquatic ascomycete *Phoma* sp. UHH 5-1-03. *Applied Microbiology and Biotechnology, 100*, 2381–2399.

Hom-Diaz, A., Jaén-Gil, A., Bello-Laserna, I., Rodríguez-Mozaz, S., Vicent, T., Barceló, D., & Blánquez, P. (2017). Performance of a microalgal photobioreactor treating toilet wastewater: Pharmaceutically active compound removal and biomass harvesting. *Science of the Total Environment, 592*, 1–11.

Hulak, M., Psenicka, M., Gela, D., Rodina, M., & Linhart, O. (2010). Morphological sex change upon treatment by endocrine modulators in meiogynogenetic tench (*Tinca tinca* L.). *Aquaculture Research, 41*, 233–239.

Ikehata, K., Naghashkar, N. J., & El-Din, M. G. (2006). Degradation of aqueous pharmaceuticals by ozonation and advanced oxidation processes: A review. *Ozone: Science & Engineering, 28*, 353–414.

Jebapriya, G. R., & Gnanadoss, J. J. (2013). Bioremediation of textile dye using white-rot fungi: A review. *International Journal of Current Research and Review, 5*, 1–13.

Jelic, A., Cruz-Morato, C., Marco-Urrea, E., Sarra, M., Perez, S., Vicent, T., Petrovic, M., & Barceló, D. (2012). Degradation of carbamazepine by *Trametes versicolor* in an air pulsed fluidized bed bioreactor and identification of intermediates. *Water Research, 46*, 955–964.

Ji, K., Kim, S., Han, S., Seo, J., Lee, S., Park, Y., Choi, K., Kho, Y. L., Kim, P. G., & Park, J. (2012). Risk assessment of chlortetracycline, oxytetracycline, sulfamethazine, sulfathiazole, and erythromycin in aquatic environments. Are the current environmental concentrations safe? *Ecotoxicology, 21*, 2031–2050.

Jjemba, P. K. (2006). Excretion and ecotoxicity of pharmaceutical and personal care products in the environment. *Ecotoxicology and Environmental Safety, 63*(1), 113–130.

Johnson, D. J., Sanderson, H., Brain, R. A., Wilson, C. J., & Solomon, K. R. (2007). Toxicity and hazard of selective serotonin reuptake inhibitor antidepressants fluoxetine, fluvoxamine, and sertraline to algae. *Ecotoxicology and Environmental Safety, 67*, 128–139.

Joss, A., Siegrist, H., & Ternes, T. A. (2008). Are we about to upgrade wastewater treatment for removing organic micropollutants? *Water Science and Technology, 57*, 251–255.

Joss, A., Zabczynski, S., Gobel, A., Hoffmann, B., Loffler, D., McArdell, C. S., Ternes, T. A., Thomsen, A., & Siegrist, H. (2006). Biological degradation of pharmaceuticals in municipal wastewater treatment: Proposing a classification scheme. *Water Research, 40*, 1686–1696.

Kallmeyer, J., Pockalny, R., Adhikari, R. R., Smith, D. C., & D'Hondt, S. (2012). Global distribution of microbial abundance and biomass in subseafloor sediment. *Proceedings of the National Academy of Sciences of the United States of America, 109*(40), 16213–16216.

Kasonga, T. K., Coetzee, M. A. A., Kamika, I., & Momba, M. N. B. (2020). Data on the degradation of pharmaceuticals and their metabolites by a fungal consortium in a non-sterile stirred fluidized bioreactor. *Data in Brief, 28*, 105057.

Khunjar, W. O., Mackintosh, S. A., Skotnicka-Pitak, J., Baik, S., Aga, D. S., & Love, N. G. (2011). Elucidating the relative roles of ammonia oxidizing and heterotrophic bacteria during the biotransformation of 17α-ethinylestradiol and trimethoprim. *Environmental Science & Technology, 45*, 3605–3612.

Kim, Y. J., & Nicell, J. A. (2006). Impact of reaction conditions on the laccase-catalyzed conversion of bisphenol A. *Bioresource Technology, 97*, 1431–1442.

Kong, L., Kadokami, K., Wang, S., Duong, H. T., & Chau, H. T. C. (2015). Monitoring of 1300 organic micro-pollutants in surface waters from Tianjin, North China. *Chemosphere, 122*, 125–130.

Kramer, V. J., Miles-Richardson, S., Pierens, S. L., & Giesy, J. P. (1998). Reproductive impairment and induction of alkaline-labile phosphate, a biomarker of estrogen exposure, in fathead minnows (*Pimephales promelas*) exposed to waterborne 17β-estradiol. *Aquatic Toxicology, 40*, 335–360.

Kumar, M. S., Kabra, A. N., Min, B., El-Dalatony, M. M., Xiong, J., Thajuddin, N., Lee, D. S., & Jeon, B. H. (2016). Insecticides induced biochemical changes in freshwater microalga *Chlamydomonas mexicana*. *Environmental Science and Pollution Research, 23*(2), 1091–1099.

Kumari, J., & Sahoo, P. K. (2005). Effects of cyclophosphamide on the immune system and disease resistance of Asian catfish *Clarias batrachus*. *Fish and Shellfish Immunology, 19*, 307–316.

Kumirska, J., Migowska, N., Caban, M., Łukaszewicz, P., & Stepnowski, P. (2015). Simultaneous determination of nonsteroidal anti-inflammatory drugs and oestrogenic hormones in environmental solid samples. *Science of the Total Environment, 508*, 498–505.

Kümmerer, K. (2009). Antibiotics in the aquatic environment: A review–Part I. *Chemosphere, 75*(4), 417–434.

Kümmerer, K., & Al-Ahmad, A. (1997). Biodegradability of the anti-tumor agents 5-fluorouracil, cytarabine, and gemcitabine: Impact of the chemical structure and synergistic toxicity with hospital effluent. *Acta Hydrochimica et Hydrobiologica, 25*, 166–172.

Kümmerer, K., Haiß, A., Schuster, A., Hein, A., & Ebert, I. (2016). Antineoplastic compounds in the environment—Substances of special concern. *Environmental Science and Pollution Research, 23*(15), 14791–14804.

Kwak, K., Ji, K., Kho, Y., Kim, P., Lee, J., Ryu, J., & Choi, K. (2018). Chronic toxicity and endocrine disruption of naproxen in freshwater waterfleas and fish, and steroidogenic alteration using H295R cell assay. *Chemosphere, 204*, 156–162.

Lahti, M., & Oikari, A. (2011). Microbial transformation of pharmaceuticals naproxen, bisoprolol, and diclofenac in aerobic and anaerobic environments. *Archives of Environmental Contamination and Toxicology, 61*, 202–210.

Lai, K. M., Scrimshaw, M. D., & Lester, J. N. (2002). Biotransformation and bioconcentration of steroid estrogens by *Chlorella vulgaris*. *Applied and Environmental Microbiology, 68*, 859–864.

Li, Q., Wang, P., Chen, L., Gao, H., & Wu, L. (2016). Acute toxicity and histopathological effects of naproxen in zebrafish (*Danio rerio*) early life stages. *Environmental Science and Pollution Research, 23*, 18832–18841.

Li, X., de Toledo, R. A., Wang, S., & Shim, H. (2015). Removal of carbamazepine and naproxen by immobilized *Phanerochaete chrysosporium* under non-sterile conditions. *New Biotechnology, 32*(2), 282–289.

Liu, J., Wu, Y., Wu, C., Muylaert, K., Vyverman, W., Yu, H. Q., Muñoz, R., & Rittmann, B. (2017). Advanced nutrient removal from surface water by a consortium of attached microalgae and bacteria: A review. *Bioresource Technology, 241*, 1127–1137.

Liu, Y., & Liu, Y. J. (2008). Biosorption isotherms, kinetics and thermodynamics. *Separation and Purification Technology, 61*, 229–242.

Liu, Y., Zhang, J., Gao, B., & Feng, S. (2014). Combined effects of two antibiotic contaminants on *Microcystis aeruginosa*. *Journal of Hazardous Materials, 279*, 148–155.

López-Serna, R., Jurado, A., Vázquez-Suñé, E., Carrera, J., Petrović, M., & Barceló, D. (2013). Occurrence of 95 pharmaceuticals and transformation products in urban groundwaters underlying the metropolis of Barcelona, Spain. *Environmental Pollution, 174*, 305–315.

Lu, T., Zhang, Q. L., & Yao, S. J. (2016). Application of biosorption and biodegradation functions of fungi in wastewater and sludge treatment. In D. Purchase (Ed.), *Fungal Applications in Sustainable Environmental Biotechnology* (pp. 85–105). Springer.

Maes, H. M., Maletz, S. X., Ratte, H. T., Hollender, J., & Schaeffer, A. (2014). Uptake, elimination, and biotransformation of 17α-ethinylestradiol by the freshwater alga *Desmodesmus subspicatus*. *Environmental Science & Technology, 48*, 12354–12361.

Mahmood, A. R., Al-Haideri, H. H., & Hassan, F. M. (2019). Detection of antibiotics in drinking water treatment plants in Baghdad City, Iraq. *Advances in Public Health, 2019*, 1–10.

Maradonna, F., Batti, S., Marino, M., Mita, D. G., & Carnevali, O. (2009). Tamoxifen as an emerging endocrine disruptor: Effects on fish reproduction and detoxification target genes. *Annals of the New York Academy of Sciences, 1163*, 457–459.

Marco-Urrea, E., Pérez-Trujillo, M., Vicent, T., & Caminal, G. (2009). Ability of white-rot fungi to remove selected pharmaceuticals and identification of degradation products of ibuprofen by *Trametes versicolor*. *Chemosphere, 74*(6), 765–772.

Martín, J., Camacho-Muñoz, D., Santos, J. L., Aparicio, I., & Alonso, E. (2011). Simultaneous determination of a selected group of cytostatic drugs in water using high-performance liquid chromatography-triple-quadrupole mass spectrometry. *Journal of Separation Science, 34*(22), 3166–3177.

Mashi, B. H. (2013). Biorremediation: Issues and challenges. *Journal of Research in Industrial Engineering, 11*, 1596–8303.

Matamoros, V., Uggetti, E., García, J., & Bayona, J. M. (2016). Assessment of the mechanisms involved in the removal of emerging contaminants by microalgae from wastewater: A laboratory scale study. *Journal of Hazardous Materials, 301*, 197–205.

Mathias, F. T., Fockink, D. H., Disner, G. R., Prodocimo, V., Ribas, J. L. C., Ramos, L. P., Cestari, M. M., & Silva de Assis, H. C. (2018). Effects of low concentrations of ibuprofen on freshwater fish *Rhamdia quelen*. *Environmental Toxicology and Pharmacology, 59*, 105–113.

Megharaj, M., Ramakrishnan, B., Venkateswarlu, K., Sethunathan, N., & Naidu, R. (2011). Bioremediation approaches for organic pollutants: A critical perspective. *Environmental International, 37*, 1362–1375.

Michalak, I., Chojnacka, K., & Witek-Krowiak, A. (2013). State of the art for the biosorption process—A review. *Applied Biochemistry and Biotechnology, 170*, 1389–1416.

Mijangos, L., Ziarrusta, H., Ros, O., Kortazar, L., Fernández, L. A., Olivares, M., Zuloaga, O., Prieto, A., & Etxebarria, N. (2018). Occurrence of emerging pollutants in estuaries of the Basque Country: Analysis of sources and distribution, and assessment of the environmental risk. *Water Research, 147*, 152–163.

Milledge, J. J., & Heaven, S. A. (2013). A review of the harvesting of micro-algae for biofuel production. *Reviews in Environmental Science and Biotechnology, 12*, 165–178.

Mir-Tutusaus, J. A., Baccar, R., Caminal, G., & Sarra, M. (2018). Can white-rot fungi be a real wastewater treatment alternative for organic micropollutants removal? A review. *Water Research, 138*, 137–151.

Misal, S. A., Lingojwar, D. P., Shinde, R. M., & Gawai, K. R. (2011). Purification and characterization of azoreductase from alkaliphilic strain *Bacillus badius*. *Process Biochemistry, 46*, 1264–1269.

Mishra, S., Zhang, W., Lin, Z., Pang, S., Huang, Y., Bhatt, P., & Chen, S. (2020). Carbofuran toxicity and its microbial degradation in contaminated environments. *Chemosphere, 259*, 127419.

Moreira, I. S., Bessa, V. S., Murgolo, S., Piccirillo, C., Mascolo, G., & Castro, P. M. L. (2018). Biodegradation of diclofenac by the bacterial strain *Labrys portucalensis* F11. *Ecotoxicology and Environmental Safety, 152*, 104–113.

Moreira, I. S., Ribeiro, A. R., Afonso, C. M., Tiritan, M. E., & Castro, P. M. L. (2014). Enantioselective biodegradation of fluoxetine by the bacterial strain *Labrys portucalensis* F11. *Chemosphere, 111*, 103–111.

Mulla, S. I., Hu, A., Sun, Q., Li, J., Suanon, F., Ashfaq, M., & Yu, C. P. (2018). Biodegradation of sulfamethoxazole in bacteria from three different origins. *Journal of Environmental Management, 206*, 93–102.

Muñoz, R., & Guieysse, B. (2006). Algal–bacterial processes for the treatment of hazardous contaminants: A review. *Water Research, 40*, 2799–2815.

Muter, O., Erkons, P., Selga, T., Berzins, A., Gudra, D., Radovica-Spalvina, I., Fridmanis, D., & Bartkevics, V. (2017). Removal of pharmaceuticals from municipal wastewaters at laboratory scale by treatment with activated sludge and bio stimulation. *Science of the Total Environment, 584–585*, 402–413.

Nassour, C., Barton, S. J., Nabhani-Gebara, S., Saab, Y., & Barker, J. (2020). Occurrence of anticancer drugs in the aquatic environment: A systematic review. *Environmental Science and Pollution Research, 27*(2), 1339–1347.

Navarro, A. E., Lim, H., Chang, E., Lee, Y., & Manrique, A. S. (2014). Uptake of sulfa drugs from aqueous solutions by marine algae. *Separation Science and Technology, 49*, 2175–2181.

Nguyen, L., Hai, F. I., Yang, S., Kang, J., Leusch, F., Roddick, F., Price, W. E., & Nghiem, L. D. (2014). Removal of pharmaceuticals, steroid hormones, phytoestrogens, UV-filters, industrial chemicals and pesticides by *Trametes versicolor*: Role of biosorption and biodegradation. *International Biodeterioration & Biodegradation, 88*, 169–175.

Niemuth, N. J., Jordan, R., Crago, J., Blanksma, C., Johnson, R., & Klaper, R. D. (2015). Metformin exposure at environmentally relevant concentrations causes potential endocrine disruption in adult male fish. *Environmental Toxicology and Chemistry, 34*(2), 291–296.

Nowakowska, K., Giebułtowicz, J., Kamaszewski, M., Adamski, A., Szudrowicz, H., Ostaszewska, T., Solarska-Dzięciołowska, U., Nałęcz-Jawecki, G., Wroczyński, P., & Drobniewska, A. (2020). Acute exposure of zebrafish (*Danio rerio*) larvae to environmental concentrations of selected antidepressants: Bioaccumulation, physiological and histological changes. *Comparative Biochemistry and Physiology Part C: Toxicology & Pharmacology, 229*, 108670.

Onesios, K. M., & Bouwer, E. J. (2012). Biological removal of pharmaceuticals and personal care products during laboratory soil aquifer treatment simulation with different primary substrate concentrations. *Water Research, 46*, 2365–2375.

Onesios, K. M., Yu, J. T., & Bouwer, E. J. (2009). Biodegradation and removal of pharmaceuticals and personal care products in treatment systems: A review. *Biodegradation, 20*, 441–466.

Osorio, V., Larrañaga, A., Aceña, J., Pérez, S., & Barceló, D. (2016). Concentration and risk of pharmaceuticals in freshwater systems are related to the population density and the livestock units in Iberian Rivers. *Science of the Total Environment, 540*, 267–277.

Otto, B., & Schlosser, D. (2014). First laccase in green algae: Purification and characterization of an extracellular phenol oxidase from *Tetracystis aeria*. *Planta, 240*, 1225–1236.

Oulton, R. L., Kohn, T., & Cwiertny, D. M. (2010). Pharmaceuticals and personal care products in effluent matrices: A survey of transformation and removal during wastewater treatment and implications for wastewater management. *Journal of Environmental Monitoring, 12*, 1956–1978.

Padhye, L. P., Yao, H., Kung'u, F. T., & Huang, C. H. (2014). Year-long evaluation on the occurrence and fate of pharmaceuticals, personal care products, and endocrine disrupting chemicals in an urban drinking water treatment plant. *Water Research, 51*, 266–276.

Paíga, P., Santos, L. H., Amorim, C. G., Araújo, A. N., Montenegro, M. C., Pena, A., & Delerue-Matos, C. (2013). Pilot monitoring study of ibuprofen in surface waters of north of Portugal. *Environmental Science and Pollution Research, 20*, 2410–2420.

Palli, L., Castellet-Rovira, F., Pérez-Trujillo, M., Caniani, D., Sarrá-Adroguer, M., & Gori, R. (2017). Preliminary evaluation of *Pleurotus ostreatus* for the removal of selected pharmaceuticals from hospital wastewater. *Biotechnology Progress, 33*, 1529–1537.

Pan, L., Li, J., Li, C., Tang, X., Yu, G., & Wang, Y. (2018). Study of ciprofloxacin biodegradation by a *Thermus* sp. isolated from pharmaceutical sludge. *Journal of Hazardous Materials, 343*, 59–67.

Pan, L., Tang, X., Li, C., Yu, G., & Wang, Y. (2017). Biodegradation of sulfamethazine by an isolated thermophile–*Geobacillus* sp. S-07. *World Journal of Microbiology and Biotechnology, 33*, 85.

Papageorgiou, M., Zioris, I., Danis, T., Bikiaris, D., & Lambropoulou, D. (2019). Comprehensive investigation of a wide range of pharmaceuticals and personal care products in urban and hospital wastewaters in Greece. *Science of the Total Environment, 694*, 133565.

Park, D., Yun, Y. S., & Park, J. M. (2010). The past, present, and future trends of biosorption. *Biotechnology and Bioprocess Engineering, 15*, 86–102.

Paul, D., Pandey, G., Pandey, J., & Jain, R. K. (2005). Accessing microbial diversity for bioremediation and environmental restoration. *Trends in Biotechnology, 23*, 135–142.

Peng, F. Q., Ying, G. G., Yang, B., Liu, S., Lai, H. J., Liu, Y. S., Chen, Z. F., & Zhou, G. J. (2014). Biotransformation of progesterone and norgestrel by two freshwater microalgae (*Scenedesmus obliquus* and *Chlorella pyrenoidosa*): Transformation kinetics and products identification. *Chemosphere, 95*, 581–588.

Petrović, M., Škrbić, B., Živančev, J., Ferrando-Climent, L., & Barceló, D. (2014). Determination of 81 pharmaceutical drugs by high performance liquid chromatography coupled to mass spectrometry with hybrid triple quadrupole–Linear ion trap in different types of water in Serbia. *Science of the Total Environment, 468*, 415–428.

Pienkos, P., & Darzins, A. (2009). The promise and challenges of microalgal-derived biofuels. *Biofuels, Bioproducts and Biorefining, 3*, 431–440.

Pinu, F. R., Beale, D. J., Paten, A. M., Kouremenos, K., Swarup, S., Schirra, H. J., & Wishart, D. (2019). Systems biology and multi-omics integration: Viewpoints from the metabolomics research community. *Metabolites, 9*(4), 76.

Pointing, S. B. (2001). Feasibility of bioremediation by white-rot fungi. *Applied Microbiology and Biotechnology, 57*, 20–33.

Polprasert, C., Dissanayake, M., & Thanh, N. (1983). Bacterial die-off kinetics in waste stabilization ponds. *Water Pollution Control Federation, 55*, 285–296.

Puckett, S. P., Samples, R. M., Schloss, P. D., & Balunas, M. J. (2020). Metabolomics and the microbiome: Characterizing molecular diversity in complex microbial communities. In *Comprehensive Natural Products III: Chemistry and Biology* (pp. 502–518). Elsevier.

Rabiet, M., Togola, A., Brissaud, F., Seidel, J. L., Budzinski, H., & Elbaz-Poulichet, F. (2006). Consequences of treated water recycling as regards pharmaceuticals and drugs in surface and ground waters of a medium-sized Mediterranean catchment. *Environmental Science & Technology, 40*(17), 5282–5288.

Radjenović, J., Petrović, M., & Barceló, D. (2009). Fate and distribution of pharmaceuticals in wastewater and sewage sludge of the conventional activated sludge (CAS) and advanced membrane bioreactor (MBR) treatment. *Water Research, 43*(3), 831–841.

Rahman, R. A., Molla, A. H., & Fakhru'l-Razi, A. (2014). Assessment of sewage sludge bioremediation at different hydraulic retention times using mixed fungal inoculation by liquid-state bioconversion. *Environmental Science and Pollution Research, 21*(2), 1178–1187.

Ramanan, R., Kim, B. H., Cho, D. H., Oh, H. M., & Kim, H. S. (2016). Algae-bacteria interactions: Evolution, ecology and emerging applications. *Biotechnology Advances, 34*(1), 14–29.

Ramírez-Durán, N., Moreno-Pérez, P. A., & Sandoval-Trujillo, A. H. (2017). Bacterial treatment of pharmaceutical industry effluents. In L. Gómez-Oliván (Ed.), *Ecopharmacovigilance* (Vol. 66, pp. 1–25). Springer.

Ramírez-Morales, D., Masís-Mora, M., Montiel-Mora, J. R., Cambronero-Heinrichs, J. C., Briceño-Guevara, S., Rojas-Sánchez, C. E., Méndez-Rivera, M., Arias-Mora, V., Tormo-Budowski, R., Brenes-Alfaro, L., & Rodríguez-Rodríguez, C. E. (2020). Occurrence of pharmaceuticals, hazard assessment and ecotoxicological evaluation of wastewater treatment plants in Costa Rica. *Science of the Total Environment, 746*, 141200.

Rhee, J. S., Kim, R. O., Chang, H. H., Lee, J., Lee, Y. M., & Lee, J. S. (2011). Endocrine disrupting chemicals modulate expression of O6-methylguanine DNA methyltransferase

(O6-MGMT) gene in the hermaphroditic fish, *Kryptolebias marmoratus*. *Comparative Biochemistry and Physiology C: Toxicology & Pharmacology, 153*, 141–149.

Rivera-Utrilla, J., Sánchez-Polo, M., Ferro-García, M. Á., Prados-Joya, G., & Ocampo-Pérez, R. (2013). Pharmaceuticals as emerging contaminants and their removal from water: A review. *Chemosphere, 93*(7), 1268–1287.

Rodarte-Morales, A. I., Feijoo, G., Moreira, M. T., & Lema, J. M. (2011). Degradation of selected pharmaceutical and personal care products (PPCPs) by white-rot fungi. *World Journal of Microbiology and Biotechnology, 27*(8), 1839–1846.

Rodriguez-Rodriguez, C. E., García-Galán, M. J., Blánquez, P., Díaz-Cruz, M. S., Barceló, D., Caminal, G., & Vicent, T. (2012). Continuous degradation of a mixture of sulfonamides by *Trametes versicolor* and identification of metabolites from sulfapyridine and sulfathiazole. *Journal of Hazardous Materials, 213–214*, 347–354.

Rodríguez-Rodríguez, C. E., Marco-Urrea, E., & Caminal, G. (2010). Degradation of naproxen and carbamazepine in spiked sludge by slurry and solid-phase *Trametes versicolor* systems. *Bioresource Technology, 101*(7), 2259–2266.

Roepke, T. A., Snyder, M. J., & Cherr, G. N. (2005). Estradiol and endocrine disrupting compounds adversely affect development of sea urchin embryos at environmentally relevant concentrations. *Aquatic Toxicology, 71*, 155–173.

Rosal, R., Rodríguez, A., Perdigón-Melón, J. A., Petre, A., García-Calvo, E., Gómez, M. J., Agüera, A., & Fernández-Alba, A. R. (2010). Occurrence of emerging pollutants in urban wastewater and their removal through biological treatment followed by ozonation. *Water Research, 44*(2), 578–588.

Rossello-Mora, R., & Amann, R. (2001). The species concept for prokaryotes. *FEMS Microbiology Reviews, 25*, 39–67.

Saeed, A., & Iqbal, M. (2013). Loofa (*Luffa cylindrica*) sponge: Review of development of the biomatrix as a tool for biotechnological applications. *Biotechnology Progress, 29*(3), 573–600.

Sahu, O. (2014). Reduction of organic and inorganic pollutants from wastewater by algae. *International Letters of Natural Sciences, 8*, 1–8.

Salim, S., Kosterink, N. R., Tchetkoua Wacka, N. D., Vermuë, M. H., & Wijffels, R. H. (2014). Mechanism behind autoflocculation of unicellular green microalgae *Ettlia texensis*. *Journal of Biotechnology, 174*, 34–38.

Salyers, A. A., Gupta, A., & Wang, Y. (2004). Human intestinal bacteria as reservoirs for antibiotic resistance genes. *Trends in Microbiology, 12*, 412–416.

Santos, I. J. S., Grossman, M., Sartoratto, A., Ponezi, A., & Durranta, L. (2012). Degradation of the recalcitrant pharmaceuticals carbamazepine and 17α-ethinylestradiol by ligninolytic fungi. *Chemical Engineering Transactions, 27*, 169–174.

Sapkota, A., Sapkota, A. R., Kucharski, M., Burke, J., McKenzie, S., Walker, P., & Lawrence, R. (2008). Aquaculture practices and potential human health risks: Current knowledge and future priorities. *Environmental International, 34*, 1215–1226.

Saussereau, E., Lacroix, C., Guerbet, M., Cellier, D., Spiroux, J., & Goullé, J. P. (2013). Determination of levels of current drugs in hospital and urban wastewater. *Bulletin of Environmental Contamination and Toxicology, 91*, 171–176.

Sayadi, S., & Ellouz, R. (1995). Roles of lignin peroxidase and manganese peroxidase from *Phanerochaete chrysosporium* in the decolorization of olive mill wastewaters. *Applied and Environmental Microbiology, 61*, 1098–1103.

Sehonova, P., Hodkovicova, N., Urbanova, M., Örn, S., Blahova, J., Svobodova, Z., Faldyna, M., Chloupek, P., Briedikova, K., & Carlsson, G. (2019). Effects of antidepressants with different modes of action on early life stages of fish and amphibians. *Environmental Pollution, 254*, 112999.

Seo, Y., Lee, W. H., Sorial, G., & Bishop, P. L. (2009). The application of a mulch biofilm barrier for surfactant enhanced polycyclic aromatic hydrocarbon bioremediation. *Environmental Pollution, 157*, 95–101.

Sgroi, M., Roccaro, P., Korshin, G. V., & Vagliasindi, F. G. A. (2017). Monitoring the behavior of emerging contaminants in wastewater-impacted rivers based on the use of fluorescence Excitation Emission Matrixes (EEM). *Environmental Science & Technology, 51*(8), 4306–4316.

Shakya, M., Lo, C. C., & Chain, P. S. G. (2019). Advances and challenges in metatranscriptomic analysis. *Frontiers in Genetics, 10*, 904.

Shekhar, S. K., Godheja, J., & Modi, D. R. (2020). Molecular technologies for assessment of bioremediation and characterization of microbial communities of pollutant-contaminated sites. In R. N. Bharagava & G. Saxena (Eds.), *Bioremediation of Industrial Waste for Environmental Safety* (pp. 447–474). Springer Nature.

Shi, W., Wang, L., Rousseau, D. P. L., & Lens, P. N. L. (2010). Removal of estrone, 17α-ethinylestradiol, and 17β-estradiol in algae and duckweed-based wastewater treatment systems. *Environmental Science and Pollution Research, 17*, 824–833.

Shishir, T. A., Mahbub, N., & Kamal, N. E. (2019). Review on bioremediation: A tool to resurrect the polluted rivers. *Pollution, 5*, 555–568.

Silva, A., Delerue-Matos, C., Figueiredo, S. A., & Freitas, O. M. (2019). The use of algae and fungi for removal of pharmaceuticals by bioremediation and biosorption processes: A review. *Water, 11*, 1555.

Silva, B., Costa, F., Neves, I. C., & Tavares, T. (2015). Psychiatric pharmaceuticals as emerging contaminants in wastewater. *SpringerBriefs in Molecular Science*.

Spina, F., Anastasi, A., Prigione, V., Tigini, V., & Varese, G. C. (2012). Biological treatment of industrial wastewaters: A fungal approach. *Chemical Engineering Transactions, 27*, 175–180.

Su, Y., Mennerich, A., & Urban, B. (2011). Municipal wastewater treatment and biomass accumulation with a wastewater-born and settleable algal-bacterial culture. *Water Research, 45*, 3351–3358.

Subashchandrabose, S. R., Ramakrishnan, B., Megharaj, M., Venkateswarlu, K., & Naidu, R. (2013). Mixotrophic cyanobacteria and microalgae as distinctive biological agents for organic pollutant degradation. *Environmental International, 51*, 59–72.

Sui, Q., Huang, J., Deng, S., Chen, W., & Yu, G. (2011). Seasonal variation in the occurrence and removal of pharmaceuticals and personal care products in different biological wastewater treatment processes. *Environmental Science & Technology, 45*, 3341–3348.

Sung, H. H., Chiu, Y. W., Wang, S. Y., Chen, C. M., & Huang, D. J. (2014). Acute toxicity of a mixture of acetaminophen and ibuprofen to green neon shrimp, *Neocaridina denticulate*. *Environmental Toxicology and Pharmacology, 38*(1), 8–13.

Taylor, R. L., Rand, J. D., & Caldwell, G. S. (2012). Treatment with algae extracts promotes flocculation, and enhances growth and neutral lipid content in *Nannochloropsis oculata*—A candidate for biofuel production. *Marine Biotechnology, 6*, 774–781.

Ternes, T., & Joss, A. (2015). Human pharmaceuticals, hormones and fragrances—The challenge of micropollutants in urban water management. *IWA Publishing*.

Topp, E., Hendel, J. G., Lapen, D. R., & Chapman, R. (2008). Fate of the nonsteroidal anti-inflammatory drug naproxen in agricultural soil receiving liquid municipal biosolids. *Environmental Toxicology and Chemistry, 27*, 2005.

Tortella, G., Diez, M., & Durán, N. (2005). Fungal diversity and use in decomposition of environmental pollutants. *Critical Reviews in Microbiology, 31*, 197–212.

Tran, N. H., Reinhard, M., & Gin, K. Y. H. (2018). Occurrence and fate of emerging contaminants in municipal wastewater treatment plants from different geographical regions: A review. *Water Research, 133*, 182–207.

Tran, N. H., Reinhard, M., Khan, E., Chen, H., Nguyen, V. T., Li, Y., Goh, S. G., Nguyen, Q. B., Saeidi, N., & Gin, K. Y. H. (2019). Emerging contaminants in wastewater, stormwater runoff, and surface water: Application as chemical markers for diffuse sources. *Science of The Total Environment, 676*, 252–267.

Tsezos, M., & Bell, J. P. (1989). Comparison of the biosorption and desorption of hazardous organic pollutants by live and dead biomass. *Water Research, 23*, 561–568.

Urbachka, R., Bottero, S., Mandich, A., Lutz, I., & Kloas, W. (2007). Endocrine disrupters with (anti)estrogenic and (anti)androgenic modes of action affecting reproductive biology of *Xenopus laevis*: I. Effects on sex steroid levels and biomarker expression. *Comparative Biochemistry and Physiology C: Toxicology & Pharmacology, 144*, 310–318.

Vazquez-Roig, P., Segarra, R., Blasco, C., Andreu, V., & Picó, Y. (2010). Determination of pharmaceuticals in soils and sediments by pressurized liquid extraction and liquid chromatography tandem mass spectrometry. *Journal of Chromatography A, 1217*, 2471–2483.

Verlicchi, P., Al Aukidy, M., & Zambello, E. (2012). Occurrence of pharmaceutical compounds in urban wastewater: Removal, mass load and environmental risk after secondary treatment—A review. *Science of The Total Environment, 429*, 123–155.

Vulliet, E., & Cren-Olivé, C. (2011). Screening of pharmaceuticals and hormones at the regional scale, in surface and groundwaters intended for human consumption. *Environmental Pollution, 159*, 2929–2934.

Wang, C., Shi, H., Adams, C. D., Gamagedara, S., Stayton, I., Timmons, T., & Ma, Y. (2011). Investigation of pharmaceuticals in Missouri natural and drinking water using high performance liquid chromatography-tandem mass spectrometry. *Water Research, 45*, 1818–1828.

Wang, H., Jin, M., Mao, W., Chen, C., Fu, L., Li, Z., Du, S., & Liu, H. (2020). Photosynthetic toxicity of non-steroidal anti-inflammatory drugs (NSAIDs) on green algae *Scenedesmus obliquus*. *Science of The Total Environment, 707*, 136176.

Wang, J., & Wang, S. (2016). Removal of pharmaceuticals and personal care products (PPCPs) from wastewater: A review. *Journal of Environmental Management, 182*, 620–640.

Wang, M., Yang, H., Ergas, S. J., & van der Steen, P. (2015). A novel shortcut nitrogen removal process using an algal-bacterial consortium in a photo-sequencing batch reactor (PSBR). *Water Research, 87*, 38–48.

Wesenberg, D., Kyriakides, I., & Agathos, S. N. (2003). White-rot fungi and their enzymes for the treatment of industrial dye effluents. *Biotechnology Advances, 22*, 161–187.

Wijffels, R. H., Kruse, O., & Hellingwerf, K. J. (2013). Potential of industrial biotechnology with cyanobacteria and eukaryotic microalgae. *Current Opinion in Biotechnology, 24*(3), 405–413.

Wittenberg, K. M. (1991). Preservation of high-moisture hay in storage through the use of forage additives. *Canadian Journal of Animal Science, 71*, 429–437.

Wohde, M., Berkner, S., Junker, T., Konradi, S., Schwarz, L., & Düring, R. A. (2016). Occurrence and transformation of veterinary pharmaceuticals and biocides in manure: A literature review. *Environmental Sciences Europe, 1*, 28.

Wolf, D. C., Cryder, Z., Khoury, R., Carlan, C., & Gan, J. (2020). Bioremediation of PAH-contaminated shooting range soil using integrated approaches. *Science of The Total Environment, 726*, 13844.

Wu, S., Zhang, L., & Chen, J. (2012). Paracetamol in the environment and its degradation by microorganisms. *Applied Microbiology and Biotechnology, 96*, 875–884.

Wu, X. L., Xiang, L., Yan, Q. Y., Jiang, Y. N., Li, Y. W., Huang, X. P., Li, H., Cai, Q. Y., & Mo, C. H. (2014). Distribution and risk assessment of quinolone antibiotics in the soils from organic vegetable farms of a subtropical city, Southern China. *Science of The Total Environment, 487*, 399–406.

Xia, L., Zheng, L., & Zhou, J. L. (2017). Effects of ibuprofen, diclofenac and paracetamol on hatch and motor behavior in developing zebra fish (*Danio rerio*). *Chemosphere, 182*, 416–425.

Xiao, R., & Zheng, Y. (2016). Overview of microalgal extracellular polymeric substances (EPS) and their applications. *Biotechnology Advances, 34*, 1225–1244.

Xiong, J. Q., Kurade, M. B., & Jeon, B. H. (2017). Biodegradation of levofloxacin by an acclimated freshwater alga *Chlorella vulgaris*. *Chemical Engineering Journal, 313*, 1251–1257.

Xiong, J. Q., Kurade, M. B., & Jeon, B. H. (2018). Can microalgae remove pharmaceutical contaminants from water? *Trends in Biotechnology, 36*(1), 30–44.

Xiong, J. Q., Kurade, M. B., Abou-Shanab, R. A. I., Ji, M. K., Choi, J. J. O., & Jeon, B. H. (2016). Biodegradation of carbamazepine using freshwater microalgae *Chlamydomonas mexicana* and *Scenedesmus obliquus* and the determination of its metabolic fate. *Bioresource Technology, 205*, 183–190.

Yan, C., Wang, F., Geng, H., Liu, H., Pu, S., Tian, Z., Chen, H., Zhou, B., Yuan, R., & Yao, J. (2020). Integrating high-throughput sequencing and metagenome analysis to reveal the characteristic and resistance mechanism of microbial community in metal contaminated sediments. *Science of The Total Environment, 707*, 136116.

Yu, K. S. H., Wong, A. H. Y., Yau, K. W. Y., Wong, Y. S., & Tam, N. F. Y. (2005). Natural attenuation, biostimulation and bioaugmentation on biodegradation of polycyclic aromatic hydrocarbons (PAHs) in mangrove sediments. *Marine Pollution Bulletin, 51*, 1071–1077.

Yuan, S. F., Liu, Z. H., Yin, H., Dang, Z., Wu, P. X., Zhu, N. W., & Lin, Z. (2019). Trace determination of sulfonamide antibiotics and their acetylated metabolites via SPE-LC-MS/MS in wastewater and insights from their occurrence in a municipal wastewater treatment plant. *Science of The Total Environment, 653*, 815–821.

Zhang, D., Gersberg, R. M., Ng, W. J., & Tan, S. K. (2014). Removal of pharmaceuticals and personal care products in aquatic plant-based systems: A review. *Environmental Pollution, 184*, 620–639.

Zhang, J. G., & Hu, B. (2012). A novel method to harvest microalgae via co-culture of filamentous fungi to form cell pellets. *Bioresource Technology, 114*, 529–535.

Zhang, L., Hu, J., Zhu, R., Zhou, Q., & Chen, J. (2013). Degradation of paracetamol by pure bacterial cultures and their microbial consortium. *Applied Microbiology and Biotechnology, 97*, 3687–3698.

Zhang, Y., & Geißen, S. U. (2012). Elimination of carbamazepine in a nonsterile fungal bioreactor. *Bioresource Technology, 112*, 221–227.

Zhao, X., Chen, Z., Wang, X., Li, J., Shen, J., & Xu, H. (2015). Remediation of pharmaceuticals and personal care products using an aerobic granular sludge sequencing bioreactor and microbial community profiling using Solexa sequencing technology analysis. *Bioresource Technology, 179*, 104–112.

Zhao, X., Ma, F., Feng, C., Bai, S., Yang, J., & Wang, L. (2017). Complete genome sequence of *Arthrobacter* sp. ZXY-2 associated with effective atrazine degradation and salt adaptation. *Journal of Biotechnology, 248*, 43–47.

Zhou, G. J., Ying, G. G., Liu, S., Zhou, L. J., Chen, Z. F., & Peng, F. Q. (2014). Simultaneous removal of inorganic and organic compounds in wastewater by freshwater green microalgae. *Environmental Sciences, 16*, 2018–2027.

Zhou, W., Cheng, Y., Li, Y., Wan, Y., Liu, Y., Lin, X., & Ruan, R. (2012). Novel fungal palletization-assisted technology for algae harvesting and wastewater treatment. *Applied Biochemistry and Biotechnology, 167*, 214–228.

CHAPTER 8

Detoxification and Biodegradation of Pharmaceutics by Common Microorganisms

NISHAT BHATIA,[1,2] ASHA KUMARI,[1,2] and RAHUL SHARMA[1,2*]

[1]*Department of Chemistry, Career Point University, Bhoranj (Tikker–Kharwarian), Hamirpur, Himachal Pradesh, India*

[2]*Center for Nano-Science and Technology, CPU, Hamirpur, Himachal Pradesh, India*

*Corresponding author.

ABSTRACT

Therapeutically active substances and organic pollutants are emerging into a significant environmental concern that poses a severe hazard to the earthly ecosystem and aquatic bodies. Subsequently, microorganisms are proficient in self-purification; microbial degradation progression is seen to be an efficacious way to contest detrimental and resistant contaminants. Pharmacological contaminants in aquatic organizations can be detached using normal wastewater treatment methods, however, hindered sludge settling, the existence of pharmaceutical mixtures, and persistent substances frequently constitute a possible ecological danger. The natural microbial population consciously prejudices the ecological progressions of pharmaceutical and natural chemicals such as nonsteroidal anti-inflammatory medicines, analgesics, and additional contaminants. Exceptional bacterial separates may function in the form of biodegraders. These microbes translate the toxins through co-metabolic pathways using the pollutants as their only supply of carbon.

Biomedical Waste Management: Recycling and Remediation Strategies. Younis Ahmad Hajam, Javid Ahmad Malik, Anu Pandita, and Deepu Pandita (Eds.)

The mechanisms used to degrade the target pollutant are biostimulation, bioaugmentation (BA), and natural attenuation by inborn microorganisms. Microorganisms include medicines, significant xenobiotics, and organic matter degraders that produce compounds that serve as nourishment for other organisms in the food chain. The communities of microbes play a significant role in sustaining ecosystem function. Variations in ecological procedures are connected with changes in microbial functional groupings. Thus, using microorganisms to detoxify contaminants likely increases their sustainable capacity for biodegradation (BD) and ensures an environmentally benevolent alternative bioremediation technique.

8.1 INTRODUCTION

The term "biodegradation (BD)" refers to the organically arbitrated decrease in the complexity of various chemical particles (Alexander, 1994). During the BD process, microorganisms (mainly aerobic bacteria) break down organic chemicals into more basic components like water, ammonia, and carbon dioxide.

Indeed, the procedure of organic materials being fragmented down into trivial molecules via active microorganisms is known as BD (Marinescu et al., 2009). "Mineralization" refers to the method after BD is finished. However, "biodegradation" is normally used to refer to almost any alteration in a substrate that is caused by a biological process (Bennet et al., 2002). Understanding the microorganisms that drive BD hence needs knowledge of the process itself. The material is changed by the micro-organism's enzymatic reactions or via metabolic pathways. It is composed of two processes: growth and metabolism. The major basis of energy and carbon for growth is organic contamination. As a result of this process, the organic pollutants are destroyed (mineralized). Cometabolism is the metabolic process of any material, which works as a main foundation of energy, is present (Fritsche et al., 2008). Bacteria, fungi, and yeast are just a few of the microorganisms that contribute to BD. There is little information on the involvement of algae and protozoa in BD (Das et al., 2008). Although BD methods vary greatly, the outcome is often carbon dioxide (Pramila et al., 2012). Anaerobic (without oxygen) or aerobic (with oxygen) degradation of organic material are both possible (Mrozik et al., 2003).

The saying "renewable matter" denotes material that bacteria may degrade, which is often organic matter such as animal and plant matter, as well as other substances produced by living creatures. Some microbes have an amazing, naturally occurring variety of catabolic enzymes that allow them to disrupt further, change, or collect a variety of substances, polyaromatic hydrocarbons, hydrocarbons, polychlorinated biphenyls (PCBs), metals, and radionuclides are examples of such substances (Leitão, 2009).

When addressing ecology, environmental remediation, and management of waste, the term "biodegradation" is commonly used (also known as "bioremediation"). Three stages or phases make up the bioremediation process shown in Figure 8.1. First, without human intervention, local microorganisms remove pollutants through a process called natural attenuation. The second strategy, known as biostimulation, involves supplying oxygen and essential nutrients to the ecosystems in order to increase their performance and speed up BD. Lastly, the systems are supplemented with microorganisms throughout the process of bioaugmentation (BA). These extra organisms must be well-organized than natural flora in humiliating the objective pollutant (Diez, 2009). Microorganisms that can swiftly adapt to and utilize pollutants of interest in a specific context are required for a viable remedial technique (Seo et al., 2009).

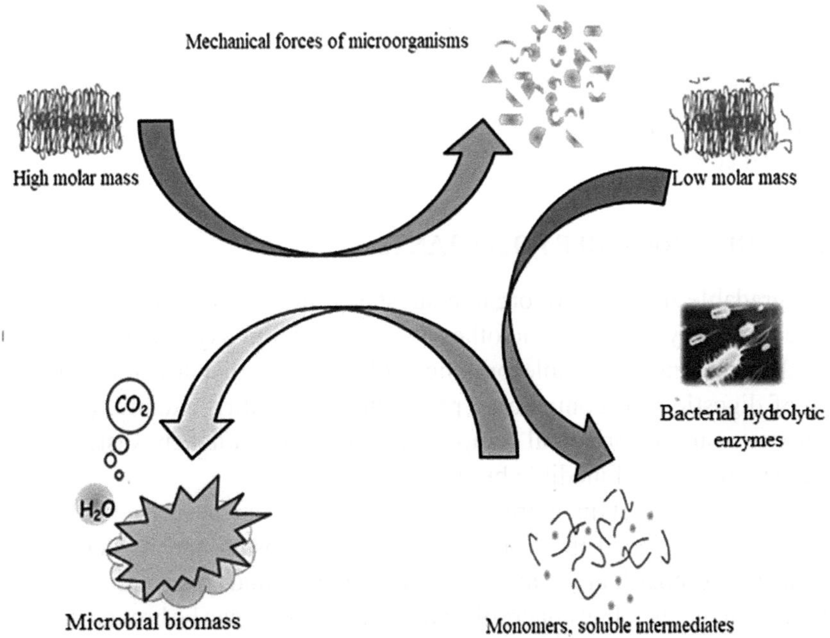

FIGURE 8.1 Stages or phases make up the bioremediation process.

The hereditary impending and specific ecological parameters like pH, temperature, and the availability of calcium and nitrogen bases then appear to impact the pace and degree of degradation. Microorganisms can utilize contaminants as substrates or cometabolize them. Consequently, the usage of genetically modified microorganisms (GMM) in the process of bioremediation has drawn a keen interest. These GEMs have been successfully employed to degrade a range of contaminants in controlled environments and have a higher degradative capacity. The main barriers to field testing GEM, however, are governmental restrictions and ecological and environmental issues (Menn et al., 2008). By attempting to cover every type of microbe involved in the breakdown of various contaminants, we have attempted to nurture a detailed grasp of the process of BD in this chapter.

Furthermore, while the term "biodegradation" is frequently used in relation to conservation, waste disposal, bioengineering, and the natural environment (bioremediation), and is now typically associated with environmentally friendly products, this chapter will focus on "biodegradation" in relation to "bioremediation" by describing processes that use microorganisms' degradation abilities (natural attenuation, biostimulation, and bioaugmentation (BA)). Microorganisms may be genetically engineered to perform a wide range of activities. The effective degradation of contaminants is one such goal. As a result, the second area of this section is to illustrate the significance of certain GEMs in the concerned procedure and to discuss challenges that must be solved before GEM can offer an efficient cleanup procedure at a reduced cost.

8.2 BIODEGRADABLE POLLUTANTS

Biodegradable trash is any organic substance in rubbish that may be fragmented down by microbes and other living things into oxygen, water, carbon dioxide, methane, or simple organic molecules by disintegration, aerobic way of digestion, or comparable procedures. Ash, dirt, manure, and other plant matter are the principal components, along with kitchen waste (spoiled food, trimmings, and inedible bits).

Additionally, certain inorganic elements that can be broken down by microbes are included in waste management. Gypsum and its derivatives, such as plasterboard, are examples of these compounds, as are other simple sulfates that, when left in anaerobic landfill conditions, can be broken down by sulfate-reducing bacteria to create hydrogen sulfide. Some other chemicals, such as nuclides of radioactive elements and metals, are much less likely to be biodegraded by local flora than normally occurring biomolecules, which may be effortlessly broken down after their introduction to the environment.

8.2.1 HYDROCARBONS

A biological substance known as a hydrocarbon is composed only of carbon and hydrogen atoms. It is feasible for carbon atoms to create double or triple bonds as well as much larger structures like rings. Apart from aromatic the long-chained aliphatic one comes in various forms such as alkanes, alkenes, and alkynes, while the first one has benzene (McMurry, 2000).

8.2.2 POLYCYCLIC AROMATIC HYDROCARBONS (PAHS)

These are also recognized as aromatic compounds with condensed rings aromatic compounds with fused rings or polycyclic aromatic hydrocarbons (PAHs). They have at least two fused aromatic rings.

The bulk of polycyclic aromatic hydrocarbons (PAHs) present in the environment contain two (naphthalene) to seven (coronene) fused benzene rings; however certain PAHs have more rings. Anthropogenic (pollution) linked PAH inputs can produce PAH complexes and accumulations that are similar to those of ordinary sources, but not exactly the same (Boehm, 1964). An appealing approach for restoring polluted places appears to be the employment of microorganisms for bioremediation of PAH-contaminated settings. Fish and other aquatic species can collect PAHs, which can then be consumed by humans. They can also bind to organic-rich soils and sediment. One way to think about the BD of PAHs is as a removal of man-made contaminants from the environment, and the other is as a regular component of the carbon cycle. Bacterial breakdown of PAHs in aquatic settings is more efficient than that of other microorganisms. In soil- or sediment-based PAHs, bacterial activity is lower, and sludge-based degrading bacteria must be added. Dissolution and vaporization processes keep degradable bacteria alive longer than sorption.

8.2.3 POLYCHLORINATED BIPHENYLS (PCBS)

The stable chemical compound PCBs was utilized extensively in the 1930s and 1940s. The main cause of PCBs' environmental persistence is that native water and earth biota are unable to digest the substance at a significant pace.

To find out how the breakdown rate might be increased, several researches on PCBs BD have been carried out. Anaerobic and aerobic PCB breakdown pathways are demonstrated by studies to exist. Highly chlorinated PCBs undergo an anaerobic procedure to extract chlorine atoms, which are subsequently mineralized in aerobic conditions (Borja et al., 2005). The breakdown process is affected by the complexity of the PCB congener, the type of microorganisms utilized, and the interactions between the microorganisms. PCBs are dangerous substances that have the potential to alter hormones and cause cancer. Consequently, worry over environmental degradation caused by PCBs is growing (Seeger et al., 2010).

8.2.4 WIDE RANGE PESTICIDES

These are chemicals or combinations made with the intention of eliminating, repelling, or reducing pests. Non-persistent pesticides are ones that break down quickly, whereas persistent pesticides are those that resist degradation. Microorganisms, particularly fungus and bacteria that use pesticides as a food supply, carry out the greatest predominant kind of degradation in the soil (Vargas, 1975).

8.2.5 HEAVY METALS

Any dangerous metal can be referred to as a "heavy metal," regardless of its atomic mass or density. Heavy metals are an undefined subset of elements with metallic properties. Lanthanides, actinides, some metalloids, and transition metals are examples of these.

A heavy metal, according to one source, is one of the common transition metals, such as lead, zinc, and copper. These metals pollute the environment through a variety of means, including leaded gasoline, industrial waste, and acid rain, which leaches metal ions from the soil into lakes and rivers (Singh, 2007). Heavy metals can be interacted with by microorganisms via a range of physicochemical methods, comprising bioleaching, bioaccumulation, biosorption, biomineralization, accumulation within the cell, and transformation on the basis of enzyme catalyzed mechanism.

Heavy metals are mobilized by bioleaching acids process of methylation or by some other biological activities (Lloyd et al., 2001). Figure 8.2 provides a summary of the chief bio-based activities which influence the bioremediation of metals and metalloids.

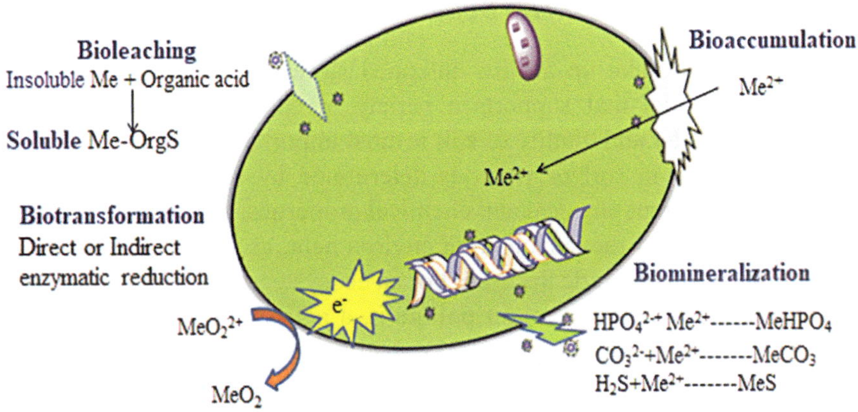

FIGURE 8.2 Microbial activities that affect the bioremediation of metals.

8.3 MICROBES MEDIATED DEGRADATION OF PHARMACEUTICS

Various microbes are employed to synthesize various medicines, other chemical substances, as well as other bioinorganic substances in the pharmaceutical industry. It also gives an idea to the pharmacists to create antibacterial drugs to tackle with various infectious diseases. In the modern era of science various microbes degrading drugs Figure 8.3 have been discovered degrading particular types of drugs.

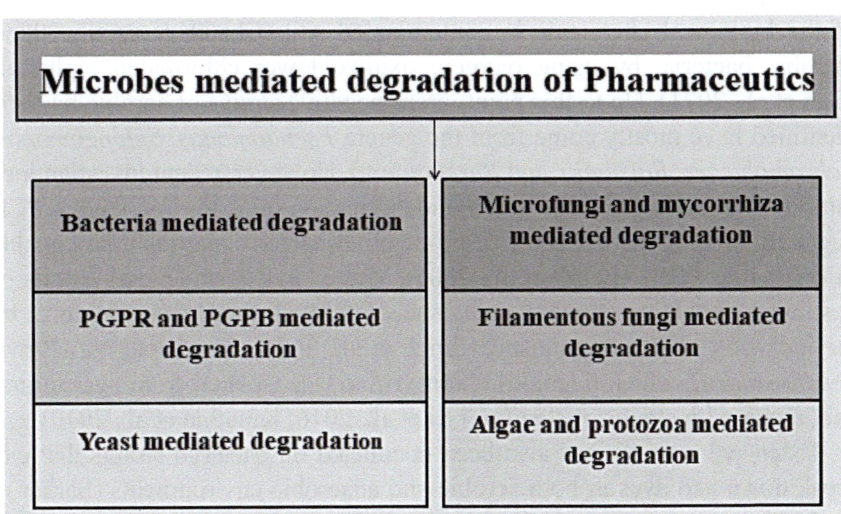

FIGURE 8.3 Microbes-mediated degradation of pharmaceutics.

8.3.1 BACTERIA MEDIATED DEGRADATION

Microorganisms make up a vast, untapped store of genetic diversity and metabolic potential that supports a variety of ecosystem functions. The protection of water and quality of soil is most important.

A drug's fate in soil or water is determined by a number of factors, including its inherent physical and chemical properties, ecological variables, and the condition of the surrounding environment, as well as the ability of microorganisms to degrade the drug. For an ecosystem to respond effectively to the different toxins that might pollute it, a native microbial community must exist (Caracciolo et al., 2015). Nitrate-reducing bacterial strains, such as *Citrobacter youngae*, *Brevibacillus* sp., *Enterobacter hormaechei*, *Pseudomonas* sp., *Arthrobacter nicotianae*, and *Pseudomonas* sp., are examples of hydrocarbons that can be biodegraded under both aerobic and anaerobic conditions (Floodgate, 1984). Among these strains of bacteria, *Bacillus* was identified as the most effective in degrading hydrocarbons (Kafilzadeh et al., 2011). These bacterial strains belonged to the following six genera: *Staphylococcus*, *Escherichia*, *Corynebacterium*, *Alcaligenes*, *Enterobacter*, and *Klebsiella*. Numerous strains of bacteria, mostly from soil, have been discovered to have the ability to break down aromatic hydrocarbons (Mrozik et al., 2003). The majority of these bacteria belong to the genus *Pseudomonas*, which are primarily gram-negative bacteria.

While many bacteria can metabolize organic contaminants, a single bacterium typically lacks the enzymatic capacity to digest all or even most of the biological chemicals in contaminated soil (Chaudhry et al., 2005). Aerobic bacteria, by using oxygen, oxidize lower chlorinated biphenyls (Cao et al., 2021). Up to this point, gram-negative strains of aerobic bacteria identified have mostly come from the genera *Pseudomonas*, *Sphingomonas*, *Achromobacter*, *Ralstonia*, and *Burkholderia*. However, recent investigations into PCB-degrading activity and the identification of genes involved in PCB degradation suggest that some gram-positive bacteria may also be capable of degrading PCBs (Leigh et al., 2006). There have been several reports of pesticides, including atrazine, being successfully removed from mixtures by the inclusion of microorganisms (Singh et al., 2004). The bacterium *Providencia stuartii*, which degrades chlorpyrifos, was isolated from agricultural soil, along with strains of *Bacillus* (Xi et al., 2016; Kinuthia et al., 2010).

Extensive research has also been conducted on bacterial strains that can break down azo dyes in both aerobic and anaerobic environments (Sarkar et al., 2017; Puvaneswari et al., 2006). According to available research, anaerobic environments are more conducive to microbial decolorization of azo dyes (Lade et al., 2015; Rau et al., 2020). For instance, Chaube et al. (2010) studied biodegradation and decolorization of dye using a mixed consortium of bacteria, including *Pseudomonas* sp., *Proteus* sp., and *Enterococcus* sp. However, several researchers have identified certain bacterial strains that are highly effective in removing azo dyes, such as *Shewanella decolorationis* (Hong et al., 2007). The use of a pure culture offers various advantages over mixed cultures, including consistent performance, a better understanding of degradation pathways, and greater confidence that, under specific environmental conditions, the catabolism of the dyes will result in harmless end products.

Another advantage is that population tracking of microbacteria over time can be achieved using culture-based or molecular approaches. During the cyclic phase, heavy metals may enter the cell via its membrane in a process commonly called active uptake (Dafale, 2010; Khalid et al., 2010). Acidophilic bacteria such as *Acidithiobacillus ferrooxidans* and sulfur-oxidizing bacteria can leach significant amounts of Ag, Co, Cd, Ca, and Mn from contaminated soils. Metals can also be precipitated as insoluble sulfides due to the catabolic and anabolic activity of sulfate-reducing bacteria (White et al., 1998). In the presence of heavy metals, the nuclear structure of an element is not "degraded" or altered; rather, it transitions between oxidation states or organic complexes.

Microorganisms are also crucial in metal bioremediation. Certain bacteria have developed the ability to protect themselves from the toxicity of heavy metals through various mechanisms, including adsorption, oxidation, absorption, and reduction. Bacteria can reduce metals by using them as terminal electron acceptors for aerobic respiration, a process known as dissimilatory metal reduction (Fourest et al., 1992; Igiri et al., 2018). Because methylated substances are typically volatile, microbial methylation plays a significant role in the bioremediation of heavy metals.

Several bacterial species are known to biomethylate mercury, Hg(II). For example, *Bacillus* sp., *Bacillus pumilus*, *Brevibacterium iodinium*, and *Pseudomonas aeruginosa* are all susceptible to gaseous methylmercury (Brierly, 1990; Sher, 2019). The bioremediation of pollutants generally employs pure microbial cultures. Under static conditions, a consortium of microbes composed of various *Pseudomonas* strains can decolorize dyes faster than individual bacteria (Jadhav et al., 2010). In a study aimed at developing a biosorbent for the removal of hazardous radioactive substances from nuclear waste streams, the *Pseudomonas* strain was identified as a strong candidate for the biosorption of thorium (IV) and uranium (VI) (Pinaki et al., 2004).

8.3.2 PGPB- AND PGPR-MEDIATED DEGRADATION

Bacteria found in endophytic regions (non-pathogenic bacteria found in naturally occurring plants) and bacteria living in rhizospheres are examples of plant-associated bacteria such as *Anabaena, Acetobacter, Arthrobacter, Achromobacter, Azotobacter, Enterobacter, Flavobacterium, Hydrogenophaga, Bacillus, Burkholderia, Clostridium, Microcoleus, Phyllobacterium, Pseudomonas*, and others.), have been publicized to subsidize to BD of lethal biological complexes in polluted soil and might have impending for civilizing phytoremediation (Divya et al., 2011). Plant growth-promoting rhizobacteria (PGPR) are naturally occurring soil bacteria that colonize plant roots aggressively and assist plants by stimulating growth. Additionally, the rhizosphere of plants in a polluted area had a greater variety of PAH-degrading bacteria, with two *Lysinibacilli* strains detected (Ma et al., 2010). A variety of plants may produce structurally similar PAHs, for example, phenols, to speed up the growth of degrading bacteria along with their decomposition of PAHs. *Pseudomonas* spp., which exhibit PGPR activity and hydrocarbon-degrading capabilities are an essential family of such kind of bacteria for such plant systems. It is fascinating to use plants along with certain microorganisms to boost the effectiveness of pollutant extraction; this approach is known as rhizoremediation. In addition, *Azospirillum lipoferum*, which is commonly found in agricultural plant rhizoplanes, was employed for Malathion breakdown, which is one of the world's greatest insecticides based upon the organic material (Hussain et al., 2009; Mishel et al., 2022). According to the literature, metals such as heavy metals may be detached from polluted soils by utilizing rhizobacteria (Dary et al., 2010).

8.3.3 MICROFUNGI- AND MYCORRHIZA-MEDIATED DEGRADATION

Microfungi are a diverse and intriguing collection of aerobic, eukaryotic microorganisms that vary from unicellular forms of yeasts to large mycelial molds (Joutey et al., 2013). Despite bacteria, fungi can flourish in low-slung moisture and optimized pH, which assistances in the degradation of organic compounds (Joutey et al., 2013). In association with mycorrhiza, the fungus colonizes the host plant's roots, either intracellularly as in arbuscular mycorrhizal fungi (AMF), or extracellularly as in ectomycorrhizal fungi. They are also important in soil life and chemistry. Mycorrhizoremediation is a kind of bioremediation that uses mycorrhiza (Kumar et al., 2021). Fungi have powerful degradative abilities, which affect the recycling of refractory polymers (such as lignin) and the elimination of dangerous contaminants from the environment (Christian et al., 2005).

8.3.4 YEAST-MEDIATED DEGRADATION

Various kinds of yeasts employ the use of various aromatic substances in the form of a growth substrate but the conversion of their ability into aromatic compound is more important (Fritsche et al., 2005). Few species of yeast like *Trichosporon cutaneum*, have aromatic absorption substrate mechanisms that are energy-dependent (Joutey et al., 2013).

Biological degradation of n-alkanes, on the other hand, with chain lengths up to n-C24 has been reported (Colombo et al., 1996). *Aureobasidion pullulans, Candida lipolytica,* and *Candida tropicalis* are examples of alkane-using yeasts. *C. ernobii* and *Rhodotorula aurantiaca* were discovered competent to decompose diesel oil.

However, these yeasts break down hydrocarbons more efficiently than bacteria (Margesin et al., 2003). In totaling to aliphatic and aromatic compounds of carbon the micro-fungi may arrange a range of other aromatic organic-based pollutants, including polycyclic aromatic hydrocarbons nitro aromatics, plasticizers, and various herbicides (Fritsche et al., 2000). Similar investigations of PCB metabolism by yeasts *C. boidinii* and *C. lipolytica* have been reported (Fritsche et al., 2005). Yeasts are well-known for their function in the elimination of harmful heavy metals. There have been several studies on the biosorption of heavy metals by yeasts. Several studies have shown that yeasts may accumulate heavy metals such as Ni (II), Cu (II), Co (II), Mg (II), and Cd (II) and are superior metal accumulators when compared to certain bacteria (Dhaliwal et al., 2020). Several yeast strains have been found to reduce Cr (VI) to Cr (III), including *P. guilliermondii, Yarrowia lipolytica,* and *Rhodotorula pilimanae* (Pandey et al., 2016). *P. anomala* can remove hexavalent chromium (Cr (VI)), and we evaluated the biosorption of Cr(V) by living and dead cells of three yeast species: *Cyberlindnera fabianii, Wickerhamomyces anomalus,* and *C. tropicalis* (Bahafid et al., 2012).

8.3.5 FILAMENTOUS FUNGI-MEDIATED DEGRADATION

The characteristics that discriminate filamentous fungi from one another living systems explain how they are better biodegraders. First, the growth of mycelium gives mycelial organisms a modest edge over solo cells like yeasts and bacteria especially when colonizing insoluble substrates. Fungi may move quickly by substrates, essentially assimilating their way through the secretion of extracellular derivative enzymes (Santos et al., 2004). Numerous researchers distribute bioremediation techniques into three basic categories: (i) the objective compound to be enzymatically attacked but not to be utilized as a carbon source; (ii) the objective compound to be used as an alternate source of carbon; and (iii) the objective composite not to be metabolized at all but needs to be taken up and to be concentrated inside the living organism (Rizwan et al., 2014). Though fungi use all these methods, commonly they are more adept at bioaccumulation than at using xenobiotics as sole carbon sources. The mechanical breakdown induced by the released enzymes is supplemented by hyphal penetration (Kirtzel et al., 2020).

In the majority, filamentous fungi are not able to mineralize the aromatic hydrocarbons totally. With the aid of bacteria, they merely turn them into indirect products with lower poisonousness and greater vulnerability to disintegration, implying that the communication between fungus and bacteria is lucrative for the process of petroleum hydrocarbon mineralization. Aspergillus and Cladosporium are filamentous fungi that degrade aliphatic hydrocarbons, whereas *Cunninghamella, Penicillinum, Fusarium*, and *Aspergillus* degrade aromatic hydrocarbons (Steliga, 2012). A fungus group of *Cephalosporium* and *Aspergillus* was also revealed to be a possible crude oil hydrocarbon degrader (Barnes et al., 2018). The role of fungi in PCB breakdown has received little attention. Several investigations have shown that filamentous fungus is capable of degrading PCBs. For the reason of their extracellular reductive enzymes, which have previously been positively employed in the breakdown of numerous pollutants, ligninolytic fungi have received considerable attention. Fungi have also been shown to degrade finished wood products, fibers, and a wide range of pseudo-cellulosic materials such as glues, gel, plastics, fuels, paints, medicines, and other human artifacts (Kirk, 1983). Metals are tolerated and detoxified by fungi through a variety of processes, including valence transformation, extracellular, active absorption, and intracellular precipitation (Saraswat et al., 2011; Zafar et al., 2007).

Zinc, copper, mercury, Cadmium, and lead may all be engrossed by spores and mycelium of many different species. Dead fungi's walls can sometimes bond better than living ones. Systems for processing uranium and thorium have been developed employing *Rhiloprzsarrhizus*. The most well-explored fungus in reference to color breakdown is the ligninolytic fungi (Tekere et al., 2001). Pesticides and their derivatives may be degraded by filamentous fungi employing two types of systems of enzymes: exocellular and intracellular (lignin-degrading system mainly consisting of lactases and peroxidases).

Pesticides have the capacity to deactivate or avoid each of these systems, hence influencing their metabolism (Chaplain et al., 2011).

8.3.6 ALGAE- AND PROTOZOA-MEDIATED DEGRADATION

Algae and protozoa are important members of the microbial community in both aquatic and terrestrial ecosystems, despite the fact that there is little known about their function in hydrocarbon BD (Joutey et al., 2013). *C. pyrenoidosa* and *C. vulgaris* degraded azo dyes utilizing dyes as nitrogen and carbon sources. General information is missing which shows that algae and protozoa play an ecologically important role in the breakdown of compounds of hydrogen and carbon in the environment (Bossert et al., 1987).

Certain fresh algae (e.g., *S. quadricauda, S. capricornutum, Chlorella vulgaris,* and *Scenedesmus platydiscus*) have been shown in studies to be capable of absorbing and decomposing PAHs. Protozoa, on the other hand, have been demonstrated to considerably reduce a variety of bacteria; consequently, their attendance in a BD organization may not unavoidably be helpful (Stapleton et al., 2002). Kobayashi et al. (1982) collated data on pesticide-algae interactions, demonstrating that algae not only could bioaccumulate pesticides but could also biotransform some of these environmental pollutants. Because protozoa are the primary grazers of degrading bacteria for organic pollutants, interactions between protozoa and degrading bacteria will directly alter the result of bacterial degradation. Mattison et al. (2005) developed a model of the food chain to evaluate the influence of such bacteria of the protozoa flagellate *Heteromitaglobosa* on benzene and methylbenzene BD.

They discovered that throughout the logarithmic growth stage of the flagellate population, bacteria improved their demeaning rate of methylbenzene and benzene by eight times. There seem to be various different ideas on the process by which protozoa accelerate the biological degradation of carbon-based pollutants, which mostly involve the parts listed below:

1. Nutritional mineralization, which enhances nutrient turnover;
2. Activation of bacteria, which regulates the amount or excretes the aged cells material;
3. Selective grazing, which lowers competition for resources and space, allowing degrading bacteria to flourish;
4. Physical disturbance, which can enhance the oxygen content and surface of degraded matter; and

5. Symmetabolism, which provides carbon and energy resources to various bacteria during the process of degradation (Chen et al., 2007; Joutey et al., 2013).

8.4 BIOAUGMENTATION (BA), NATURAL ATTENUATION, AND BIOSTIMULATION

Soil and groundwater bioremediation companies use bacteria that feed on harmful substances for energy, resulting in the breakdown of the targeted contaminant. Examples include junkyards, industrial spills, land development, and fertilizer. Bioremediation as a biotechnological process involving microorganisms has been a blossoming study issue in microbiology due to the rising potential of BD to answer the hazards of many pollutants. Microorganisms may be considered good pollution removal instruments in soil, water, and sediments due to their benefits over other bioremediation techniques. The most basic way to bioremediation is natural attenuation, in which soils are simply monitored for changes in pollution concentrations to ensure that the pollutant transformation is ongoing.

Bioaugmentation (BA) is typically used in situations where natural active microbial communities are present at low levels or are absent, and the addition of pollution-degrading organisms may increase the pace of transformation (El et al., 2005).

In such cases, the adaptation of foreign strains with very efficient pollutant transformation activities to unfamiliar environments is a major difficulty in implementation (El et al., 1999).

8.4.1 BIOAUGMENTATION (BA)

The addition of microorganisms with the potential to biodegrade refractory complexes in polluted environments is recognized as BA. This method is less costly and more ecologically friendly than physicochemical procedures. Uniting BA and biostimulation may be an operative way to accelerate bioremediation.

8.4.2 NATURAL ATTENUATION

Natural attenuation or bioattenuation is the decrease of contaminant concentrations in the environment via various biological processes (aerobic and anaerobic BD, plant, and animal uptake), physical phenomena, and chemical reactions.

The notion of natural attenuation includes terms like intrinsic remediation and biotransformation (Army et al., 1995). If natural attenuation is insufficiently swift or thorough, bioremediation will be aided by BA or biostimulation.

8.4.3 BIOSTIMULATION

Biostimulation is the process of changing the environment to activate existing microorganisms capable of bioremediation. Biostimulation by means of soil nutrients, electron acceptors, trace minerals, or electron donors increases soil pollutant biotransformation. Here are numerous examples of BD of contaminants by indigenous microorganisms. Liliane et al. (2012) exposed that the process of biostimulation was more operative than spontaneous biodiesel attenuation in polluted soils. BA had the largest potential for degradation and spontaneous attenuation was more successful than biostimulation of diesel-contaminated soils (Fatima et al., 2005).

Natural attenuation by Yu et al. (2005) seemed to be the better technique to cure phenanthrene and fluorene-polluted mangrove sediments, whereas biostimulation was better accomplished in degrading pyrene-contaminated sediments. When lactate is injected during biostimulation, microorganisms completely convert trichloroethene and perchloroethene to ethane in a short duration.

8.5 BIOREMEDIATION OR BIO-DEGRADATION AND FACTORS INFLUENCING MICROBIAL DEGRADATION

Microorganisms may eliminate a widespread range of biological pollutants due to their capacity to adapt to harsh environments. Due to this microorganisms are critical in site cleansing. However, their effectiveness is determined by a number of factors, comprising the chemical composition and amount of pollutants, their obtainability to microorganisms, and the physical and chemical-based features of the environment. As a result, the variables that influence the rate of pollutant breakdown by microorganisms are either related to their requirements for nutrition or related to the environment (environmental factors).

8.5.1 ORGANIC FACTORS

The metabolic capacity of the bacteria is a biotic component. The factors that affect the degradation of organic compounds by microbes are straight suppression of enzyme activities and the rate of growth of the processes of degradation by bacteria. The concentration of the contaminant and the amount of "catalyst" present are often used to calculate the rate of contamination degradation. Temperature, moisture, and pH are the factors that affect the rate of BD. Moisture affects the amount of elements available that are of soluble nature, also the osmotic pressure, to govern contaminant metabolism. Indeed, every 10°C change in temperature reduces the percentage of BD by about half.

BD may occur in a wide range of temperatures and pH; however, in most aquatic and terrestrial systems, a pH of 6 to 8 is regarded as optimum for BD. BD may occur in a wide range of pH; however, in most aquatic and terrestrial systems, a pH of 6.4 to 8.2 is regarded as optimum for BD (Abatenh et al., 2017).

8.5.2 ENVIRONMENTAL FACTORS

The potential for organic compound adsorption on a solid's surface is influenced by soil type and carbon-based matter concentration (Chiou et al., 2000; Rutherford et al., 1992). A similar process occurs when a pollutant enters the bulk mass of the soil matrix. Adsorption and absorption both reduce the contaminant's availability to maximum bacteria besides the pace at which the chemical degrades. The capacity of the matrix to transmit gases such as oxygen, methane, and carbon dioxide is reduced in fine-grained sediments and as soils become more saturated with water. This can affect the rate and kind of BD that happens.

8.6 MECHANISM OF BIODEGRADATION (BD)

The two main BD forms are known to be hydrolysis and biological oxidation. Based on the sources used, such as enzymes or bacteria, these mechanisms are further divided. The transformation of xenobiotics by microorganisms occurs by growth and co-metabolic processes. The growth process uses biological contaminants by way of the solitary carbon vitality for the complete mineralization of organic contaminants. In the presence of growth substrates like carbon, the organic pollutants are metabolized, which is termed co-metabolism (Jørgensen, 2008). Bacteria, protozoa, yeast, and certain fungi are engaged in the degradation of active pollutants. The core advantage of this process is the enhancement of complete mineralization of the organic compounds into water and carbon dioxide, which occurs either aerobically or anaerobically (Fritsche and Hofrichter, 2008). There are three phases of BD: (i) Natural attenuation involves the reduction of pollutants by natural microorganisms; (ii) biostimulation is a technique to improve the biodegrading efficiency by providing nutrients and oxygen to the process; and (iii) bioaugmentation (BA), where more efficient supplementary microorganisms are added to the system in addition to natural microflora to target the specific contaminants. Microorganisms are affected by a number of factors to use organic pollutants as a substrate or cometabolize them, such as pH, phosphorus, nitrogen, and temperature content which affects the extent and rate of degradation (Fritsche and Hofrichter, 2008). BD methods are essentially a vital process using natural flora to transform active therapeutic pollutants like drugs and other medical wastes.

8.7 DEGRADATION OF VARIOUS DRUGS BY MICROORGANISMS

8.7.1 MICROBIAL DEGRADATION OF ANTIBIOTICS

Antibiotics are substances that are pharmaceutically active and destroy microorganisms in the host body. Many of these chemicals are often used as over-the-counter drugs and are excreted in the urine or feces in the form of active metabolites or substances. Because these antibiotics are active biologically, they can damage the environment when they assemble in places like groundwater, sewage systems, and effluents. The microbial transformation of current antibiotics has been carried out in order to generate new, updated, and upgraded antibiotics with features such as minimum toxicity, broad antibacterial spectrum, increased oral adsorption, and less conferring resistance effects. Secondary metabolites of bacterial and fungal origin, as well as semi-synthetic natural and synthetic compounds (Fair and Toy, 2014). Another key issue connected with antibiotic survival is the absence or unwillingness of organic matter and chemical catalytic bacteria in wastewater treatment facilities and soil (Polianciuc et al., 2020). Sulfonamide antibiotics have bacteriostatic properties, and sulfamethazine concentrations of 0.6 ng/L have been found in groundwater. Sulfamethazine in soils can disrupt soil respiration and native microbial populations (Kotzerke et al., 2008). The research found that using Gram-positive *microbacteria* sp. strains in the field was capable of mineralizing benzylic and pyrimidine molecule of sulfamethazine. Sulfadiazine is a sulfonamide antibiotic that is partly mineralized by the strain of *Microbacterium lacus* (Tappe et al., 2013).

Reis et al. (2014) worked on *Rhodococcus rhodochrous* and demonstrated a 20% drop in concentrations of sulfamethizole and sulfamethoxazole (Reis et al., 2014). Kim et al. (2011) were the first to discover bacteria capable of biotransforming a fluoroquinolone antibiotic.

Figure 8.4 shows how a *Microbacterium* sp. that was resistant to norfloxacin was improved by employing wastewater as an inoculum and N-phenylpiperazine. Although norfloxacin did not allow the bacteria to grow, it was biotransformed into different metabolites: 6-defluoro-6-hydroxynorfloxacin, desethylenenorfloxacin, and N-acetylnorfloxacin. The final need was the occurrence of casamino acids in the concerned medium, and a glutamine synthase that catalyzed this process was later identified (Kim et al., 2013).

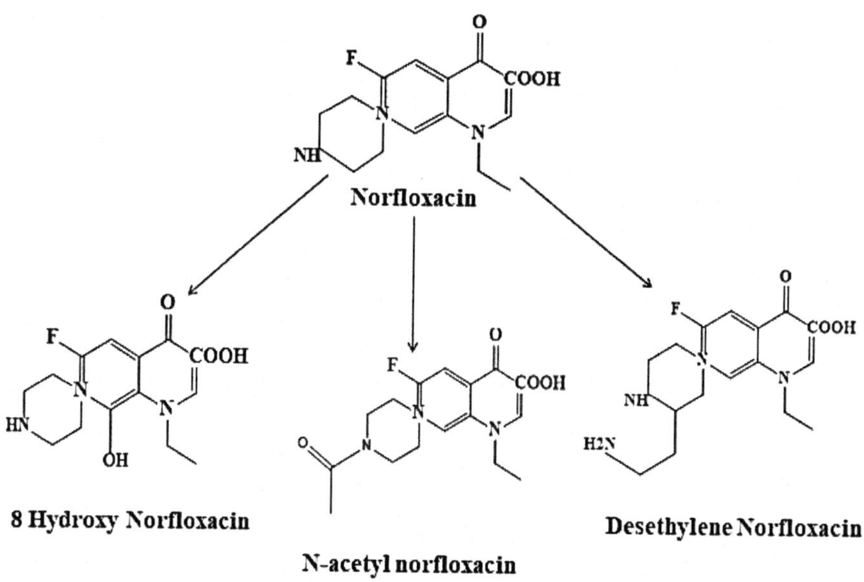

FIGURE 8.4 *Microbacterium* sp. degrading norfloxacin.

Non-steroidal anti-inflammatory drugs (NSAIDs) are pain-relief medications that contain narcotics and are used to alleviate pain in nearly all diseases. Naproxen, ibuprofen, diclofenac, ketoprofen, and aspirin are some of the pharmacologically active NSAIDs Flurbiprofen, a fluorobiphenyl-containing medicine, may be biotransformed into several oxidized fluorometabolites by *Cunninghamella* fungus and Streptomyces bacteria (Amadio et al., 2010; Bright et al., 2011). These species are known models of mammalian xenobiotic metabolism and generate cytochromes P450 (Murphy, 2015).

At a relatively low (5%) yield, *E. coli* articulating cyanobacterial cytochrome P450 converted flurbiprofen methyl ester into two hydroxylated compounds (Makino et al., 2012). Figure 8.5 represents the degradation of diclofenac by *T. versicolor* bacteria in the presence of its enzyme LAC.

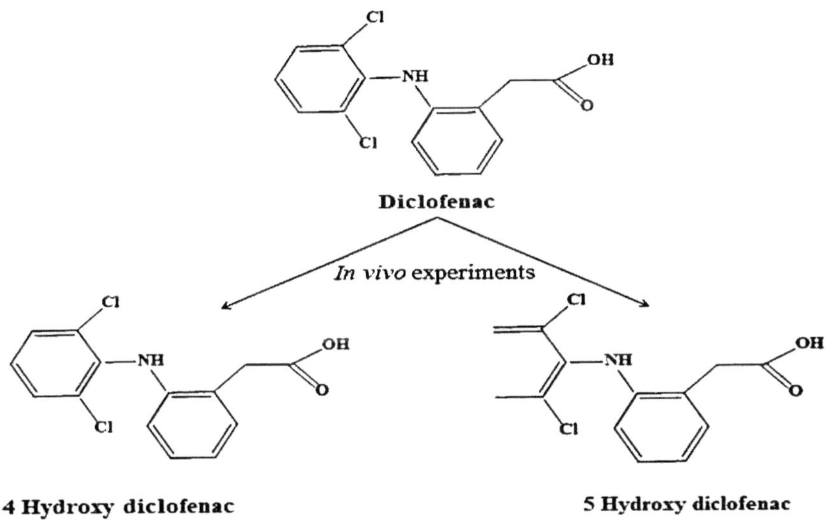

FIGURE 8.5 Bacteria-mediated degradation of diclofenac by *T. versicolor* microbacterium.

This type of biotransformation may be valuable in finding labile regions on medicinal molecules, allowing for the production of more stable/potent derivatives. Nonsteroidal anti-inflammatory medicines (NSAIDs) are narcotic-containing pain relievers that are used to treat practically all disorders (Figure 8.6).

Studies indicate that ketoprofen was seen to be more active than the parent molecule and the first concentration of BD utilizing river water microbial biofilms was found to be 100 μg/L and the breakdown occurs within 5 to 10 days after the initial addition (Marco-Urrea et al., 2010).

8.7.2 MICROBIAL DEGRADATION OF ANTI-CANCEROUS DRUGS

Degradation of Flutamide is a popular nonsteroidal antiandrogen medication used to treat prostate cancer. It is extensively converted *in vivo* to bioactive metabolites, most notably phenylalanine. These investigations have observed the biotransformation by microbes with this medication. Herath

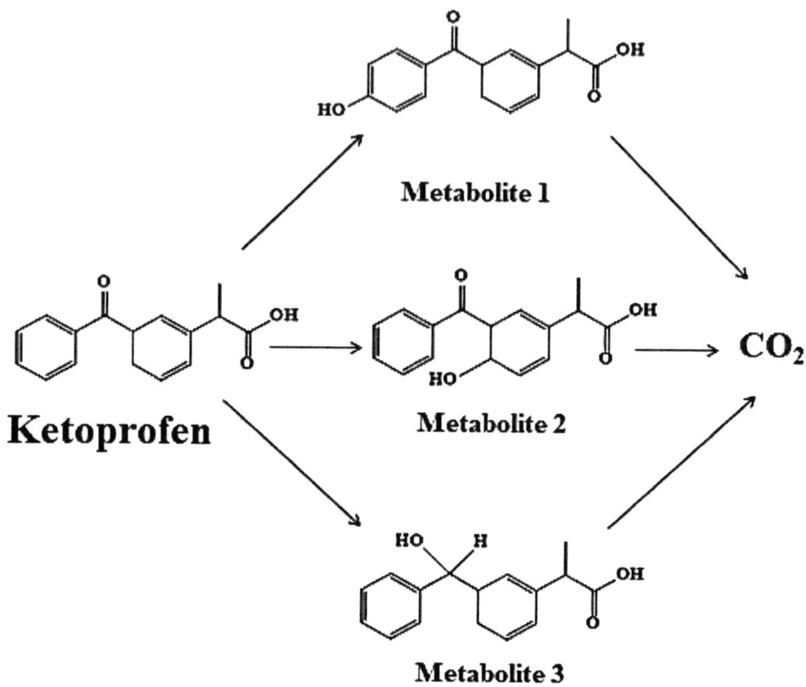

FIGURE 8.6 Pathway for the degradation of ketoprofen by *T. versicolor* bacteria.

and Khan (2010) worked on the fungus *Beauveri abassiana* and *Rhodotorula mucilaginosa* (ATCC 20129) as well as the bacteria *Nocardia* sp. (NRRL 5646) and created three metabolites via nitroreduction, isobutanoyl group hydrolysis, and N-acetylation. Amadio and Murphy (2011) established the creation of metabolites in hydroxial form, from flutamide incubated with bacterial species *C. elegans*, which possesses the mandatory cytochrome P450 activity to catalyze biotransformations. When this drug was incubated with *C. elegans* biofilms, a noteworthy decrease in the biomass was detected, representing that the compound is extremely poisonous and might be focused on the biofilm (Quinn et al., 2015).

8.7.3 MICROBIAL DEGRADATION OF DEGRADATION OF ANTIDEPRESSANTS

Antidepressant medication relieves anxiety, dysthymia, sleeplessness, and other related problems. Susceptibility of antidepressants to microorganisms for drugs Prozac (fluoxetine HCl), Valium, and its foremost human metabolites (temazepam, and oxazepam) in bacterial liquid cultures from sewage sludge (SS)-soil amendments studies could give an insight into antidepressant drug degradation. These drugs were proven to be unaffected by BD in a 60-day test culture, and even after prolonged exposure to SS-soil amendment (fluoxetine HCl). Oxazepam was found to be the only 1,4-benzodiazepine that could undergo a biotic transition in test culture, with an 80% loss documented under both abiotic and biotic conditions (Redshaw et al., 2008).

Fluoxetine is a commonly recommended antidepressant with a trifluoromethylphenyl, and citalopram (another fluorinated antidepressant) can be degraded by *Bjerkandera* sp. R1 and *Phanerochaete chrysosporium, Bjerkandera adjusta* (Rodarte-Morales et al., 2012). These three species were able to entirely degrade citalopram (1 mg/l) for 14 14-day treatment period, although fluoxetine was only weakly degraded, reaching a limit of 46% when cultured in a similar variant in the culture of *Bjerkandera* sp. The bacterium *L. portucalensis* can breakdown fluoxetine in the presence of modest concentrations (2 M) of substrate, succeeding in molar amounts of fluoride ion (Moreira et al., 2014), and in the attendance of a carbon source, acetate, a considerable quantity of fluoxetine may be degraded. Fluoxetine (Prozac) is a popular antidepressant with a trifluoromethylphenyl complex. It is intractable and has been found in rivers where it has been shown to have measurable biological impacts on aquatic species (Silva et al., 2015).

Rodarte-Morales et al. (2011) investigated the degradation of citalopram and fluoxetine by white rot molds: an anamorph of *Bjerkandera adjusta, Bjerkandera* sp., and *Phanerochaete chrysosporium.* While all three species completely decomposed citalopram (1 mg/l) after two weeks, fluoxetine was rather weakly degraded, reaching a percentage of 46 in the culture of the anamorph of *Bjerkandera* sp. R1. The investigation did not find any intermediates, and while MnP movement was measured, no *in vitro* tests were done.

8.7.4 MICROBIAL DEGRADATION OF ANTIHISTAMINES

For several years, antihistamines have been a breakthrough in the treatment of allergic diseases. The oxidative deamination activity of certain microorganisms degrades histamine by catalyzing histamine oxidase or histamine dehydrogenase. *Bacillus polymyxa, Rummeliibacillus stabekisii, Agrobacterium tumefaciens, Bacillus licheniformis, Bacillus cereus, Bacillus amyloliquefaciens, Bacillus licheniformis, Bacillus subtilis,* were the 8-bacteria isolated from salted fish products to degrade histamines. However, in contrast to the other isolates, *B. polymyxa* took the advantage of highest activity of histamine degradation in liquid broth media by producing histamine dehydrogenase. *Lactobacillus plantarum* D-103 is a histamine-degrading bacteria reported to have 100% degradation (Kandhasamy et al., 2022) *Histamine* and *tyramine* could get degraded by *Staphylococcus xylosus*.

8.7.5 MICROBIAL DEGRADATIONS OF ANTIHYPERTENSIVE OR β BLOCKERS

Beta-blockers are a type of medicine that is used to treat abnormal heart rhythms and are known as beta-adrenergic blocking agents. White-rot fungus may destroy propranolol and atenolol via peroxidases, ligninolytic enzymes, intracellular enzymes, and lactases (Asgher et al., 2008), as well as formylation, hydroxylation, dehalogenation, and deamination processes.

Trametes versicolor decomposed blockers in sterile and nonsterile wastewater throughout eight days within a fluidized bed bioreactor (Cruz-Morató et al., 2013). Atorvastatin (AT) is a cholesterol-lowering medication that is one of the most popular medications in history. Despite its widespread use, there have been few investigations on its BD. Although no research on degradation by bacteria from such chemicals has been documented, Rodrguez et al. (2011) investigated the use of the white-rot fungus *T. versicolor* including AT for pharmaceutical clearance. They showed a decrease from 38 to 5 ng/g AT when the fungus was cultivated on sterile sewage sludge for 42 days.

8.7.6 MICROBIAL DEGRADATION OF ANTISEPTICS

Antiseptics are biocides that are used to limit the growth of bacteria, but they are not antibiotics. Triclosan is found in antiseptics used in hospital surroundings, paints, cosmetics, cleaning goods, toys, materials of plastic, and other consumer items, and it has also been linked to changes in normal endocrine function in humans (Carr et al., 2011). *Nitrosomonas europaea*, an activated sludge bacterium, and *Methylobacillus* sp., a proteobacterial, were discovered to biodegrade triclosan in liquid culture conditions (Murdoch and Hay, 2005). Triclosan resistance is high in *Pseudomonas putida* TriRY and Al*caligenes xylosoxidans* subsp. denitrificans TR1, which may utilize it as their sole carbon source (0.4 mg/L; Meade et al., 2001).

8.7.7 MICROBIAL DEGRADATION OF ANTICONVULSANT OR ANTI-EPILEPTIC DRUG

Carbamazepine and Gabapentin, used for treating seizures are described as recurring medicines in surface and groundwater and the degradation of such compounds is a typical process. In the presence of glucose, *Rhodococcus rhodochrous* and *Aspergillus niger* were shown to be important in the breakdown of carbamazepine.

Cunninghamella elegans was also reported as a bio-degrader of carbamazepine in a liquid medium and 60–80% elimination of carbamazepine was obtained by *Phanerochaete chrysosporium* in a non-sterile bioreactor when utilized for over 100 days (Zhang and Geißen, 2012). A study showed that *Trichoderma harzianum*, which degrades carbamazepine by a co-metabolic process, resulted in 57% removal and aerobic degradation by *Trametes versicolor*. Figure 8.7 shows the complete mineralization of carbamazepine (94%) in six days, and the rate of degradation increased up to 95.6% in the bioreactor with acridine, acridone, 11-epoxy-carbamazepine, and 10,11-dihydro-10,11-dihydroxycarbamazepine (Kang et al., 2008) as the major end products.

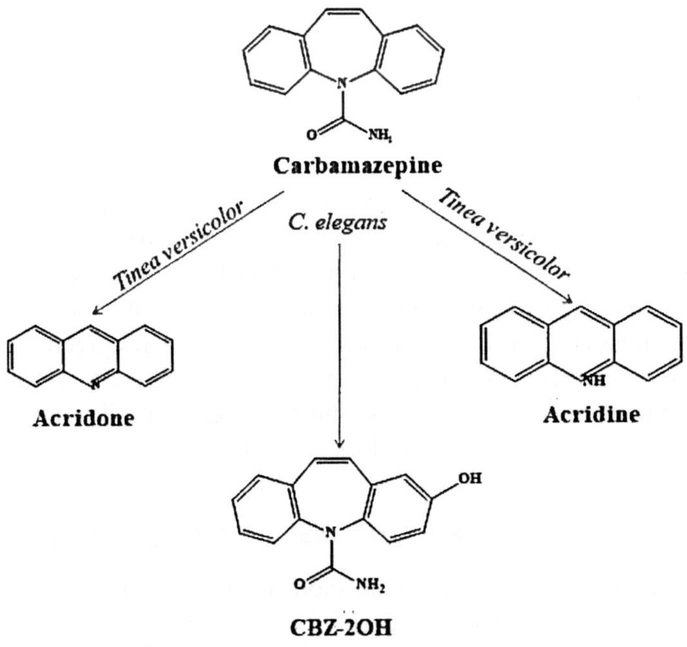

FIGURE 8.7 Degradation of carbamazepine by various fungi.

8.8 DEGRADATION BY GENETICALLY ENGINEERED MICROORGANISMS

Likewise, previously stated, biostimulation and BA are two ways that may be used to quicken the reclamation of polluted settings. Bacterial genes that produce biocatalysts for recalcitrant complexes were duplicated and reported for the first time in the late 1970's and early 1980's. A modified microorganism (GMM) or genetically engineered microorganism (GEM) is a microorganism whose DNA has been distorted by using genetic engineering performances enthused by natural gene exchange between various microorganisms. Such methods are commonly referred to as recombinant DNA technology. These genetically engineered microorganisms have shown potential for bioremediation of groundwater, soil, and triggered sludge, with healthier degrading proficiencies of an extensive variety of chemical pollutants. *Pseudomonas putida* containing the NAH and XYL plasmids, as well as a hybrid plasmid derived by conjugating parts of CAM and OCT, could degrade camphor, and naphthalene and grow rapidly on crude oil because it could absorb hydrocarbons more competently than any other solitary plasmid. Even though numerous projections in the field of bioremediation were predicted to succeed in the detection of recombinant DNA technology, downstream processing could not have occurred so far.

Genetic engineering of endophytic and rhizospheric bacteria for use in plant-associated degradation of hazardous substances in soil is one of the most promising developing ways for restoring damaged environmental regions (Divya et al., 2011; McGuinness and Dowling, 2009).

Many microorganisms in the rhizosphere demonstrate relatively little capacity to digest organic contaminants (Afzal et al., 2014; Chaudhary et al., 2005). Underneath the direction of a bacteroid-based specific promoter, the nifH gene, the gene was expressed to make phytochelatins and accumulate Cd^{2+} (Sussman et al., 1988). However, there is much disagreement near the discharge of such genetically altered bacteria into the ecosystem, therefore ground challenges of such organisms must be overdue until safety issues and the prospective for environmental loss are addressed.

8.9 GAPS IN KNOWLEDGE AND RESEARCH REQUIREMENTS

For obtaining desired results, gene modification in microorganisms may play an effective role. So, considering their effectiveness lots of attention is needed to degrade various kinds of drugs which can be degraded with gene-modified microorganisms.

8.10 CONCLUSIONS

The amount of medicines that are eliminated from the environment relies on the availability of natural microbes that can break them down. The majority of drugs are not particularly durable once in the environment and may experience biotransformation via metabolic and co-metabolic pathways. Both the comprehensive cycle of carbon and the renewal of our environment rely greatly on microbial activity. The term "biodegradation" refers to a variety of processes. Microorganisms have a special aptitude to degrade or change a wide range of manufactured chemicals and other molecules, including hydrocarbons and heavy metals, which have ecotoxicological consequences. This phrase, however, typically denotes to prospective abilities of degradation assessed in the lab using chosen cultures and proper growth conditions.

It is necessary to enhance the activity of native microorganisms in polluted biotopes and to boost their capacity for degradation by BA or biostimulation. Another technique is to improve the biodegradability of microorganisms through genetic engineering. Nonetheless, the use of GEM in the field is laden with hazards. Whether or if these technologies eventually succeed in eliminating industrial pollutants via bioremediation may have an influence on our ability to reduce waste, eliminate pollution, and enjoy a more sustainable future. These microbes convert the toxins via co-metabolic pathways and rely only on the pollutants for carbon.

Biostimulation, BA, and natural attenuation via means of inherent microorganisms are the techniques engaged to break down the marked contaminant. Microorganisms can also be genetically modified to progress the effectiveness of neutralization and help in the oxidation of pollutants. As a result, using microorganisms to detoxify contaminants likely increases the sustainability possible biodegradability and ensures an environmentally acceptable substitute method of bioremediation.

KEYWORDS

- **biodegradation**
- **bioremediation**
- **biostimulation**
- **detoxification**
- **genetic engineering**
- **natural attenuation**

REFERENCES

Abatenh, E., Gizaw, B., Tsegaye, Z., & Wassie, M. (2017). The role of microorganisms in bioremediation: A review. *Journal of Environmental Biology, 2*, 038–046.

Afzal, M., Khan, Q. M., & Sessitsch, A. (2014). Endophytic bacteria: Prospects and applications for the phytoremediation of organic pollutants. *Chemosphere, 117*, 232–242.

Alexander, M. (1999). *Biodegradation and Bioremediation*. Gulf Professional Publishing.

Amadio, J., & Murphy, C. D. (2010). Biotransformation of fluorobiphenyl by *Cunninghamella elegans*. *Applied Microbiology and Biotechnology, 86*, 345–351.

Anthony, I. O. (2006). Biodegradation alternative in the cleanup of petroleum hydrocarbon pollutants. *Biotechnology and Molecular Biology Reviews, 1*, 38–50.

Asgher, M., Bhatti, H. N., Ashraf, M., & Legge, R. L. (2008). Recent developments in biodegradation of industrial pollutants by white rot fungi and their enzyme system. *Biodegradation, 19*, 771–783.

Atlas, R. M., & Bartha, R. (1992). Hydrocarbon biodegradation and oil spill bioremediation. In *Microbial Ecology* (pp. 1287–1338). Springer, Boston, MA.

Bahafid, W., Joutey, N. T., Sayel, H., Iraqui-Houssaini, M., & El Ghachtouli, N. (2013). Chromium adsorption by three yeast strains isolated from sediments in Morocco. *Geomicrobiology Journal, 30*, 422–429.

Barnes, N. M., Khodse, V. B., Lotlikar, N. P., Meena, R. M., & Damare, S. R. (2018). Bioremediation potential of hydrocarbon-utilizing fungi from select marine niches of India. *Biotechnology Reports, 8*, 1–10.

Bennett, J. W., Wunch, K. G., & Faison, B. D. (2002). Use of fungi for biodegradation. *Environmental Microbiology*.

Bento, F. M., Camargo, F. A., Okeke, B. C., & Frankenberger, W. T. (2005). Comparative bioremediation of soils contaminated with diesel oil by natural attenuation, biostimulation, and bioaugmentation. *Bioresource Technology, 96*, 1049–1055.

Boehm, P. D. (1964). Polycyclic aromatic hydrocarbons (PAHs). In *Environmental Forensics* (pp. 313–337). Academic Press.

Borja, J., Taleon, D. M., Auresenia, J., & Gallardo, S. (2005). Polychlorinated biphenyls and their biodegradation. *Process Biochemistry, 40*, 1999–2013.

Brierley, C. L. (1990). Bioremediation of metal-contaminated surface and groundwaters. *Geomicrobiology Journal, 8*, 201–223.

Bright, T. V., Clark, B. R., O'Brien, E., & Murphy, C. D. (2011). Bacterial production of hydroxylated and amidated metabolites of flurbiprofen. *Journal of Molecular Catalysis Enzymatic, 72*, 116–121.

Cao, S., Davis, A., & Kjellerup, B. V. (2021). Presence of bacteria capable of PCB biotransformation in stormwater bioretention cells. *Microbial Ecology, 97*, 159.

Caracciolo, A. B., Topp, E., & Grenni, P. (2015). Pharmaceuticals in the environment: Biodegradation and effects on natural microbial communities. A review. *Journal of Pharmaceutical and Biomedical Analysis, 106*, 25–36.

Chaplain, V., Mamy, L., Vieublé, L., Mougin, C., Benoit, P., Barriuso, E., & Nélieu, S. (2011). The fate of pesticides in soils: Toward an integrated approach of influential factors. *Microbial Ecology, 9*, 157.

Chaube, P., Indurkar, H., & Moghe, S. (2010). Biodegradation and decolorization of dye by mixed consortia of bacteria and study of toxicity on *Phaseolus mungo* and *Triticum aestivum*. *Asiatic Journal of Biotechnology Research, 1*, 45–56.

Chaudhry, Q., Blom-Zandstra, M., Gupta, S. K., & Joner, E. (2005). Utilizing the synergy between plants and rhizosphere microorganisms to enhance the breakdown of organic pollutants in the environment. *Environmental Science and Pollution Research, 12*, 34–48.

Chen, X., Liu, M., Hu, F., Mao, X., & Li, H. (2007). Contributions of soil micro-fauna (protozoa and nematodes) to rhizosphere ecological functions. *Acta Ecologica Sinica, 27*, 3132–3143.

Chiou, C. T., Kile, D. E., Rutherford, D. W., Sheng, G., & Boyd, S. A. (2000). Sorption of selected organic compounds from water to a peat soil and its humic-acid and humin fractions: Potential sources of the sorption nonlinearity. *Environmental Science & Technology, 34*, 1254–1258.

Christian, V., Shrivastava, R., Shukla, D., Modi, H. A., & Vyas, B. R. M. (2005). Degradation of xenobiotic compounds by lignin-degrading white-rot fungi: Enzymology and mechanisms involved. *Indian Journal of Experimental Biology, 43*, 301–312.

Colombo, J. C., Cabello, M., & Arambarri, A. M. (1996). Biodegradation of aliphatic and aromatic hydrocarbons by natural soil microflora and pure cultures of imperfect and lignolytic fungi. *Environmental Pollution, 94*, 355–362.

Cruz-Morató, C., Ferrando-Climent, L., Rodriguez-Mozaz, S., Barceló, D., Marco-Urrea, E., Vicent, T., & Sarrà, M. (2013). Degradation of pharmaceuticals in non-sterile urban wastewater by *Trametes versicolor* in a fluidized bed bioreactor. *Water Research, 47*, 5200–5210.

Dafale, N. A. (2011). Exploration of genetic information from dynamic microbial populations for enhancing the efficiency of azo-dye-degrading systems. *Environmental Reviews, 19*, 310–332.

Das, N., & Chandran, P. (2011). Microbial degradation of petroleum hydrocarbon contaminants: An overview. *Biotechnology Research International, 2011*.

Dhaliwal, S. S., Singh, J., Taneja, P. K., & Mandal, A. (2020). Remediation techniques for removal of heavy metals from soil contaminated through different sources: A review. *Environmental Science and Pollution Research, 27*, 1319–1333.

Diez, M. C. (2010). Biological aspects involved in the degradation of organic pollutants. *Journal of Soil Science and Plant Nutrition, 10*, 0718–9516.

Divya, B., & Deepak Kumar, M. (2011). Plant-microbe interaction with enhanced bioremediation. *Research Journal of Biotechnology, 6*, 4.

El Fantroussi, S., & Agathos, S. N. (2005). Is bioaugmentation a feasible strategy for pollutant removal and site remediation? *Current Opinion in Microbiology, 8*, 268–275.

El Fantroussi, S., Belkacemi, M., Top, E. M., Mahillon, J., Naveau, H., & Agathos, S. N. (1999). Bioaugmentation of a soil bioreactor designed for pilot-scale anaerobic bioremediation studies. *Environmental Science & Technology, 33*, 2992–3001.

Fair, R. J., & Tor, Y. (2014). Antibiotics and bacterial resistance in the 21st century. *Perspectives in Medicinal Chemistry, 6*, PMC-S14459.

Farshid, K., Parvaneh, S., Hooshang, J., & Yaghoob, T. (2011). Isolation and identification of hydrocarbons degrading bacteria in soil around Shiraz Refinery. *African Journal of Microbiology Research, 5*, 3084–3089.

Floodgate, G. D., & Atlas, R. M. (1984). The fate of petroleum in marine environments. In *Petroleum Microbiology* (pp. 35–60).

Fourest, E., & Roux, J. C. (1992). Heavy metal biosorption by fungal mycelial by-products: Mechanisms and influence of pH. *Applied Microbiology and Biotechnology, 37*, 399–403.

Fritsche, W., & Hofrichter, M. (2001). Aerobic degradation by microorganisms. In *Biotechnology* (pp. 144–167).

Fritsche, W., & Hofrichter, M. (2005). Aerobic degradation of recalcitrant organic compounds by microorganisms. *Environmental Biotechnology*.

Herath, W., & Khan, I. A. (2010). Microbial metabolism of flutamide. *Chemical and Pharmaceutical Bulletin, 58*, 562–564.

Hong, Y., Xu, M., Guo, J., Xu, Z., Chen, X., & Sun, G. (2007). Respiration and growth of *Shewanella decoloration is* S12 with an azo compound as the sole electron acceptor. *Applied and Environmental Microbiology, 73*, 64–72.

Hussain, S., Siddique, T., Saleem, M., Arshad, M., & Khalid, A. (2009). Impact of pesticides on soil microbial diversity, enzymes, and biochemical reactions. *Advances in Agronomy, 102*, 159–200.

Igiri, B. E., Okoduwa, S. I., Idoko, G. O., Akabuogu, E. P., Adeyi, A. O., & Ejiogu, I. K. (2018). Toxicity and bioremediation of heavy metals contaminated ecosystem from tannery wastewater: A review. *Journal of Toxicology*, 2018.

Jadhav, J. P., Kalyani, D. C., Telke, A. A., Phugare, S. S., & Govindwar, S. P. (2010). Evaluation of the efficacy of a bacterial consortium for the removal of color, reduction of heavy metals, and toxicity from textile dye effluent. *Bioresource Technology, 101*, 165–173.

Jong-Su, S., Young-Soo, K., & Qing, X. L. (2009). Bacterial degradation of aromatic compounds. *International Journal of Environmental Research and Public Health, 6*, 278–309.

Joutey, N. T., Bahafid, W., Sayel, H., & El Ghachtouli, N. (2013). Biodegradation: Involved microorganisms and genetically engineered microorganisms. *International Biodeterioration & Biodegradation, 1*, 289–320.

Kandhasamy, S., Balraj, S., Kanagaraj, S., & Sundaravadivelu, S. (2022). Microbial degradation of pharmaceuticals. *Microbial Degradation of Pharmaceuticals*, 183–210.

Kang, S. I., Kang, S. Y., & Hur, H. G. (2008). Identification of fungal metabolites of anticonvulsant drug carbamazepine. *Applied Microbiology and Biotechnology, 79*, 663–669.

Kesavaram, P., & Krishnan, M. (2012). *Brevibacillus parabrevis*, *Acinetobacter baumannii*, and *Pseudomonas citronellolis*: Potential candidates for biodegradation of low-density polyethylene (LDPE). *African Journal of Microbiology Research, 4*, 9–14.

Khalid, A., Arshad, M., & Crowley, D. (2010). Bioaugmentation of azo dyes. *Biodegradation of Azo Dyes, 9*, 1–37.

Kim, D. W., Feng, J., Chen, H., Kweon, O., Gao, Y., Yu, L. R., & Sutherland, J. B. (2013). Identification of the enzyme responsible for N-acetylation of norfloxacin by *Microbacterium* sp. strain 4N2-2. *Applied and Environmental Microbiology, 79*, 314–321.

Kinuthia, M., Hamadi, I. B., Anne, W. M., Ciira, K., & Muniru, K. T. (2010). Degradation of dichlorodiphenyltrichloroethane (DDT) by bacterial isolates from cultivated and uncultivated soil. *African Journal of Microbiology Research, 4*, 185–196.

Kirk, T. K. (1983). Degradation and conversion of lignocelluloses. In *The Filamentous Fungi* (Vol. 4, pp. 266–295).

Kirtzel, J., Ueberschaar, N., Deckert-Gaudig, T., Krause, K., Deckert, V., Gadd, G. M., & Kothe, E. (2020). Organic acids, siderophores, enzymes, and mechanical pressure for black slate bioweathering with the basidiomycete *Schizophyllum commune*. *Environmental Microbiology, 22*, 1535–1546.

Kobayashi, H., & Rittmann, B. E. (1982). Microbial removal of hazardous organic compounds. *Environmental Science & Technology, 16*, 170A–183.

Kumar, A., Kumar, A., Singh, R., Singh, R., Pandey, S., Rai, A., & Rahul, B. (2020). Genetically engineered bacteria for the degradation of dye and other organic compounds. *Environmental Pollution and Bioavailability, 33*, 331–350.

Kumar, V. (Ed.). (2021). *Rhizomicrobiome Dynamics in Bioremediation.* Boca Raton, FL, USA: CRC Press, Taylor & Francis Group.

Lade, H., Kadam, A., Paul, D., & Govindwar, S. (2015). Biodegradation and detoxification of textile azo dyes by a bacterial consortium under sequential microaerophilic/aerobic processes. *Journal of Experimental and Clinical Research, 14*, 158.

Leigh, M. B., Prouzová, P., Macková, M., Macek, T., Nagle, D. P., & Fletcher, J. S. (2006). Polychlorinated biphenyl (PCB)-degrading bacteria associated with trees in a PCB-contaminated site. *Applied and Environmental Microbiology, 72*, 2331–2342.

Leitão, A. L. (2009). Potential of *Penicillium* species in the bioremediation field. *International Journal of Environmental Research and Public Health, 6*, 1393–1417.

Ma, B., Chen, H. H., He, Y., & Xu, J. M. (2010). Isolations and consortia of PAH-degrading bacteria from the rhizosphere of four crops in PAH-contaminated field. In *19th World Congress of Soil Science*.

Makino, T., Otomatsu, T., Shindo, K., Kitamura, E., Sandmann, G., Harada, H., & Misawa, N. (2012). Biocatalytic synthesis of flavones and hydroxyl-small molecules by recombinant *Escherichia coli* cells expressing the cyanobacterial CYP110E1 gene. *Microbial Cell Factories, 11*, 1–13.

Margesin, R., Gander, S., Zacke, G., Gounot, A. M., & Schinner, F. (2003). Hydrocarbon degradation and enzyme activities of cold-adapted bacteria and yeasts. *Extremophiles, 7*(6), 451–458.

Marinescu, M., Dumitru, M., & Lăcătuşu, A. R. (2009). Biodegradation of petroleum hydrocarbons in artificially polluted soil. *International Journal of Environmental Research, 41*, 157–162.

Mattison, R. G., Taki, H., & Harayama, S. (2005). The soil flagellate *Heteromita globosa* accelerates bacterial degradation of alkylbenzenes through grazing and acetate excretion in batch culture. *Microbial Ecology, 49*, 142–150.

McGuinness, M., & Dowling, D. (2009). Plant-associated bacterial degradation of toxic organic compounds in soil. *International Journal of Environmental Research, 6*, 2226–2247.

McMurry, J. (2000). Aromatic hydrocarbons. In *Organic Chemistry* (5th ed., pp. 120–180). Brooks/Cole: Thomas Learning.

Meade, M. J., Waddell, R. L., & Callahan, T. M. (2001). Soil bacteria *Pseudomonas putida* and *Alcaligenes xylosoxidans* subsp. *denitrificans* inactivate triclosan in liquid and solid substrates. *Microbiology Letters, 204*, 45–48.

Menn, F. M., Easter, J. P., & Sayler, G. S. (2008). Genetically engineered microorganisms and bioremediation. In *Biotechnology: Environmental Progress, 14*, 441–463.

Mishel, P. F., Steffi, P. F., Thirumalaiyammal, B., & Anburaj, R. (2022). Bioremediation: A green technology for environmental cleanup. In *Metagenomics to Bioremediation* (pp. 3–21). Academic Press.

Moreira, I. S., Ribeiro, A. R., Afonso, C. M., Tiritan, M. E., & Castro, P. M. (2014). Enantioselective biodegradation of fluoxetine by the bacterial strain *Labrys portucalensis* F11. *Chemosphere, 111*, 103–111.

Mrozik, A., Piotrowska-Seget, Z., & Labuzek, S. (2003). Bacterial degradation and bioremediation of polycyclic aromatic hydrocarbons. *Polish Journal of Environmental Studies, 12*, 11–15.

Murdoch, R. W., & Hay, A. G. (2005). Formation of catechols via removal of acid side chains from ibuprofen and related aromatic acids. *Applied and Environmental Microbiology, 71*, 6121–6125.

Murphy, C. D., Quirke, S., & Balogun, O. (2008). Degradation of fluorobiphenyl by *Pseudomonas pseudoalcaligenes* KF707. *Microbiology Letters, 286*, 45–49.

Pan, X., Lin, D., Zheng, Y., Zhang, Q., Yin, Y., Cai, L., & Yu, Y. (2016). Biodegradation of DDT by *Stenotrophomonas* sp. DDT-1: Characterization and genome functional analysis. *Scientific Reports, 6*, 1–10.

Pandey, P., Pathak, H., & Dave, S. (2016). Microbial ecology of hydrocarbon degradation in the soil: A review. *Research Journal of Environmental Toxicology, 10*, 1.

Polianciuc, S. I., Gurzău, A. E., Kiss, B., Ştefan, M. G., & Loghin, F. (2020). Antibiotics in the environment: Causes and consequences. *Medical and Pharmaceutical Reports, 93*, 231.

Puvaneswari, N., Muthukrishnan, J., & Gunasekaran, P. (2006). Toxicity assessment and microbial degradation of azo dyes. *Indian Journal of Experimental Biology, 44*(8), 618–626.

Quinn, L., Dempsey, R., Casey, E., Kane, A., & Murphy, C. D. (2015). Production of drug metabolites by immobilized *Cunninghamella elegans*: From screening to scale up. *Journal of Industrial Microbiology & Biotechnology, 42*, 799–806.

Rafii, F., Hall, J. D., & Cerniglia, C. E. (1997). Mutagenicity of azo dyes used in foods, drugs, and cosmetics before and after reduction by *Clostridium* species from the human intestinal tract. *Food and Chemical Toxicology, 35*, 897–901.

Rau, J., Knackmuss, H. J., & Stolz, A. (2002). Effects of different quinoid redox mediators on the anaerobic reduction of azo dyes by bacteria. *Environmental Science & Technology, 36*, 1497–1504.

Redshaw, C. H., Cooke, M. P., Talbot, H. M., McGrath, S., & Rowland, S. J. (2008). Low biodegradability of fluoxetine HCl, diazepam, and their human metabolites in sewage sludge-amended soil. *Journal of Soils and Sediments, 8*, 217–230.

Reis, P. J., Reis, A. C., Ricken, B., Kolvenbach, B. A., Manaia, C. M., Corvini, P. F., & Nunes, O. C. (2014). Biodegradation of sulfamethoxazole and other sulfonamides by *Achromobacter denitrificans*. *Journal of Hazardous Materials, 280*, 741–749.

Rizwan, M., Singh, M., Mitra, C. K., & Morve, R. K. (2014). Ecofriendly application of nanomaterials: Nanobioremediation. *Journal of Nanoparticles*, 2014.

Rodarte-Morales, A. I., Feijoo, G., Moreira, M. T., & Lema, J. M. (2011). Degradation of selected pharmaceutical and personal care products (PPCPs) by white-rot fungi. *Journal of Microbiology and Biotechnology, 27*, 1839–1846.

Rodarte-Morales, A. I., Feijoo, G., Moreira, M. T., & Lema, J. M. (2012). Operation of stirred tank reactors (STRs) and fixed-bed reactors (FBRs) with free and immobilized *Phanerochaete chrysosporium* for the continuous removal of pharmaceutical compounds. *Biochemical Engineering Journal, 66*, 38–45.

Rodríguez, J. B., & Comin, M. J. (2003). New progresses in the enantioselective synthesis and biological properties of carbocyclic nucleosides. *Mini-Reviews in Medicinal Chemistry, 3*, 95–114.

Rutherford, D. W., Chiou, C. T., & Kile, D. E. (1992). Influence of soil organic matter composition on the partition of organic compounds. *Environmental Science & Technology, 26*, 336–340.

Samuels, R. I., Santos, A. V., & Silva, C. P. (2011). Enzymology of entomopathogenic fungi. In *Microbial Insecticides: Principles and Applications* (pp. 71–92). Nova Science Publishers, Inc.

Sar, P., Kazy, S. K., & D'Souza, S. F. (2004). Radionuclide remediation using a bacterial biosorbent. *International Biodeterioration & Biodegradation, 54*, 193–202.

Saraswat, S., & Rai, J. P. N. (2011). Mechanism of metal tolerance and detoxification in mycorrhizal fungi. In *Biomanagement of Metal-Contaminated Soils* (pp. 225–240). Springer, Dordrecht.

Sarkar, S., Banerjee, A., Halder, U., Biswas, R., & Bandopadhyay, R. (2017). Degradation of synthetic azo dyes of the textile industry: A sustainable approach using microbial enzymes. *Water Conservation Science and Engineering, 2*, 121–131.

Seeger, M., Hernández, M., Méndez, V., Ponce, B., Córdova, M., & González, M. (2010). Bacterial degradation and bioremediation of chlorinated herbicides and biphenyls. *Journal of Soil Science and Plant Nutrition, 10*, 320–332.

Sher, S., & Rehman, A. (2019). Use of heavy metals resistant bacteria—A strategy for arsenic bioremediation. *Applied Microbiology and Biotechnology, 103*, 6007–6021.

Silva, L. J., Pereira, A. M., Meisel, L. M., Lino, C. M., & Pena, A. (2015). Reviewing the serotonin reuptake inhibitors (SSRIs) footprint in the aquatic biota: Uptake, bioaccumulation, and ecotoxicology. *Environmental Pollution, 197*, 127–143.

Singh, B. K., Walker, A., Morgan, J. A. W., & Wright, D. J. (2004). Biodegradation of chlorpyrifos by *Enterobacter* strain B-14 and its use in bioremediation of contaminated soils. *Applied and Environmental Microbiology, 70*, 4855–4863.

Singh, M. R. (2007). Impurities-heavy metals: IR perspective. *International Journal of Physical Sciences, 5*, 1045–1058.

Stapleton Jr, R. D., & Singh, V. P. (2002). *Biotransformation's: Bioremediation Technology for Health and Environmental Protection*. Elsevier.

Steliga, T. (2012). Role of fungi in biodegradation of petroleum hydrocarbons in drill waste. *Polish Journal of Environmental Studies, 21*, 471–479.

Sussman, M., Collins, C. H., Skinner, F. A., & Stewart-Tull, D. E. (1988). *Release of Genetically-Engineered Microorganisms*. Academic Press, London.

Tappe, W., Herbst, M., Hofmann, D., Koeppchen, S., Kummer, S., Thiele, B., & Groeneweg, J. (2013). Degradation of sulfadiazine by *Microbacterium lacus* strain SDZm4, isolated from lysimeters previously manured with slurry from sulfadiazine-medicated pigs. *Applied and Environmental Microbiology, 79*, 2572–2577.

Tekere, M., Mswaka, A. Y., Zvauya, R., & Read, J. S. (2001). Growth, dye degradation, and ligninolytic activity studies on Zimbabwean white rot fungi. *Enzyme and Microbial Technology, 28*, 420–426.

U.S. Army. (1995). *Interim Army Policy on Natural Attenuation for Environmental Restoration*. Washington, DC: Department of the Army, Assistant Chief of Staff for Installation Management. DAIM-ED.

Vargas, J. M. (1975). Pesticide degradation. *Journal of Arboriculture, 1*, 232–233.

Vickers, N. J. (2017). Animal communication: When I'm calling you, will you answer too? *Current Biology, 27*, 713–715.

White, C., Shaman, A. K., & Gadd, G. M. (1998). An integrated microbial process for the bioremediation of soil contaminated with toxic metals. *Nature Biotechnology, 16*, 572–575.

Yu, K. S. H., Wong, A. H. Y., Yau, K. W. Y., Wong, Y. S., & Tam, N. F. Y. (2005). Natural attenuation, biostimulation, and bioaugmentation on biodegradation of polycyclic aromatic hydrocarbons (PAHs) in mangrove sediments. *Marine Pollution Bulletin, 51*, 1071–1077.

Zafar, S., Aqil, F., & Ahmad, I. (2007). Metal tolerance and biosorption potential of filamentous fungi isolated from metal contaminated agricultural soil. *Bioresource Technology, 98*, 2557–2561.

Zhang, Y., & Geißen, S. U. (2012). Elimination of carbamazepine in a non-sterile fungal bioreactor. *Bioresource Technology, 112*, 221–227.

CHAPTER 9

Bioremediation of Pharmaceuticals with Cow Dung and Earthworms

AMOL D. GHOLAP,* YASH P. DADARKAR, and ASHISH S. DUBEY

Department of Pharmaceutics, St. John Institute of Pharmacy and Research, Palghar, Maharashtra, India

*Corresponding author.

ABSTRACT

The human advancement and innovations into the pharmaceutical industry helping mankind with beneficial medicines for treatment of several disease and health benefits. But the pharmaceutical wastewater and sludge are one of the effluents along with wastes induced from pharmaceutical manufacturing operations. Different management strategies involved in the disposal of these wastes including open dumping, incineration along with landfilling which further creates the issue of groundwater contamination, esthetic loss, and the release of the harmful contaminants into the environment. Several compounds along with heavy metals from the industrial waste can induces the genotoxic, phytotoxic as well as zootoxic. Bioremediation methods by using cow dung (CD) and earthworms are better alternative to address these issues in effective way. The easily available cow dung slurry into the rural areas can be able to bioremediate the harmful pollutants. Earthworms can help into the reduction of the toxicity induced from heavy metal concentration. They are also used for the stabilization of the organic waste with faster degradation with enhancement of plant growth promoting substances. But still there is more exploration into these sectors for effective utilization of this methods for human benefits.

Biomedical Waste Management: Recycling and Remediation Strategies. Younis Ahmad Hajam, Javid Ahmad Malik, Anu Pandita, and Deepu Pandita (Eds.)

9.1 INTRODUCTION

Human evolution and consistent technological advancement have given a boost to industrialization in the last three decades for the betterment of mankind in all aspects. To seek this approach the rapid growth in global industrialization has also given major issues like pollution. Human pollution into the environment, soil along with fresh water has become a major issue over a period of time ultimately affecting water quality and giving more hazardous health problems. The major activities which have led to the worsening of the water quality are mining. The intense type of mining will result into the generation of different toxic pollutants. Several industries engaged for battery, petrochemicals, metals modifications, polymers processing induces harmful toxin which later on produces the issues for environments. These toxins remain in the environment for longer time and hence are more dangerous more their toxic effects production for many decades inducted due to human and animal encounters. Most of the pollutants show the property of mutagenicity to both humans along with the surrounding. When such heavy metals get accumulated in the organs including the brain, liver, along with kidney. Other harmful effects which are commonly seen in animals include cancer, stunted growth, and damage to the nervous system, which may ultimately lead to death. Th soil containing heavy metals, they affect the soil in such a way that the quality and quantity of the food are decreased. Several heavy impacts are inducing on the plant are getting inhibited due to these heavy metals including plant growth and various types of metabolic processes. The nutrient absorption pattern for plants is also get disturbed due to accumulated heavy effects. The application of physicochemical, methods induce a hug quantity of waste along with pollution into the soil. So, such methods are not implemented for the same.

The bioremediation technology is seeming to be cost-effective and one of the best practical solutions for removing the contaminants present in the environment. The key functions of the soil microorganisms include plant growth promotion. They are also implemented for insect control, nutrient recycling, soil conversation and pollutant reduction. In context with efficiency, cost, and social acceptability, are present into bioremediation have come a long way. The main focus area of bioremediation research is bacterial processes, which have immense applications. In many applications of bioremediation, Archaea is given major contribution for the same. The various other advantages of microbes involve the elimination of the pollutants given from various sources like hyperthermal, and hypersaline along with the acidic and basic industrial waste. Many researchers are currently going on, and it has been suggested that the efficiency and results can be highly improved by using more than one living organisms. Such type of process is possible due to various microbial diversity applicable for bioremediations.

Bioremediation technology is employed by many researchers for the pollutant's removal induced from organic along with inorganic sources. A study was performed, by using bioremediation technology for the treatment of various pollutants by using fungi like *Aspergillus sydowii.* Another study was performed, in which *Cymbella* sp. has been employed for the detoxification of naproxen-polluted water which showed to have effective reach of 97.1%. The bioremediation techniques mainly require the microbial enzymes which play their role by breaking down hydrocarbons. Such products act as less harmful compounds. A modified step in bioremediation technology involves the use of microorganisms which are modified by using genetic techniques. These are also known to eliminate certain xenobiotic chemicals. For obtaining better outcomes and results, various factors are to be considered which include the temperature of the surrounding environment, availability of nutrients, and aerobic or anaerobic conditions employed. The various environmental pollutants including persistent heavy metals and air pollutants which includes either synthetic or natural origin have the ability to reach the ecosystems by using anthropogenic activities. They are quite dangerous to lifeforms like humans, animals, and plants. Bioremediation is thus considered to be one of the better favorable innovations in the field of biotechnology. Waste management also relies on bioremediation technology.

With the use of bioremediation, persistent organic pollutants can be removed effectively, which are usually difficult to break down. This review enlightens about the newer advances and updated information on bioremediation and utilizing the bioremediation strategies. Such strategies are definitely working for eco-friendly detoxification along with the effective degradation of several contaminants from the organic and inorganic origin so as to control environmental pollution. Another potential contaminant that can pollute the aquatic zone of environment. Along with this they are also able to pollute terrestrial and atmospheric environment includes synthetic and semisynthetic pharmaceuticals and pesticides. These pharmaceuticals and pesticides get into the drinking water as a dilute cocktail containing several drugs in varied concentrations. Several drugs such as antibiotics, painkillers, tranquilizers, antiepileptic, chemotherapeutics, inflammatory, hypolipidemic drugs in trace amounts are involved in this process. The potential to cause toxicity. They are also able to induce mutagenicity, carcinogenicity. The occurrence of genotoxicity by the presence of organo-xenobiotics have led to alarming state and is a great matter of significance for the public domain. These pollutants have severely affected the ecosystem through their negative impacts. The excretion after human use, petroleum waste, disposal of expired medicines, agriculture and veterinary use, and manufacturing waste has given a path for the entry of nonbiodegradable inside the environment. They also facilitated the entry of synthetic along with semisynthetic compounds into the environment. Pharmaceutical companies do not focus on the biodegradation (BD) aspects of the new drugs. As many are busy into the manufacturing of more efficacious drugs applicable for various diseases. The drugs that have the ability to persist in the environment for more decades are those that have a longer duration of action in the body. With the help of various modifications in the structure of the drug, the drug can be embedded with the property to remain inside the body for a prolong period of time, that is, longer duration of action. These drugs which get accumulate and reach into the food chain which can affect humans. Apart from humans, their accumulation can also affect other living systems including non-targeted organisms, animals, and plants (Boricha and Fulekar, 2009; Carlsson et al., 2006; Daughton and Ternes, 1999; Hirsch et al., 1999; Jones et al., 2005; Kobayashi et al., 2006; Kolpin et al., 2002; Lee et al., 2008; Marinovich et al., 1996; Singh et al., 2022; Sutar and Das, 2012).

9.2 EFFECT OF CONTAMINANTS

Many chemicals are used in combinations to give toxic effects. Also compounds with various action mechanisms can give toxicity induced from interactions between the chemicals. Many chemicals are able to induces the toxic effects into the non-target species as in several condition monotherapy is replaced by multidrug therapy then the intensity of adverse effects get worsened. The combination of drugs induces pile up concentration that is quite dangerous and impacting many ecosystem pillars. However, in terms of change in behavior, this change is subtle, while the documentation becomes difficult. Due to chronic exposure to eco-damaging chemicals in nature, the incidence of cancers has tremendously increased. Some European research projects have established the environmental impact of these pharmaceutical and veterinary medicine products for both human and animal use, respectively. Much evidence suggested that the some of the pharmaceutical compounds have the ability to withstand the processes occurring in sewage treatment plants (STP) without getting degraded. Additionally, these are non-biodegradable. The removal rates of the different pharmaceuticals differ in the STPs. The removal rates vary from 0% (Carbamazepine) to 30–60% (amoxicillin (AMX), ciprofloxacin, and so forth) only. One of the utmost important tasks is the rejuvenation of the environment and by using the physicochemical process. However, by using biological processes the pollutants can be transformed into harmless products. Therefore, there is need to have a low-cost bioremediation technique. The bioremediation technique usually uses the microbes to diminish the harmful components. The application bioremediation in its raw version through different methods are routinely utilized in many India as per the detail shared from Vedic literature (Freer and Detroy, 1982; Gupta et al., 2016; Randhawa and Kullar, 2011; Seiler, 2002; Strickland and Rousk, 2010; Ternes, 1998; Thiele-Bruhn, 2003; Thorpe et al., 2001; Wetzstein et al., 1997).

9.3 OVERVIEW OF BIOREMEDIATION

The bioremediation involves the use of microorganisms or even the microorganisms that are genetically tailored are used in order to detoxify man-made pollutants. For example, earthworms, especially chloragocytes, have the capability to bioaccumulate the several heavy metals in the body and by using their microflora they are used to detoxify the pesticides. In order to enhance the activity of composts and thus provide better detoxification of the environment, earthworms serve as a good additive. Both, chemically and physically these microorganisms can interact with compounds occurring both naturally as well as man-made resulting in complete degradation or structural changes of the target molecule. For increases, in the production of crops farmers and growing industries have used a wide range of xenobiotic compounds and released them into the environment. And more amount of pollutant compound causes toxic soil which affects crop production and pollutes clean water. Bioremediation is used to clean polluted water and toxic soil by the application of biological agents, mostly microorganisms like fungi or bacteria, and yeast. And these methods are useful for the increases the development of the micro-flora which makes some desired activities. This micro-flora helps in increasing the yield of nutrients, through the management of temperature and by the addition of electrons on the terminal side. In this process, the bacteria and fungi use the pollutant for the generation of energy sources and nutrients. In the last, some years population gets increased worldwide which causes an increase in the growth of polluted water and soil. And the increase in the people of number which generate a burden on natural sources like land, air, and water sources. To the outfit of the people, the increases the industries, cars, vehicle, food, and health it is required for people. But we know that in this growing phase it is very difficult to maintain the quality of our life which are troublesome in the ecosystem in which we live when management is not applied. Bacteria, fungi, and microorganisms that are present in nature break the organic compound. And also, global attention is focused on the prevent the environment from polluted compounds. This attempt is made feasible through the use of biotechnology. Chlorinated solvents, petroleum hydrocarbons, polynuclear aromatic hydrocarbons, ketones, TNT, inorganic nitrogen (NO_3, NH_4), Tt, Pu, Np, Cr, U, and other heavy metals are among the pollutants studied by environmental Biotechnology researchers. Bioremediation is a word used to describe biological processes that employ biological components to heal damaged environments. In the event of oil spills, the procedure takes use of the catabolic capacity of microorganisms that feed on oil. Several employees have studied the different applications of microorganisms.

9.3.1 PRINCIPLE OF BIOREMEDIATION

The current studies of ecology and molecular biology perform well and create some effective biological processes, and the significance of this study is the clean the contaminated water and soil areas. Bioremediation is the process where chemical waste is degraded by the application of biological agents under the guidance of regulatory agencies and the control of the level below the concentration range of regulatory standards. As stated by definition bioremediation is used for the degradation of the ecosystem pollutant into the minimum level of toxic form by the application of microorganisms. And it uses bacteria and fungi or plants to detoxify the compound which creates hazards to the health of humans and the ecosystem. The microorganisms might be native to a polluted region, or they might be separated from somewhere else and brought to the polluted site. Toxin compounds are changed by residing organic entities through responses that occur as a part of their metabolic cycles. The BD of a compound is many times a consequence of the activities of a different organism. At the point when microorganisms are imported to a polluted site to increase bioconversion, we have a cycle known as bioaugmentation (BA). Microorganisms must enzymatically assault the contaminations and convert them to dangerous items. The bioremediation method is effective in the only ecosystem, and it participates in the changes in the environmental limits to accept the growth and degradation of proceeding at a quicker rate, that is, other technologies, which has its limitations. However, aromatic hydrocarbon and chlorinated organic are inhibiting the attack of microorganisms. These compounds are degrading in slow amounts and not at all, from this rate not easy to expect some clean-up and there is no specific rule for the prediction of pollutants (Allard and Neilson, 1997; Mueller et al., 2009; United States Environmental Protection Agency, 2012).

9.3.2 *TYPES OF BIOREMEDIATIONS*

There are two major classes of bioremediations namely in situ bioremediation and ex situ remediations. The subclasses of them are presented in Table 9.1.

9.3.2.1 *IMPACT OF IN SITU BIOREMEDIATION*

Due to the presence of minimum cost and less disturbance, these methods are most commonly used, and they provide the treatment set up staying away from uncovering and transporting impurities. And these treatments are restricted by the profundity of the dirt that can be treated. In many soils, compelling oxy-gen dispersion for helpful paces of bioremediation reaches out to a limit of some centimeters to around 30 cm into the dirt, even though deep of 60 cm and more prominent have been successfully treated at times.

9.3.2.2 *BIOVENTING*

It is the most well-known in situ treatment and includes providing air and supplements through wells to degrade soil to trigger the native microbes and low flow rates of air provide for the BD. These treatment methods work on land surfaces with simple hydrocarbon. It is included in the providing the nutrients by the supply of oxygen in the polluted land to trigger the bacteria for degradation of chemicals. These methods are only used for the land and groundwater. Also, these methods involve various forms like infiltration of water-containing oxygen, and nutrients or the acceptance of electrons.

9.3.2.3 BIOSPARING

It works by the application of air under pressure injection for water to enhance the oxygen limit in the groundwater to increase the rate of degradation of pollution by bacteria. It enhances the link between land and groundwater. This method is conducted at a lower cost.

9.3.2.4 BIOAUGMENTATION (BA)

It is included in both exogenous polluted sites and the addition of micro-organisms. Two factors range the addition of microbial in a soil treatment unit: (i) mostly soil preventive unit is well managed by the long-term contact with microorganisms; and (ii) the development of nonindigenous microbes' culture for increases the population level of microbes.

9.3.2.5 IMPACTS OF EX-SITU BIOREMEDIATION

These techniques work on the elimination or reduction of polluted soil from the land surface. Farming is the common method in which polluted land is spread on the prepared bed and tilled regularly until contaminants are degraded. This method aims to trigger the microorganism and provide the degradation of pollutants.

9.3.3 USE OF COMPOSTING

It is a method that includes a combination of polluted land with manure. The availability of this composition helps increase the growth of microbial population levels at a specific temperature.

9.3.3.1 BIOPILES

These are a cross between land cultivation and waste. Manufactured cells are essentially built as aerated composting piles. They are a modified form of landfarming that tends to reduce physiological losses of pollutants via leaking and evaporation. They are often employed for the treatment of surface contamination with petroleum hydrocarbons. Biopiles promote the growth of indigenous aerobic and anaerobic bacteria.

9.3.3.2 BIOREACTORS

The bioreactor uses slurry reactors which are employed to remove polluted soil and water pumped up from a polluted plume. In reactors, bioremediation entails passing polluted solid material (soil, sediment, sludge) or water via a designed vessel. A wet bioreactor is a containment tank and device used to establish three phases (solid, liquid, and gas) combining state to boost the bioremediation rate of soil-bound and water-soluble toxins as a water slurry of polluted soils and biomass capable of decomposing specific toxins. In overall, the pace and amount of digestion in a bioreactor system are larger than in situ or solid-phase systems because the enclosed atmosphere is much more maintainable, controlled, and reliable. Despite the benefits of reactor systems, there are certain drawbacks. Before placing the polluted soil in a bioreactor, it must be pre-treated (e.g., excavated) or the pollutant can be extracted from the soil by soil washing or physical extraction (e.g., vacuum extraction) as presented in Figure 9.1. There are two types of bioremediation methods (Fahnestock and Wickramanayake, n.d.; United States Environmental Protection Agency, 1990; Table 9.1).

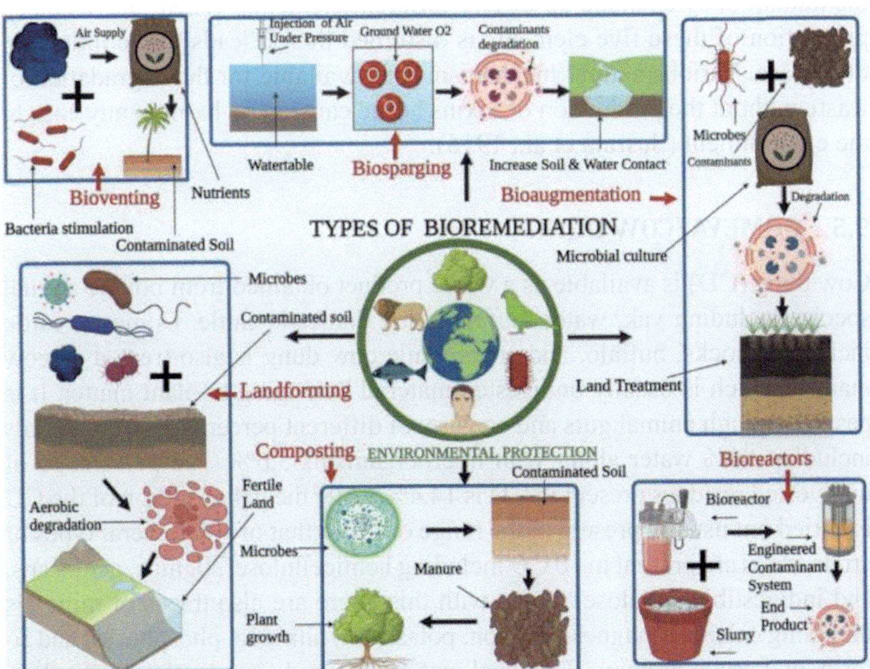

FIGURE 9.1 Types of bioremediations include bioventing, biosparging, bioaugmentation, bioreactors, landfarming, and composting.

TABLE 9.1 Types of Bioremediations

Sl. No.	Types	Method
1.	*In situ* bioremediation	• Bioventing • Biosparging • Bioaugmentation
2.	*Ex-situ* bioremediation	• Composting • Biopiles • Bioreactor

9.4 VEDIC LITERATURE

Ayurveda (AY) is one of the oldest sciences to give information about methods of utilization for cow dung. The most widely used therapy where in cow products such as milk, curd, and dung is basically used. This part of AY is coined as Panchgavya Chikitsa. This majorly works on the principle that the World is made up of a combination of air, water, and fire along with a harmonious blend of the human body. Sushruta Samhita also described the human body as made up of five different elements. When this balanced proportion of these five elements is disturbed then it leads to the induction of disease. Various natural mechanisms are available for the degradation of waste without the production of toxins hence causing no harm in any case to the environment (Susruta et al., 1916).

9.5 GOMEYA/COW DUNG

Cow dung (CD) is available as a waste product obtained from bovine animal species including yak, water buffalo along domestic cattle. Domestic cattle include bullocks, buffalo, and cows while cow dung is also treated as cow manure which is usually undigested material belonging to plant matter. It is passed through animal guts and consists of different percentages of materials including 80% water along with microorganisms 5.6%. The percentage of undigested residues present in CD is 14.4% while the pH condition of the CD is varied but usually present in the range of 7.1 to that of 7.4. Several types of crude fibers are present in 10 CD including hemicelluloses, lignins, pentosans, and indigestible cellulose. Along with this, there are also traces of minerals including calcium, magnesium, iron, potassium, nitrogen, phosphorus, and so forth, and crude proteins. The fecal matter obtained from cattle rumen plays an important role in the enrichment of CD by providing mucus, intestinal bacteria along bile pigments like biliverdin. From ancient times the use of CD was done in the Asian and African communities as an organic fertilizer for the betterment of crops. The CD plays many important roles in the development of plant growth by facilitating phosphorus s and sulfur solubilization, stimulating plant growth. It also helps plants combat with disease and pests by improving their resistance against it. The microorganism is required to nurture the soil through a biogeochemical process and such microbes are harbored by CD which ultimately enriches the soil through the same process. The medicinal properties of cow products are well known in India and in fact as per AY several different products of obtained including curd, urine, milk, and ghee. And dung is in medicinal preparations. In the last two decades, the use of CD as a bioresource has been intensified and is now playing a great role in sustainable development. With the advancement in technology now the utilization of CD is implemented in major sectors like bioprocess, pharmaceuticals, agriculture, and so forth (Gupta et al., 2016; Mandavgane and Kulkarni, 2020; Radha and Rao, 2014; Sawatdeenarunat et al., 2016).

The CD has very useful benefits mentioned in the AY including natural waste purifier. It gradually degrades the waste. It's usually used in many parts of India including urban along rural areas in the solution form of a certain concentration (1:10–1:25 solution). It is termed a gold mine due to its various applications in many sectors including agriculture as biofertilizers, therapeutic applications, and so forth. One of the products obtained d from it is cow manure containing a high concentration of microflora. As per the Indian Vedic traditions, cow dung from indigenous Indian cows, such as the Bos indicus and Zebu varieties, is superior to that from other more modern types. According to AY, the best cow dung comes from a healthy Zebu cow that has been raised clean and hygienic and is fed a nutritious diet of grasslands that includes a variety of natural plants. The Rig Veda recommends man allow cows to graze unrestrictedly in the meadows in all four directions every day (Wikipedia, 2022).

It contains approximately a 3:1 blend of dung and urine. It includes 24 minerals, including cobalt, calcium, traces of sulfur, magnesium, iron, potassium, manganese, nitrogen, and more. It also contains fiber content (cellulose plus lignin), crude protein, cellulose, and hemicelluloses. Around 60 species of bacteria (species of *Lactobacillus* and *Corynebacterium*), two fungi especially *Trichoderma* and *Aspergillus*, and 100 species of yeast (*Candida* and *Saccharomyces*), with some protozoa make up the microbial composition of cow dung. The major contributing bacteria are pectin and hemicellulose fermenters. Undigested fiber-shredded intestinal epithelium, certain expelled bile products (pigments), intestinal bacteria, and mucus are all components of cow manure. Cow dung (an herbivore) gets its greenish color due to biliverdin, which is predominantly found in it. The presence of bile salts is very important as they provide emulsifying properties through coating for hydrophobic droplets from that of hydrophilic ones. CD sludge is generally applied in a proportion of 1:10 or 1:25 to hospital, rural, and urban trash as well as oil spills to naturally decompose them into the basic five elements. *Streptococcus, Faecal streptococcus, Mucor* spp., *Sarcina, Pseudomonas* species, and *Penicillium* microbes are found in cow dung slurry. Composting (bioaugmentation) is a bioremediation technique that induces the BD and stabilization of biodegradable constituents when performed under restricted conditions that include the presence of oxygen. The composting process consists of four major stages which are sequenced as the initial phase followed by the thermophilic phase and mesophilic phase. The maturation phase is the last phase which results in good quality biofertilizer. The extent of organic material humification is broadly acknowledged as a set of criteria for compost maturity or preservation. As a result, of the humification process, functional groups are formed. The humification process used to generate functional groups, along with elevated oxidation of organic matter results in an increase in cation exchange capacities. As a result, compost with a high exchange of cations potential is perceived as a maturity marker. The attainment of the maturity phase of compost is very important as in the immature phase the compost can be able to give toxic effects including phototoxic substances including phenolic acids along volatile fatty acids. It is also risky for the germination of seeds and impacts the root length. Composting is progressively being used to speed further breakdown events and conversion of pollutants, along with pesticides, to stabilize heavy metals available in the soil. Such interactions confirm that contaminants are subjected to a diverse range of microorganisms in the atmosphere, increasing the likelihood of breakdown or conversion by various microorganisms. Furthermore, a few organic molecules established during composting can bind certain metals, preventing their complete removal and therefore translocation from fragile habitats (Balakrishna and Pisupati, 2010; Garg and Mudgal, 2007; Geetha and Fulekar, 2008).

9.6 BIOREMEDIATION FOR PHARMACEUTICALS AND PESTICIDE WASTE BY USING COW DUNG

9.6.1 ANTIMICROBIAL AGENTS

Antimicrobial agents constitute a major part of the medications used in healthcare. It contributes up to 6% of all prescriptions in human health and around

70% of veterinary medicines. It is therefore important to notice that if the antibiotics are not degraded, there's a possibility of the development of multi-drug resistance. This in turn can affect the morbidity and mortality of humans. This brings us to find a correct solution for the degradation of these antimicrobials. Manures like cow dung act as an efficient model of the ecosystem to ensure the degradation of the drug. Few of the coprophilous basidiomycetes are able well degrade enrofloxacin. Wicklow and his colleagues isolated two such basidiomycetes from aged cow dung around 20 years ago. These strains namely, Strain NRRL 6464 along with Strain *Cyathus stercoreus* exhibited high concentration capabilities for lignocellulose and studied via *in vitro* mode. Enrofloxacin has been degraded by *C. stercoreus.*

9.6.2 BIOMEDICAL WASTE (BMW) DEGRADATION

In India, disposing of biomedical waste (BMW) is an upcoming concern. Looking at the economy and population of our country, catering to a cheap and proper method of medical waste treatment and disposal is difficult. Currently used practices include the utilization of incinerators. Such processes are very expensive as well as not good for climates' fate as they produce toxic gases such as polychlorinated dibenzofurans. In these, the dioxins are very harmful to human health. They can cause genetic aberrations along with hormonal imbalances. They are also able to damage immunity and the reproductive system. The role of CD comes in here, the microbial flora present in the CD is useful in this process. The *periconiella* species of fungus that was found in CD was evaluated to be an excellent BMW degrader. This species degraded the BMW such as used bandages, and cotton which was kept in the culture plates by the 50th day of commencement. This turned out to be an effective, economical, and environmentally friendly way to treat BMW.

9.6.3 PESTICIDES

The pesticide production that happens in India is one of the largest in the world and Asia. Indian pesticide industry ranks second in Asia, according to the statistics generated in 2005–2006 with an overall production of around 90,000 tons. When a particular pesticide is used, only 2–3% is utilized, while the rest of it just deposits into the soil causing environmental pollution and toxicity. The soil water environment also gets degraded due to the deposition of pesticide residues into the soil. India has been an agriculture-driven country, and such a pesticide problem is a major concern for India. The majority of India's population is engaged in agricultural practices; thus, pesticides affect the maximum population directly. The current practices used in India to degrade pesticide waste are the physicochemical methods. These are not as useful and effective. Thus, it leads to pesticide residues still prevailing in the soil environment and eventually being present in the food chain. The pesticides deposited in the soil are eventually eaten up by the animals. When animal products are consumed by humans the toxic pesticide tends to accumulate in the adipose tissue along with lymphoid organs of humans. It is also accumulated in the blood. Morbidity cases increase due to the presence of pesticides in the environment which ultimately affect the health due to the pesticide. The pesticides in the environment also affect immunology. Immunodeficiency, immunosuppression or autoimmune diseases, or hypersensitivity reactions as some of the immunological effects of pesticides in animals. Some of the listed reactions are eczema, dermatitis, and allergic respiratory diseases. Carcinoma of the liver and lung is also seen in some of the pesticide Residue which could cause mutations in the chromosomes. Such mutation is present in both humans and animals. These pesticides that are deposited in the environment are teratogenic along with mutagenic properties. They can cause neuropathy along with reproductive disorders. A particular study suggested cow dung as the bioremediation of pesticides such as chlorpyrifos along with cypermethrin. Some of the studies showed that the bioremediation process initiated by the microorganisms tends to degrade a wide range of anthropogenic chemicals. To decontaminate an environment, it is very important that proper techniques of bioremediation of the pesticides in the soil are implemented. Bioremediation of harmful compounds including cypermethrin and chlorpyrifos which are majorly related to organophosphate insecticide through *Pseudomonas plecoglossicida*, a novel organism. This microorganism has not only beneficial effects in the control laboratory step but also is effective in the soils that are contaminated due to the pesticides in the environment. Fenvalerate, a synthetic pyrethroid, contaminates the soil and makes it toxic, thus increasing the toxicity level in the soil water. It is used as an extensive agriculture tool. Some of the harmful byproducts are acclimatized in the environment for a greater period of time then they could degrade into inorganic biomass eventually. While the use of pesticides can also be avoided by the implementation of organic farming. Cow dung is one of the better manures for organic farming (Geetha and Fulekar, 2010; Ingerslev and Halling-Sørensen, 2000; John, 2017; Khan et al., 2015; Kumar et al., 2020; Thiele-Bruhn, 2003; Wetzstein et al., 1999; Wicklow et al., 1980).

9.7 BIOREMEDIATION OF PHARMACEUTICAL WASTE BY USING EARTHWORMS

The environmental process degrades the heavy metals that are prevalent in the environment. Although they are attenuated, they are still found in different water sources like surface water, groundwater along drinking water. The medication residues that are in the environment have a great biological oxygen requirement along with chemical oxygen necessity. Many pharmaceutically active compounds such as hormones and antibiotics are also found in the environment which are hazards to the biosphere. The residues of these substances are very toxic to the environment, and they have complex recalcitrant components that are very difficult for the environmental degraders to degrade. This increase is a health concern in human health services. The variety of BMWs that are being generated has complicated the treatment plan and the management of waste. The pharmaceutical waste that is being disposed of, has a few techniques. The techniques include open dumping, landfilling, incineration, and various other physiochemical processes that do not degrade the pharmaceutical sludge properly. These techniques have a downside and are also hazardous to the environment due to the compromising esthetic value of the release of harmful chemicals into the environment, the high cost, and ineffectiveness. These also degrade the aesthetic value of the environment. All the worsening conditions have brought us to find a way that is cost-effective and environment-friendly. At this point, biological methods of treatment of BMW are proven old-aged treatment processes. The earthworms are cultured and with the help of vermitechnology which is the product of vermiculture, we can convert the waste into vermicompost. Vermicompost can be used as an excellent nutrient-rich manure for agriculture which consists of nitrogen, phosphorus, and potassium to a large extent. It is an excellent soil fertilizer too. Vermicomposting with earthworms is a very feasible option as it converts the BMW into vermicompost which is eco-friendly, reduces pollution, and also increases the fertility of the soil. The process used in bioremediation by earthworms and cow dung is presented in Figure 9.2.

9.7.1 ENVIRONMENTAL AND HUMAN HEALTH EFFECTS OF PHARMACEUTICAL WASTE

Inappropriate disposal techniques and ineffective treatment plants have caused pharmaceutical waste to have an impact on the environment along

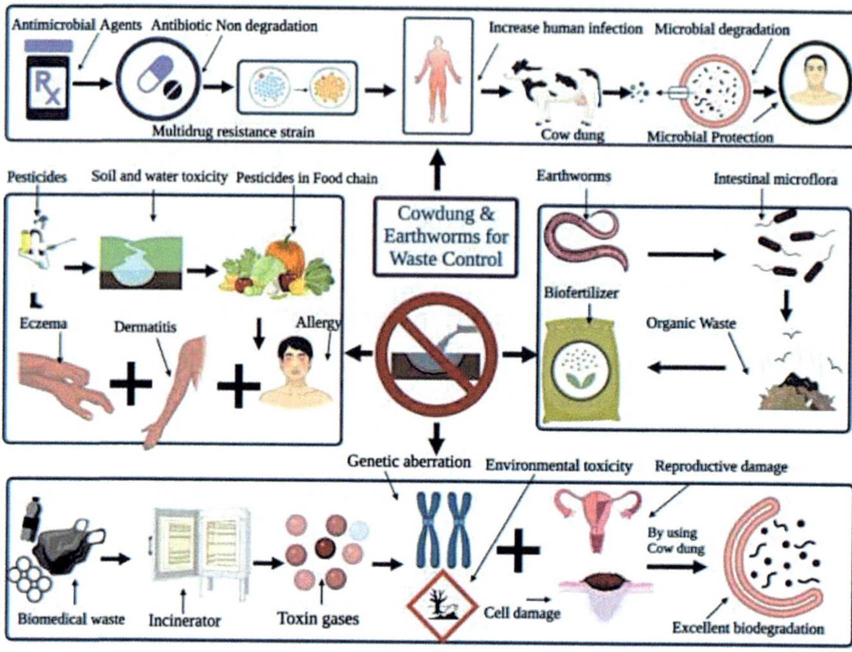

FIGURE 9.2 Solution of cow dung and earthworms for pharmaceutical waste, including antimicrobial agents, pesticides, and biomedical waste.

with human health. Many biomedical bases or pharmaceutical sludge that are found in the environment are refractory organic compounds that are non-biodegradable, and they have a high risk of biomagnification. The pharmaceutical waste generated is having a very high organic load of oxygen content and bioactive compounds including hormones, antibiotics, toxic chemicals, and surfactants. Pharmaceuticals have proven to be very hazardous to the environment if not treated properly. They have the capacity to react with biological processes even in small quantities which is a huge risk to the environment. Many drugs are found to be semi-persistent which means that they do not degrade easily when they are introduced into the water body. Thus, they cause potential harm to human health as well as the water ecosystem. These quantities are lesser than the usual organic pollutants, but the long-term effect of these environmental pollutants is to be considered. Pharmaceuticals are typically lipophilic and thus, make them non-biodegradable. Pharmaceutical activity is a bigger harm to the environment as compared to other organic pollutants. According to a research study, the chances of the environment getting polluted due to the medication in the ecosystem is high in underdeveloped countries or developing countries as compared to developed countries because of the differences in BMW generation and their treatment plants. According to some surveys, it is concluded that the people living near pharmaceutical industries have become sick due to waterborne diseases caused by contaminated water. The disorders caused by drugs in water are gastroenteritis, hypertension, fetomaternal mortality, cardiac problems, diabetes mellitus, and impaired neurobehavioral consequences. According to Sharif et al., the inorganic and organic pollutants in the environment are one of the reasons for the increase in diabetes mellitus cases. Pharmaceutical industries can release such constituents into the ecosystem which causes pollution. Heavy metals like Lead (Pb), Cadmium (Cd), Chromium (Cr), Nickel (Ni), and Mercury (Hg) are fear of the components of these pollutants. Other organic compounds and phenolic compounds are also present in the environment. These compounds eventually degrade the quality of the ground and surface water. These pollutants become a major concern when we consider the food chain in the environment. This has led to a lot of research projects. The World Health Organization (WHO) directives and some studies, set up a permissible limit for the pollutants or these constituents that must be present in the environment, it also gives us an idea about the health hazard of the exposed active pharmaceutical ingredients even at an extremely low quantity. According to a 2013 report by the WHO and the United Nations Environment Program (UNEP), many countries especially developing countries are facing problems with health hazards due to the API getting injected into the water resources. According to Green et al. (2004), the dramatic decline in the population of vultures in the Indian subcontinent is because of the biomagnification of an API named Diclofenac in the food chain of vulture and its ecosystem (Sanchez-Hernandez, 2020; Singh and Suthar, 2012).

9.7.2 VERMICOMPOSTING

Vermicomposting has been proven of great help in maintaining soil quality and the ecosystem as it offers a lot of benefits such as chemical compound breakdown and transformation of organic waste into biofertilization and also providing aeration to the soil. Some of the researchers mentioned how vermicomposting can be used beneficially to remediate herbal pharmaceutical waste. They conducted an experiment under laboratory control conditions. The earthworm species that were used for vermicomposting was *Eisenia fetida*. This species along with cow dung in different combinations and varying amounts treated herbal pharmaceutical industry waste. Significant chemical changes were observed during vermicomposting. The pH level also drought by a considerable amount when the residual combination was analyzed. The decreased pH made it very useful as fertilizer. The residual product was also more stable, odor-free, and content with better-fertilizing capacity.

9.7.3 VERMICOMPOSTING OF DIFFERENT INDUSTRIAL WASTE AND SLUDGES

The process of vermicomposting is getting more popular day by day due to its property to enhance pharmaceutical waste. Does increasing the value of the waste. When it comes to bio-conservation of waste vermicomposting has having faster processing speed than traditional composting making it both efficient and economical. Vermicomposting has a lot of advantages over a traditional composting system such as lesser processing time and improved nutrient recovery. Vermicomposting has been gaining popularity and several research projects are being done on it. Studies on biological waste like animal waste, sewage sludge industrial sludge/waste, and agriculture waste are getting popular nowadays. Many research studies have proven that vermicomposting is an efficient and effective waste management strategy for industries. According to a recent study, the degradation of industrial sludge was a result of the proteolytic enzymes identified in the earthworms' gastrointestinal tract.

9.7.4 VERMIREMEDIATION OF PHARMACEUTICAL WASTE AND SLUDGES

The earthworms are popularly known as the "Farmer's friend" and "Intestine of the Earth" because of the variety of services they provide to decompose a number of organic substances. The mechanical and chemical chewing done by the muscular gizzard present in the intestine of the earthworm helps it to break several organic materials into smaller sizes. Earthworms' intestine contains a variety of microbes enzymes and hormones. These factors altogether can biodegrade a partially digested food resulting in a nutrient-rich vermicompost that is developed in a short period. Biodegradable materials when passing through the earthworm elementary canal get crushed into small pieces or fine powder which are then acted upon by the fermenting enzymes leading to the BD of these products and synthesis of vermicompost with the aid of bacteria and enzymes that are present in the intestinal flora of the earthworm. Depending on the natural habitat and feeding habits of the earthworms they are classified into the following classes: epigeic, anecic, and endogeic. The earthworm is an abundantly encountered organism. There are more than 420 species and subspecies known to have been discovered in India. Vermicomposting has gained a lot of popularity in the world and is a commonly used method to manage a variety of organic residues and wastes. Vermicomposting is among the most researched and the most efficient way of managing organic waste. Vermicomposting is a process in which microorganisms along with earthworms in the intestine of the earthworm break down and degrade trash. This is a speeded-up process. The earthworm species with the help of their microflora and chloragocyte cells can detoxify most of chemicals and heavy metals. According to some recent studies, industrial sludge and municipal waste with a good amount of moisture proved to be an excellent combination for the nutrition of the earthworm. Some of the most commonly used earthworm species in vermicomposting are *Dendrobaena rubida*, *Lumbricus rubellus*, *Eisenia fetida*, *Pheretima elongata*, *Perionyx excavatus*, *Eudrilus eugeniae*, and *Dendrobaena veneta*. By using *Eisenia fetida* vermicomposting was performed at a very successful rate in the reprocessing of plant-available nutrients by utilizing herbal pharmaceutical industrial waste. According to the study, the trash combinations were divided into five categories by varying the amount of cow dung and herbal pharmaceutical industry waste and one of the subjects of the studies was waste-free (control). The study duration was 60 days with an interval of 10 days to procure a sample. All the vermi-beds reflected a high rate of cocoon generation. This above study proved one that vermicomposting could be used to manage pharmaceutical waste. As a result, vermicomposting is a viable and efficient option for treating pharmaceutical sludge in the future. In the near future, we could generate a vermicomposting method that is 100% efficient in producing nutrient-rich and detoxified manure. This method will also be considered effective and environment-friendly technology in future (Bhat et al., 2016, 2018; Fernández-Gómez et al., 2013; Garg et al., 2012; Karwal and Kaushik, 2020; Lee et al., 2018; Lim et al., 2016; Malafaia et al., 2015; Mohan, 2017; Raj et al., 2017; Sharma and Garg, 2019; Singh et al., 2020; Usmani et al., 2017).

KEYWORDS

- **bioremediation**
- **earthworms**
- **municipal waste**
- **organic waste**
- **pharmaceutical waste**
- **pharmaceuticals**
- **vermi-beds**
- **vermicomposting**

REFERENCES

Allard, A. S., & Neilson, A. (1997). Bioremediation of organic waste sites: A critical review of microbiological aspects. *International Biodeterioration & Biodegradation, 39*(4), 253–285.

Balakrishna, S. M. S., & Pisupati. (2010). Traditional knowledge in policy and practice: Approaches to development and human well-being.

Bhat, S. A., Singh, J., & Vig, A. P. (2016). Genotoxicity reduction in bagasse waste of sugar industry by earthworm technology. *SpringerPlus, 5*(1), 1186. https://doi.org/10.1186/s40064-016-2882-1.

Bhat, S. A., Singh, S., Singh, J., Kumar, S., Bhawana, & Vig, A. P. (2018). Bioremediation and detoxification of industrial wastes by earthworms: Vermicompost as powerful crop nutrient in sustainable agriculture. *Bioresource Technology, 252*, 172–179. https://doi.org/10.1016/j.biortech.2018.01.003.

Boricha, H., & Fulekar, M. (2009). *Pseudomonas plecoglossicida* is a novel organism for the bioremediation of cypermethrin. *Biology and Medicine, 1*.

Carlsson, C., Johansson, A. K., Alvan, G., Bergman, K., & Kühler, T. (2006). Are pharmaceuticals potent environmental pollutants? *Science of The Total Environment, 364*(1–3), 67–87. https://doi.org/10.1016/j.scitotenv.2005.06.035.

Daughton, C. G., & Ternes, T. A. (1999). Pharmaceuticals and personal care products in the environment: Agents of subtle change? *Environmental Health Perspectives, 107*(Suppl 6), 907–938. https://doi.org/10.1289/ehp.99107s6907.

Dighton, J. (2017). *The Fungal Community: Its Organization and Role in the Ecosystem* (4th ed.). CRC Press.

Fahnestock, F. M. von, & Wickramanayake, G. B. (n.d.). *Biopile Design, Operation, and Maintenance Handbook for Treating Hydrocarbon-Contaminated Soils*. https://www.osti.gov/biblio/679963 (accessed on 18 September 2024).

Fernández-Gómez, M. J., Díaz-Raviña, M., Romero, E., & Nogales, R. (2013). Recycling of environmentally problematic plant wastes generated from greenhouse tomato crops through vermicomposting. *International Journal of Environmental Science and Technology, 10*(4), 697–708. https://doi.org/10.1007/s13762-013-0239-7.

Freer, S. N., & Detroy, R. W. (1982). Biological delignification of 14C-labeled lignocelluloses by basidiomycetes: Degradation and solubilization of the lignin and cellulose components. *Mycologia, 74*(6), 943–951. https://doi.org/10.1080/00275514.1982.12021614.

Garg, A. K., & Mudgal, V. (2007). Organic and mineral composition of gomeya (cow dung) from desi and crossbred cows–A comparative study. *International Journal of Cow Science, 3*(1–2), 17–19.

Garg, V. K., Suthar, S., & Yadav, A. (2012). Management of food industry waste employing vermicomposting technology. *Bioresource Technology, 126*, 437–443. https://doi.org/10.1016/j.biortech.2011.11.116.

Geetha, M., & Fulekar, M. H. (2008). Bioremediation of pesticides in surface soil treatment unit using microbial consortia. *African Journal of Environmental Science and Technology, 2*(2), 36–45. https://doi.org/10.5897/AJEST.9000016.

Geetha, M., & Fulekar, M. H. (2010). A remediation technique for removal of fenvalerate from contaminated soil. *Asian Journal of Water, Environment and Pollution, 7*(1), 85–91.

Gupta, K. K., Aneja, K. R., & Rana, D. (2016). Current status of cow dung as a bioresource for sustainable development. *Bioresources and Bioprocessing, 3*(1), 28. https://doi.org/10.1186/s40643-016-0105-9.

Hirsch, R., Ternes, T., Haberer, K., & Kratz, K. L. (1999). Occurrence of antibiotics in the aquatic environment. *Science of The Total Environment, 225*(1–2), 109–118. https://doi.org/10.1016/S0048-9697(98)00337-4.

Ingerslev, F., & Halling-Sørensen, B. (2000). Biodegradability properties of sulfonamides in activated sludge. *Environmental Toxicology and Chemistry, 19*(10), 2467–2473. https://doi.org/10.1002/etc.5620191011.

Jones, O. A., Lester, J. N., & Voulvoulis, N. (2005). Pharmaceuticals: A threat to drinking water? *Trends in Biotechnology, 23*(4), 163–167. https://doi.org/10.1016/j.tibtech.2005.02.001.

Karwal, M., & Kaushik, A. (2020). Co-composting and vermicomposting of coal fly ash with press mud: Changes in nutrients, micro-nutrients, and enzyme activities. *Environmental Technology & Innovation, 18*, 100708. https://doi.org/10.1016/j.eti.2020.100708.

Khan, M. Y., Roy, M., Saroj, B. K., Dubey, S., & Sharma, V. K. (2015). A review: Benefits of Panchgavya therapy (Cowpathy) for the health of humans. *Asian Journal of Research in Pharmaceutical Science, 5*(2), 115. https://doi.org/10.5958/2231-5659.2015.00019.3.

Kobayashi, Y., Yasojima, M., Komori, K., Suzuki, Y., & Tanaka, H. (2006). Removal characteristics of human antibiotics during wastewater treatment in Japan. *Water Practice and Technology, 1*(3). https://doi.org/10.2166/wpt.2006.059.

Kolpin, D. W., Furlong, E. T., Meyer, M. T., Thurman, E. M., Zaugg, S. D., Barber, L. B., & Buxton, H. T. (2002). Pharmaceuticals, hormones, and other organic wastewater contaminants in U.S. streams, 1999–2000: A national reconnaissance. *Environmental Science & Technology, 36*(6), 1202–1211. https://doi.org/10.1021/es011055j.

Kumar, V., Thakur, I. S., & Shah, M. P. (2020). Bioremediation approaches for treatment of pulp and paper industry wastewater: Recent advances and challenges. In *Microbial Bioremediation & Biodegradation* (pp. 1–48). Springer Singapore. https://doi.org/10.1007/978-981-15-1812-6_1.

Lee, L. H., Wu, T. Y., Shak, K. P. Y., Lim, S. L., Ng, K. Y., Nguyen, M. N., & Teoh, W. H. (2018). A sustainable approach to biotransform industrial sludge into organic fertilizer via vermicomposting: A mini-review. *Journal of Chemical Technology & Biotechnology, 93*(4), 925–935. https://doi.org/10.1002/jctb.5490.

Lee, Y. J., Lee, S. E., Lee, D. S., & Kim, Y. H. (2008). Risk assessment of human antibiotics in Korean aquatic environment. *Environmental Toxicology and Pharmacology, 26*(2), 216–221. https://doi.org/10.1016/j.etap.2008.03.014.

Lim, S. L., Lee, L. H., & Wu, T. Y. (2016). Sustainability of using composting and vermicomposting technologies for organic solid waste biotransformation: Recent overview, greenhouse gases emissions, and economic analysis. *Journal of Cleaner Production, 111*, 262–278. https://doi.org/10.1016/j.jclepro.2015.08.083.

Malafaia, G., da Costa Estrela, D., Guimarães, A. T. B., de Araújo, F. G., Leandro, W. M., & de Lima Rodrigues, A. S. (2015). Vermicomposting of different types of tanning sludge (liming and primary) mixed with cattle dung. *Ecological Engineering, 85*, 301–306. https://doi.org/10.1016/j.ecoleng.2015.10.010.

Mandavgane, S. A., & Kulkarni, B. D. (2020). Valorization of cow urine and dung: A model biorefinery. *Waste and Biomass Valorization, 11*(3), 1191–1204. https://doi.org/10.1007/s12649-018-0406-7.

Marinovich, M., Ghilardi, F., & Galli, C. L. (1996). Effect of pesticide mixtures on *in vitro* nervous cells: Comparison with single pesticides. *Toxicology, 108*(3), 201–206. https://doi.org/10.1016/0300-483X(96)03303-3.

Mohan, S. M. (2017). Vermicomposting of paper mill sludge with *Eisenia fetida* for its conversion to nutrients using different seed materials. *Journal of The Institution of Engineers (India): Series A, 98*(4), 545–553. https://doi.org/10.1007/s40030-017-0253-8.

Mueller, J. G., Cerniglia, C. E., & Pritchard, P. H. (2009). Bioremediation of environments contaminated by polycyclic aromatic hydrocarbons. *Applied Microbiology and Biotechnology, 49*(5), 605–612.

Radha, T. K., & Rao, D. L. N. (2014). Plant growth promoting bacteria from cow dung-based biodynamic preparations. *Indian Journal of Microbiology, 54*(4), 413–418. https://doi.org/10.1007/s12088-014-0468-6.

Raj, K., Nandhivarman, M., Thongni, P., Muthulingam, P., & Gopalsamy, P. (2017). Utilization of agrowastes for vermicomposting and its impact on growth and reproduction of selected earthworm species in Puducherry, India. *Nature Environment and Pollution Technology, 16*(4), 1125–1133.

Randhawa, G. K., & Kullar, J. S. (2011). Bioremediation of pharmaceuticals, pesticides, and petrochemicals with Gomeya/cow dung. *ISRN Pharmacology, 2011*, Article 362459. https://doi.org/10.5402/2011/362459.

Sanchez-Hernandez, J. C. (2020). Vermiremediation of pharmaceutical-contaminated soils and organic amendments. https://doi.org/10.1007/698_2020_625.

Sawatdeenarunat, C., Nguyen, D., Surendra, K. C., Shrestha, S., Rajendran, K., Oechsner, H., Xie, L., & Khanal, S. K. (2016). Anaerobic biorefinery: Current status, challenges, and opportunities. *Bioresource Technology, 215*, 304–313. https://doi.org/10.1016/j.biortech.2016.03.074.

Seiler, J. P. (2002). Pharmacodynamic activity of drugs and ecotoxicology—Can the two be connected? *Toxicology Letters, 131*(1–2), 105–115. https://doi.org/10.1016/S0378-4274(02)00045-0.

Sharma, K., & Garg, V. K. (2019). Vermicomposting of waste. In *Sustainable Resource Recovery and Zero Waste Approaches* (pp. 133–164). Elsevier. https://doi.org/10.1016/B978-0-444-64200-4.00010-4.

Singh, D., & Suthar, S. (2012). Vermicomposting of herbal pharmaceutical industry waste: Earthworm growth, plant-available nutrient and microbial quality of end materials. *Bioresource Technology, 112*, 179–185. https://doi.org/10.1016/j.biortech.2012.02.101.

Singh, P., Singh, V. K., Singh, R., Borthakur, A., Madhav, S., Ahamad, A., Kumar, A., Pal, D. B., Tiwary, D., & Mishra, P. K. (2020). Bioremediation. In *Abatement of Environmental Pollutants* (pp. 1–23). Elsevier. https://doi.org/10.1016/B978-0-12-818095-2.00001-1.

Singh, S., Roshan, W., & Vig, A. (2022). Bioremediation of wastewater sludge from pharmaceutical industries through vermistabilization.

Strickland, M. S., & Rousk, J. (2010). Considering fungal dominance in soils: Methods, controls, and ecosystem implications. *Soil Biology and Biochemistry, 42*(9), 1385–1395. https://doi.org/10.1016/j.soilbio.2010.05.007.

Susruta; Bhishagratna, K. L. K., Kaviraj. (1916). *Shushruta Samhita*. https://archive.org/details/englishtranslati01susruoft (accessed on 18 September 2024).

Sutar, H., & Das, C. K. (2012). A review on Bioremediation. *International Journal of Research in Chemistry and Environment, 2*(1), 13–21.

Ternes, T. A. (1998). Occurrence of drugs in German sewage treatment plants and rivers. *Water Research, 32*(11), 3245–3260. https://doi.org/10.1016/S0043-1354(98)00099-2.

Thiele-Bruhn, S. (2003). Pharmaceutical antibiotic compounds in soils: A review. *Journal of Plant Nutrition and Soil Science, 166*(2), 145–167. https://doi.org/10.1002/jpln.200390023.

Thorpe, K. L., Hutchinson, T. H., Hetheridge, M. J., Scholze, M., Sumpter, J. P., & Tyler, C. R. (2001). Assessing the biological potency of binary mixtures of environmental estrogens using vitellogenin induction in juvenile rainbow trout (*Oncorhynchus mykiss*). *Environmental Science & Technology, 35*(12), 2476–2481. https://doi.org/10.1021/es001767u.

United States Environmental Protection Agency. (1990). *Handbook on In Situ Treatment of Hazardous Waste-Contaminated Soils*.

United States Environmental Protection Agency. (2012). *Bioremediation of Hazardous Waste Sites: Practical Approaches to Implementation* (EPA/625/K-96/001). http://cfpub.epa.gov/si/si_public_comments.cfm (accessed on 18 September 2024).

Usmani, Z., Kumar, V., & Mritunjay, S. K. (2017). Vermicomposting of coal fly ash using epigeic and epi-endogeic earthworm species: Nutrient dynamics and metal remediation. *RSC Advances, 7*(9), 4876–4890. https://doi.org/10.1039/C6RA27329G.

Wetzstein, H. G., Schmeer, N., & Karl, W. (1997). Degradation of the fluoroquinolone enrofloxacin by the brown rot fungus *Gloeophyllum striatum*: Identification of metabolites. *Applied and Environmental Microbiology, 63*(11), 4272–4281. https://doi.org/10.1128/aem.63.11.4272-4281.1997.

Wetzstein, H. G., Stadler, M., Tichy, H. V., Dalhoff, A., & Karl, W. (1999). Degradation of ciprofloxacin by basidiomycetes and identification of metabolites generated by the brown rot fungus *Gloeophyllum striatum*. *Applied and Environmental Microbiology, 65*(4), 1556–1563. https://doi.org/10.1128/AEM.65.4.1556-1563.1999.

Wicklow, D. T., Detroy, R. W., & Jessee, B. A. (1980). Decomposition of lignocellulose by *Cyathus stercoreus* (Schw.) de Toni NRRL 6473, a "white rot" fungus from cattle dung. *Applied and Environmental Microbiology, 40*(1), 169–170. https://doi.org/10.1128/aem.40.1.169-170.1980.

Wikipedia. (2022). *Rigveda*. https://en.wikipedia.org/wiki/Rigveda (accessed on 18 September 2024).

CHAPTER 10

E-Waste Management from the Point of View of Materials Engineering

RAFAEL VARGAS-BERNAL*

Departamento de Ingeniería en Materiales, Instituto Tecnológico Superior de Irapuato, Irapuato, Guanajuato, Mexico

*Corresponding author.

ABSTRACT

The accelerated and optimistic economic expansion of the globalized world, the enormous desire to be digitally connected and the enormous technological advances are causing the obsolescence of products. This is leading all countries to struggle with the massive increase in electronic waste. Within the categories of metals, polymers, glass, nanomaterials, and rare earth elements (REEs), the materials employed in electrical goods for the biomedical sector are extremely diverse. In this chapter, materials engineering-based e-waste management is analyzed to identify recycling and remediation opportunities to exploit materials considered waste and reduce the production of raw materials which increase environmental pollution and climate change on the planet. Despite the progress made so far, it is necessary to continue advancing in legislating and implementing smart and functional recycling plants that take advantage of all the waste generated by electronic products around the world.

10.1 INTRODUCTION

Electrical and electronic equipment, abbreviated as EEE, is becoming increasingly important in society, which is driving up the amount of outdated EEE that ends up in landfills (Murthy and Ramakrishna, 2022). Because of

Biomedical Waste Management: Recycling and Remediation Strategies. Younis Ahmad Hajam, Javid Ahmad Malik, Anu Pandita, and Deepu Pandita (Eds.)

its unsustainable methods of collecting, treating, and disposing of e-waste or waste EEE, which has posed considerable problems both to the environment as well as to the health of living beings. The legal sector requires technology, infrastructure, ability to collect and manage an increasing amount of electronic waste in an environmentally responsible way. Although approximately 71% of the population around the world has so far adopted e-waste regulations, requiring global legislation to be imposed on all countries. According to a 2021 report from the United Nations (UN), each human being can generate electronic waste at a rate of 7.6 kg per capita, which produces an amount of 57.4 MMT (Material Management Technologies) globally. Although multiple laws have been passed to protect the environment, exploit natural resources, as well as to provide formal job security, illegal actions have had a negative impact. Some of the actions undertaken by the informal sector are open burning, acid extraction, and illegal waste disposal.

Along with being harmful to the environment, these wastes also result in the loss of high-value-added materials such as rare earth as well as precious materials, ferrous and non-ferrous metals (Patil and Ramakrishna, 2020). In such a way that both recycling and management of this garbage become additional problems derived from the dangerous materials present including Hg, Pb, and Cd in circuit boards, computer monitors, and televisions. Per year the worldwide production of electronic waste is predicted at 50 million tons. Globally only 20% of this waste is disposed of properly, leading to most of it going unused. With a material value of $62.5 billion alone, the volume of e-waste presents a significant economic opportunity. So much less energy consumed, and carbon dioxide emission are required to extract resources from this waste instead of extracting it from the earth's crust. These wastes are both valuable and environmentally harmful due to the significant amount of metal value they contain, which negatively impacts both biotic and abiotic elements.

Due to unsustainable processing methods, e-waste is producing a challenging alternative (Murthy and Ramakrishna, 2022). Developed countries continue to expand while developing countries are struggling because of a lack of knowledge, resources, and infrastructure. Anyone can see the issue of electronic waste as a shared duty to preserve the environment as well as the safety and health of women, youth, adolescents, and children. Used equipment considered waste from high-income countries such as European countries, Australia, and North America is exported to low-income Asian, African, and Latin American nations (Baldé et al., 2017).

The demand for electronics and electrical equipment is growing daily due to changes in emerging technology and lifestyle. These devices continue to impact daily activities no matter what they have both beneficial and harmful consequences, people rarely consider what will happen after they are disposed of. Unavoidably, there will be an enormous rise in the number of electrical devices, and more and more everyday items will be using micro-processors. One of the markets that are now growing quickly is the electronic market. The application and manufacture of electronic and electrical equipment have expanded because of rapid economic growth, urbanization, and a rise in consumer products. To effectively address the e-waste problem, a multi-directional approach is necessary. It is also required to produce a balance between the promotion of the global economy and the conservation of natural resources. This is because the reimbursement of resources must consider the conservation of a clean environment as well as human health (Ahirwar and Tripathi, 2021).

This chapter was divided as follows. Section 10.2 describes the concepts and issues required to define electronic waste. To identify the composition of devices, equipment, or systems related to e-waste, the materials found in them are presented in Section 10.3. In Section 10.4, the challenges and possibilities for recycling materials as well as the function of the e-waste management structure are commented on. The last developments regarding the recycling of electronic waste materials have been highlighted in Section 10.5. In Section 10.6, the conclusions of this research chapter regarding waste management are provided.

10.2 E-WASTE BASICS

Electronic and electrical equipment (EEE) consumption has significantly increased since the turn of the 21st century grew out of economic growth and manufacturing, technological advances in information, communication, and electronics, as well as consumer convenience (Patil and Ramakrishna, 2020). Most households use several electrical machines, including air conditioners, refrigerators, washers, ovens, blenders, and televisions. The fact that over half of the world's population is online has increased demand for the use of technologies that exploit information and telecommunications such as tablets, computers, and cell phones. Multiple ICT gadgets owned by a single customer contribute significantly to the world's e-waste. Because of how quickly electronic technology is developing, these devices quickly become outdated and are replaced by newer generations of equipment, creating frightening amounts of e-waste. Electronic waste also comes from lighting equipment, consumer equipment, toys, monitoring and control instruments, power tools, medical devices, recreational and sports equipment, as well as automatic dispensers in addition to the aforementioned uses (Ahirwar and Tripathi, 2021). The most found types of e-waste are depicted in Figure 10.1.

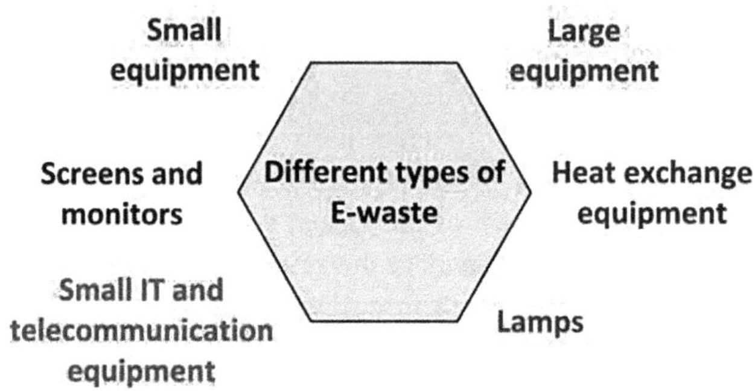

FIGURE 10.1 The different types of electronic waste.

Significant and ongoing developments in medical technology have improved patient diagnosis, care, and rehabilitation. However, the business is creating alarming amounts of electronic waste as medical equipment becomes outdated more quickly (Lubell, 2019). Most of this e-waste is sent to unidentified sites, probably landfills. Both the environment and the local community are affected by the negative role of toxic substances found in waste. In addition, healthcare providers must properly delete and destroy any patient data that has been saved digitally to comply with legal duties. Far-reaching negative effects are produced due to inappropriate biomedical e-waste disposal. Computers, lab analyzers, microscopes, and more specialist medical equipment like electrocardiograms, sphygmomanometers, and spectrophotometers are examples of e-waste in the healthcare sector. Hospitals, dental offices, toxicology labs, treatment facilities, doctor's offices, cosmetic surgery clinics, urgent care centers, and other medical facilities commonly produce this kind of healthcare e-waste. These wastes contain heavy metals which are considered hazardous substances, which makes biomedical electronic equipment regarded as a dangerous item when it is disposed of. This is only one of the reasons why responsible disposal is essential.

Electronic waste includes a host of products found in homes, consumer devices, and business applications (Mudali et al., 2021a). EEE that were completely or partially discarded, as well as manufacturing waste and rejects, is referred to as e-waste. Used EEE that is intended with the aim of reuse, resale, salvage, recycling, or disposal of materials is another expression for electronic waste.

E-waste encompasses a broad spectrum of items, from stuff for the home to consumer devices to business supplies (Mudali et al., 2021a). Heat exchange equipment, lamps, displays, little IT and telecom devices, little household equipment, and big equipment are the six main categories. In addition, this type of waste includes equipment components, among which are lead capacitors, batteries, CRTs, activated glass, circuit boards, and plastic covers. Personal devices such as computers, screens, smartphones, laptops, and televisions account for nearly half of all e-waste; the remainder consists of non-portable household equipment as well as heating and air conditioning equipment.

E-waste-related health risks can arise from coming into physical touch with dangerous objects, breathing in toxic gases that are emitted, and from the increase of dangerous chemical substances that can be deposited in water, soil, and consequently in food (Mudali et al., 2021a). E-waste includes toxins that are mutagenic, teratogenic, and carcinogenic that are persistent, transportable, and that is stored biologically in the environment. This hazardous waste negatively affects the socioeconomic, biological, and physical environments. There is severe pollution in the Earth's biosphere, hydrosphere, lithosphere, and atmosphere. People and other members of biodiversity are susceptible to lethal diseases like cancer, abnormalities of reproduction, neurological damage, endocrine disturbances, asthmatic bronchitis, and mental retardation. E-waste can have a catastrophic effect on people and the environment if recycling and disposal are not done scientifically. Human health as well as the environment are negatively impacted as a result of inappropriate care in recycling and waste management by informal sectors that use crude procedures.

E-waste has a huge amount of potential as a resource for several useful and significant materials that may be applied to create innovative goods, eliminating the requirement to exploit fresh raw materials (Mudali et al., 2021a). Because there are more valuable metals concentrated in electronic waste than in minerals, mining in cities is illegal and much more cost-effective than mining for raw materials, and recycling materials also saves much energy. Recovery of several vital resources could lessen the nation's reliance on imports.

Hazardous elements are present in high amounts of e-waste pollutants (Mudali et al., 2021a). Recycling enables safe processing and management of the stream of hazardous e-waste. E-mining makes it feasible to conserve natural resources, lessen pollution, cut back on greenhouse gas (GHG) emissions, and use less of the earth's raw materials.

Large amounts of electronic waste provide a difficult handling, storage, and disposal challenge (Mudali et al., 2021a). Because of the dangers it poses towards both life as well as the environment, the government, and the public both view the safe and sustainable disposal of these materials as a major priority. A scientific recycling method is required for the disposal of such heterogeneous complex material to safeguard the environment as well as to advance the circular economy, an economic system that aims to reduce waste and promote resource efficiency. Therefore, society needs to develop secure and scientific ways for profitably recovering embedded rare and valuable metals without putting the environment at risk. It is well acknowledged that recycling, recovering, and reusing obsolete EEE in future product cycles is a tough task. It should be noted that the foundation to improve electronic waste management and accelerate the circular economy is the adoption of the 5 R's principle (recycle, repurpose, reuse, reduce, and refuse).

Since e-waste contains both dangerous and costly metals, safe treatment of this garbage is currently a considerable concern for governments, the public, and trade openings. However, the bizarre methods of removing these priceless metals from garbage have led to environmental issues and health risks (Garlapati, 2016). Within the next 6–8 years, developing countries will yield twice as much electronic waste as developed countries. Furthermore, emerging and developed countries are expected to trash 200–300 million and 400–700 million old computers, respectively, by 2030.

10.3 MATERIALS IN E-WASTE

According to the Association of Plastics Manufacturers in Europe (APME), electronic waste is made up of materials containing metals, rare-earth elements, plastics, and ceramics (Garlapati, 2016). The components of e-waste vary according to the manufactured items and include over 1000 distinct compounds classed as dangerous or non-hazardous. Glass, plastics, and metals are the three main components of the complicated e-waste stream (Mudali et al., 2021a). A good source of certain valuable and significant metals is e-waste. It has a lot of valuable non-ferrous materials, such as palladium (Pd), platinum (Pt), copper (Cu), silver (Ag), gold (Au), and rare earth elements. There are significant amounts of base metals, including iron (Fe), aluminum (Al), and tin (Sn). There are far lesser amounts of other elements including valuable metals and rare earth elements (REEs). Flame retardants (bromine-based biphenyls) and metals including hexavalent chromium (Cr (VI)), selenium (Se), cadmium (Cd), arsenic (As), mercury (Hg), and lead (Pb) are examples of toxic materials. E-waste is full of dangerous substances that are bad for the environment and people's health, including beryllium (Be), mercury (Hg), lead (Pb), cadmium (Cd), sulfur (S), chromium (Cr), and bromine (Br) as flame retardant.

Electronic waste encompasses both ferrous and nonferrous metals, as well as polymers, rubber, glass, wood products, ceramics, concrete, boards for electronic devices (PCBs), as well as other materials (Garlapati, 2016). Electronic waste contains approximately 50% iron and its derivatives, 21% polymers, 13% nonferrous metals, as well as other materials. Nonferrous metals include specific materials like copper (Cu), and aluminum (Al), as well as valuable metals including silver (Ag), gold (Au), platinum (Pt), palladium (Pd), and others. E-waste that contains excessive amounts of chemicals including lead (Pb), mercury (Hg), arsenic (As), cadmium (Cd), selenium (Se), and hexavalent chromium (Cr (VI)), and anti-flame agents which have been deemed to be toxic waste. The application percentages of the different types of materials in electronic waste are shown in Figure 10.2. Table 10.1 lists the most common types of metals detected in e-waste.

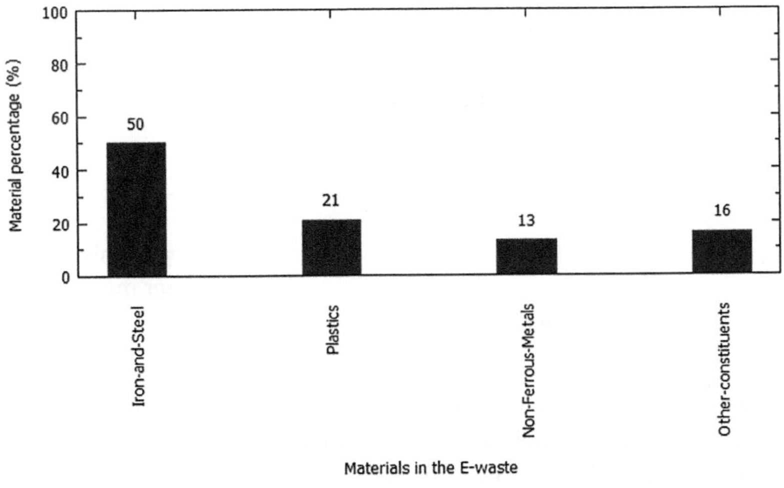

FIGURE 10.2 The percentages of various kinds of materials found in e-waste.

TABLE 10.1 The Main Types of Metals Found in E-Waste

Ferrous Materials	Nonferrous Materials	Precious Materials	Semimetals	Heavy Metal	Soft Metal
Iron (Fe)	Copper (Cu)	Silver (Ag)	Tin (Sn)	Lead (Pb)	Cadmium (Cd)
Steels (iron alloys)	Aluminum (Al)	Gold (Au)	Arsenic (As)	Mercury (Hg)	–
–	–	Platinum (Pt)	–	Beryllium (Be)	–
–	–	Palladium (Pd)	–	–	–

Figure 10.3 depicts a detailed classification of hazardous e-waste compounds. Also, this figure classifies the components found in electronic trash into four types: halogenated substances, heavy metals, radioactive elements, and other chemicals (Garlapati, 2016). The health consequences are primarily physical, affecting lung health, cell growth and function, thyroid hormone, and reproductive health. It has been demonstrated that the halogenated flame retardants included in e-waste polymers are toxic to mammals' neurological systems and interfere with reproduction. Lead causes symptoms including coma, convulsions, diarrhea, vomiting, and even death. Lead is often found in the printed circuit board solder, cable sheathing, lead-acid batteries, and Cathode Ray Tube (CRT) of obsolete televisions. Cadmium exposure can lead to acute and long-term poisoning, as well as symptoms similar to the flu.

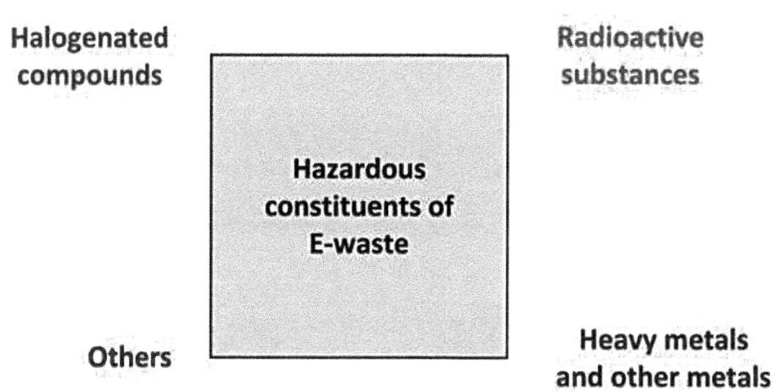

FIGURE 10.3 Hazardous components of e-waste.

E-waste from photocopier cartridges, selenium drums (copiers), batteries, capacitors, mercury switches, printed circuit board assemblies, liquid crystal displays (LCDs), and electrolytes contains a variety of contaminants and dangerous metals, including Pb, Hg, and Cr(VI) (Garlapati, 2016).

Polychlorinated biphenyls (PCBs) and mercury are toxic substances included in e-waste such as flat screen displays computer cadmium batteries, obsolete capacitors, switches, and transformers. Toxic substances such as dioxins and furans are released when bromine-based anti-flame agents in boards for printed circuits, plastic enclosures, and PVC-insulated conductors are burned. Antimony oxide, lead, and bromine-based flame retardants (included in a percentage of 5 to 10% by weight), are among the other toxic materials found in e-waste.

Numerous chemicals found in electronic trash are hazardous to both human and environmental health (Garlapati, 2016). E-waste contains persistent organic contaminants, such as polychlorinated dibenzo-p-dioxins (PCDDs), polychlorinated dibenzofurans (PCDFs), perfluoroalkyl, and polyfluoroalkyl substances (PFASs), bromine-based anti-flame agents (BFRs), polychlorinated biphenyls (PCBs), as well as polycyclic aromatic hydrocarbons (PAHs; Ahirwar and Tripathi, 2021).

Failure to recover rare-earth elements (REEs) is also causing issues in the development of future generations of technological devices. Rare earth elements are a chemically cohesive collection containing 17 chemical elements that include Lanthanides and chemically related substances such as Scandium (Sc) and Yttrium (Y) (Mudali et al., 2021a). Lanthanides are composed of Cerium (Ce), Dysprosium (Dy), Erbium (Er), Europium (Eu), Gadolinium (Gd), Holmium (Ho), Lanthanum (La), Lutetium (Lu), Neodymium (Nd), Promethium (Pm), Praseodymium (Pr), Samarium (Sm), Terbium (Tb), Thulium (Tm), and Ytterbium (Yb). REEs come in two different types: heavy REEs (HREEs) and light REEs (LREEs).

Rare earth has been exploited in a broad variety of applications and final destinations, including alloys, phosphors, permanent magnets, additives, glass polishing powders, pigments, glazes, ceramics, catalysts, and batteries (Mudali et al., 2021a). REEs are included in catalysts, co-catalysts, and electrolyte additives in various fuel cell systems. More than 75% of worldwide REE demand is accounted for by four REEs: praseodymium, neodymium, lanthanum, and cerium.

Rare earth elements are becoming more crucial in high-tech devices (Mudali et al., 2021a). Magnets, batteries, phosphors, and catalysts are only a few of the components of advanced technology that incorporate REEs. These components are used in many industries, including clean energy, computer, automotive, healthcare, consumer electronics, petroleum refining, transportation, robotics, aerospace, and defense. Rare earth elements have significant usage for treatment and diagnosis. These are included in several imaging procedures, including computed tomography scanning, imaging with positron emission tomography, magnetic resonance imaging, as well as X-rays. The enduring magnet technology as a high-tech sector requires large amounts of REEs.

There is an increasing need for the recycling of these materials from electronic waste due to scarcity, restrictions on trading internationally, and cost variations for specific types of these metals globally (Mudali et al., 2021a). Complex electronics contain about 60 different elements. Numerous e-waste may also include harmful substances like flame retardants. Most of the time, e-waste is neither properly documented nor recycled.

Printed circuit boards (PCBs) account for a major portion of electronic waste (Mudali et al., 2021a). PCBs are comprising metals (40%), ceramics (30%), and plastics (30%). PCBs typically contain copper (up to 25%), tin (up to 5%), aluminum (up to 5%), iron (up to 5%), nickel (up to 2%), zinc (up to 2%), lead (up to 2%), and deposits of valuable metals (gold (up to 0.1%), silver (up to 0.3%), palladium (up to 0.03%), as well as a few ppm of Pt). When a plastic piece of a PCB or other electrical appliance is burned, it emits halogenated flame retardants that can pose risks to the ecology and human health (Garlapati, 2016). E-waste is frequently recycled and dumped without restriction which causes heavy metal poisoning of soil, water, and plants, posing major environmental and health problems.

Recycling now recovers base metals (ferrous and nonferrous metals), polymers, valuable metals, and glass from e-waste (Mudali et al., 2021a). Rubber, ceramics, concrete, wood, and other materials can also be found. The type of electrical equipment used and other factors, such as the composition of e-waste, all have a significant role, in its model, its supplier, when it was made, and how old the junk was. E-waste is a type of garbage extremely complex and heterogeneous.

10.4 CHALLENGES AND OPPORTUNITIES TO RECOVER MATERIALS FROM THE E-WASTE MANAGEMENT STRUCTURE

Approximately 75% of the electronic equipment kept in residences, workplaces, and warehouses needs to be managed as e-waste (Garlapati, 2016). Some techniques that will eventually minimize e-waste can be used to complete this objective. Techniques for waste management include process change, recovery, and reuse. For instance, lowering the quantity of surplus raw materials in-store and the number of toxic chemicals used in the activity can both reduce the waste produced. Additionally, the recovery, as well as reuse method, seems to have the ability to do away with final disposal expenses, lower the cost of raw materials, and make money from scalable trash. One can recycle waste either off-site, on-site, or via the inter-industry interaction. A few of these techniques are roster administration, manufacturing process change, quantity decrease, as well as recovery and preservation.

E-waste is a treasure that has not been processed; it could include a number of raw and processed materials, including polymers, base metals, glass, platinum group metals, and precious metals, as well as other non-metals (Murthy and Ramakrishna, 2022). The Environmental Protection Agency (EPA) of the United States has started the sustainable materials management (SMM) electronics challenge to gather discarded electronics that should be recycled by authorized recyclers in partnership involving distributors and original equipment manufacturers (OEMs). In 1993, the European Union (EU) proposed the Waste Shipping Regulation (WSR) to ban the export of dangerous electronic trash to nations outside the OECD, which led to the restriction of hazardous substances (RoHS) directive in 2012. Its activities allowed the development of the Waste from Electrical and Electronic Equipment (WEEE) directive (2012/19/EU) which regulates the management of e-waste in their member countries. This policy enables management of electronic waste to be managed, including collection, recycling, as well as retrieval. Every nation in the globe must implement stricter e-waste management policies and processes since such unofficial approaches result in a poor recovery of e-waste as well as the dangerous compounds connected to it (Patil and Ramakrishna, 2020). The issue of handling and recycling e-waste affects both industrialized and developing countries, especially so for the latter.

Asia is the region that produces the most e-waste, and most Asian nations are subject to e-waste regulations (Baldé et al., 2017). Contrarily, Africa produces the least quantity of electronic waste because no e-waste restriction is applied in the majority of African nations. The Americas, which covers both the north and south hemispheres of the globe, has been a significant producer of e-waste. The higher-income countries such as the USA and Canada have the biggest manufacturers of electronic trash in America (Baldé et al., 2017; Patil and Ramakrishna, 2020). The basic problem with most nations' electronic waste regulations comes from inefficient and inadequate application of the law. Due to the lack of legal obligation and the availability of statistics about the removal as well as recycling of e-waste, it is hard to evaluate the effectiveness of the present electronic waste law. It is challenging to keep track of worldwide recycling of electronic trash since e-waste law varies greatly between nations. Oceania is made up of the Pacific Island area, New Zealand, and Australia (Patil and Ramakrishna, 2020). Similar to Asian nations, these nations have grasped the value of appropriate laws in regulating the management of e-waste. Australia is a unique nation in this area with a law that particularly addresses e-waste management.

Because of the contradictory environmental and good physical condition implications by part of informal e-waste processing, appropriate legislation must be enacted to create a formal and secure e-waste management system (Patil and Ramakrishna, 2020). Dangerous levels of contamination of the environment have been generated due to the incorrect disposal of both electronic garbage as well as primitive methods for obtaining valuable metals. Hazardous compounds emitted from e-waste are continuously contaminating important environmental elements like soil, air, and water. Electronic garbage is disposed of in landfills containing dangerous chemicals like heavy metals (including beryllium, lithium, cadmium, mercury, and lead) and acids which causes the leaching of such chemicals into the land, polluting similar biodegradable garbage. Additionally, the biome next to the dump is impacted, which contaminates the food chain with harmful substances. Moreover, the soil's fertility is decreased by residues from the combustion of trash and the spill of toxic acids.

Electronic waste disposal represents a significant problem everywhere in the globe. WEEE emits hazardous pollutants when disposed of with regular household garbage, putting crucial ecosystem elements in jeopardy (Garlapati, 2016). The traditional methods of electronic waste disposal, including incineration and landfilling, are also harmful to the ecosystem. Contaminated leachates from land-filled e-waste enter groundwater. Soil acidification and water resource contamination will result from a component in computer chips that melts. E-waste incineration releases poisonous gases that are dangerous, which contribute to issues with air pollution. The primary weakness of the country's uncontrolled activities in a shady e-waste recycling program is the discarded offshoot of toxins into the land, water, and atmosphere.

The groundwater table is also contaminated by hazardous substances as well as heavy metals that leach into the land from dumps (Patil and Ramakrishna, 2020). Freshwater bodies like lakes and rivers that provide drinking water to neighboring cities are contaminated with e-waste pieces and processing waste byproducts such as hazardous substances, acids, risky compounds, as well as heavy metals. Using water that is so polluted for agricultural purposes creates another pathway for the poisons to reach the food chain.

When polyvinyl chloride (PVC) cabling, wires, and PCBs from e-waste are burnt in the air to extract valuable metals, the ensuing smoke contains dangerous substances including dioxins and acid gases, which pollute the air (Patil and Ramakrishna, 2020). Mercury, lead, and cadmium are among the heavy metals that are exposed to the ecosystem when electronic waste is burned. In addition to being extremely dangerous, the leftover fly ash from burned e-waste spreads in the air around the burning sites.

Computers, phones, and other common electronic medical devices all contain dangerous components and substances that, if improperly disposed of, may have terrible impacts on both the environment as well as the health of people (Lubell, 2019). Bromine-based flame retardants, chlorinated polymers (PVC) from cables and cabling, mercury from LCDs, copper, and titanium-niobium from MRI magnets lead from CRT displays, cadmium from batteries, and polymers from resistors, are among the potentially dangerous materials that are found from computer casing to PCB in biomedical electronic equipment. Furthermore, a startling 70% of electronic waste finds itself inside an undocumented or unidentified location, most probably in a dump. These harmful materials can affect nearby ecosystems, contaminate food, and even endanger human health if they are dumped in a landfill and begin to leak into the groundwater.

Rather than focusing on production facilities, the Extended Producer Responsibility (EPR) idea of pollution control legislation places more emphasis on product systems (Garlapati, 2016). As a result, the business must be involved with the end-of-life of the product associated with the manufacturing and operating activities. Because of this, the main objective of EPR is to motivate manufacturers to minimize resource and energy consumption at every step involved in the product life cycle by altering product design as well as manufacturing technologies. Effective EPR management helps increase organization-wide awareness, increase commercial growth chances, boost firm credibility and reputation, and promote relationships and communication with important stakeholders. EPR aims to promote long-term development by promoting product development and product recovery practices that are environmentally friendly. EPR extends producers' responsibility for their goods to the point beyond consumer use. Even though there are regulations governing e-waste and EPR in around two-thirds of the world, just about 20% of electronic waste is legally recycled (Murthy and Ramakrishna, 2022). Current practices and statistics make it possible to distinguish different challenges and opportunities that could offer interesting solutions for the near future.

From a legal basis, each country has different regulations even though electronic waste has the same design and manufacturing process around the world (Murthy and Ramakrishna, 2022). It is necessary that at a global level, there is a common legal basis that is imposed in all countries to avoid errors and conflicts of interest in the required policies. Exploiting e-waste reduces costs by extracting materials directly from obsolete equipment rather than from mining or manufacturing processes, leading to more effective and sustainable processes. In addition, this preserves the natural resources of the planet and reduces environmental damage by using automated controlled systems to reduce the impact on garbage collectors directly exposed to waste, and a decarbonized circular economy can be achieved.

E-waste collection rates through existing methods are quite low, mostly because consumers are unaware of the benefits and drawbacks of recycling electronic debris and because there are no facilities for financial inducements including cost-free gathering, delivery, and exchange advantages (Ahirwar and Tripathi, 2021). Resolving these issues will encourage customers to take part in recycling campaigns. Recycling of electronic trash is done informally and is important for managing solid waste in many poor nations, is also thought to be hazardous to the environment and workers' health. Informal sectors typically cost less to create and maintain, and they have first-hand knowledge of how a city disposes of its electronic trash. Large amounts of e-waste may be collected and processed manually at low cost, frequently with little infrastructure or investment, thanks to the availability of very inexpensive labor. However, improvised methods used by informal e-waste recyclers frequently result in the release of contaminants and unextractable materials entering the environment as well as the ability to recover only a small portion of the economically valuable materials (such as gold and silver) from one of the more enriched deposits (such as PCB). Integrating such mobile recyclers and e-waste collectors into the established or cooperative sphere and ensuring remunerative returns for their work is vital to reduce the hazards to the environment and health that were produced by collection and treatment processes.

From a technological point of reference, legislation must be continuously updated not only due to the possible incorporation of new materials but also due to the updating of the infrastructure and technologies for processing electronic waste (Murthy and Ramakrishna, 2022). This means that the organizations dedicated to developing standards must continually introduce updates to those standards dedicated to the management of electronic waste. This could ensure the reuse, repair, repurposing, and recycling of materials from electronic waste.

Consumer WEEE is generally gathered by roving collectors, trash merchants, and illegal e-waste traffickers in the unorganized sector, and is then manually disassembled and recycled in small-scale facilities using subpar techniques (Ahirwar and Tripathi, 2021). The official sectors, on the other hand, gather valuable WEEE by means of a take-back system through distributors or merchants as well as transport it by using authorized recyclers for mechanical disassembly and carefully controlled treatment in metal extraction.

The extended producer responsibility (EPR) must be extended globally regardless of whether the countries belong to the OECD or not, since all companies, governments, and users must know their role and responsibility for managing electronic waste, always thinking of the well-being environment without conflicts of interest (Murthy and Ramakrishna, 2022). EPR is a common e-waste management approach utilized in many nations worldwide. The German Packaging Waste Prevention Ordinance (the GreenDot program) was the world's first mandated EPR program, requiring manufacturers to bear financial accountability for waste packaging collection and reduction (Garlapati, 2016). A law mandating producers of electrical and electronic items with the aim of offering free final disposal treatment for their gadgets using authorized handlers has been approved in Switzerland, Taiwan, Norway, and Sweden. Consumers in Norway can take old electronics like laptops, phones, and wires, and free electronic and industrial electrical supplies to vendors at local government drop-off locations, which are funded by adding a recycling fee to certain items just at the point of purchase. On a state level, the United States implemented effective EPR programs by requirements established in state statutes, rules, and policies, such as advanced distribution charges, deposit reimbursement programs like bottle bills, as well as required battery take-back programs. Japan has taken back systems in place for televisions, air conditioners, refrigerators, and washing machines, and it permits businesses to pay the customer for the service.

EPR adds post-consumer management to the manufacturer's and distributors' prioritization of existing environmental duties (Garlapati, 2016). They are required to either take back used products and control them via reuse, recovery, or energy generation, or assign this duty to a private entity called a Producer Responsibility Organization (PRO), that receives payment from the producer for spent-product administration. By compelling manufacturers, importers, and/or buyers to factor in the cost of waste management in their product pricing, EPR transfers waste management duty into the government and private sector. The Duales System Deutschland (DSD) in Germany, VALPAK in the United Kingdom, El-Kretsen AB in Sweden, and the Rechargeable Battery Recycling Corporation (RBRC) in the United States are samples belonging to these groups, which are growing quickly across several nations alongside EPR initiatives.

Electronics manufacturing companies must expand their interest in developing green technologies by choosing the chemistry of their products with their end-life in mind (Murthy and Ramakrishna, 2022). Therefore, businesses, governments, and users are responsible for the design, production, and processing of recycled electronic waste. This implies that there is financial, educational, and technological support that enables all those involved.

Users of electronic products must act responsibly in the management of electronic waste using the regulation that manufacturers must include as information when selling a product so that the product's lifespan, composition, risks associated with improper disposal, techniques for reusing, repairing, and renovating, and so forth, be known (Murthy and Ramakrishna, 2022).

E-waste management must move from informal and unsustainable methods to formal and sustainable strategies (Murthy and Ramakrishna, 2022). Women and children are regularly involved in the dismantling of electronic waste, which leads to seriously ill human beings and irreparable environmental damage. Therefore, the workers involved in this activity must be trained and protected to carry out this activity safely and with all the required protection. The primary barrier to e-waste recycling is the safety of low-wage workers (Garlapati, 2016). The recycling of e-waste (arranging erratically, disassembling, splintering, tearing apart, processing acid, and open burning) poses health risks. People working without protection in cramped spaces with limited ventilation are exposed to poisonous and slowly poisonous compounds, endangering their health. E-waste is being disposed of using crude and dangerous techniques. The majority of these methods involved recycling e-waste generated by their daily activities, which exposed them and their family to a contaminated atmosphere full of dangerous poisons. Recycling workers frequently get injuries, which, if untreated, expose the wounds to harmful substances. Additionally, cancer and endocrine, excretory, reproductive, or long-term neurodegenerative problems might emerge from the release of acid residues from precious metals recovery activities into nearby fields. Waste produced through the diagnosis, therapy, as well as vaccination of humans, animals, and research subjects is known as medical waste. Dialysis, medicine usage, insulin doses, and perhaps even animal care procedures in rural areas can all produce BMW at home. It is not just created at medical facilities, research facilities, and labs. Safe ecosystems and healthy populations depend on effective BMW treatment. The improper management of biological waste is detrimental to public health. It's necessary to handle biological waste safely and consistently. Both developed and developing nations struggle with managing biological waste. For a safer and greener environment, BMW management must be done in a way that is economical, environmentally friendly, and less polluting. The community and related employees are very interested in raising awareness about trash management. The main strategies to solve e-waste management are summarized in Figure 10.4.

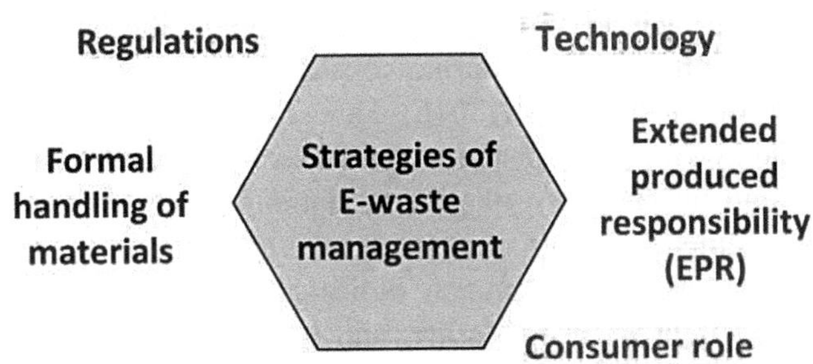

FIGURE 10.4 Main strategies for e-waste management.

Enhancing design work is another method for reducing trash generation as well as the health issues that come with it. With effective control of the chemistry as well as the purity of the raw materials applied in electronics, the amount of toxic waste created at the end of their use may be greatly decreased (Ahirwar and Tripathi, 2021). Due to limitations imposed by miniaturization, the present market trend for electronics favors lightweight, slender, and highly compact electronic devices. Recyclability should be a higher priority for device producers, as are the CPU speed, energy efficiency, and display size and performance of these modern electronic devices. One method of promoting easier and more effective recycling is to design items to make disassembling by easier reducing using glue and screws in plastic that have been thermally set. Another strategy to address the overproduction of EEE and the corresponding production of e-waste is to produce sustainable products that do not become obsolete too quickly or have a long lifespan. Such goals can be met by collaborating with industrialists, engineers, scientists, and e-waste recycling specialists.

Handling the material in a production plant is used for roster management, which lowers waste generation (Garlapati, 2016). This could be accomplished by limiting both hazards through the implementation of resource appraisal, control methods, disclosure monitoring systems, and excess natural resources in stock. Setting performance evaluations for every piece of equipment acquired is the first stage in developing an entry management program. A suitable mechanism should be in place to review all materials before purchasing. For the approval route, both harmful and safe unusual components are filtered through all production resources. A stringent register monitoring system necessitates another way that just needed the quantity of supplies to be acquired. Based on the present time restrictions, materials management events must be put into action.

Production of waste is reduced by employing alternative materials, better-trained laborers, or both during construction and by the alteration of the industrial process (Garlapati, 2016). Three methods can be used to implement this strategy: improved performance and continuity procedures, change of material, and modification of process equipment. Another approach to e-waste management consists of reducing the quantity of electronic waste (quantity reduction). This approach indicates that electrical components must only be bought when necessary, and prompt maintenance should be carried out to increase the service life of the product. Old electronic items should be given to charity causes so that they can be refurbished and used again, as opposed to being thrown away. To save expenses associated with disposal, quantity could also be decreased by substituting non-hazardous trash for hazardous waste during system operation. Techniques for trash concentration and resource recovery can be used to achieve this. The most economical method for managing e-waste is metal separation and possible treatments for retrieved metal value. Precious metals can be recovered and reused from e-waste using techniques including freeze vaporization, reverse osmosis, ultra-filtering, and gravity and vacuum filtering.

Revenue from discharged garbage is feasible through the e-waste management strategy of recovery and restoration (Garlapati, 2016). Inter-industry trash recycling at off-site or on-site treatment centers makes this possible and eventually lowers the costs of sanitary landfills and raw materials. Centrifugation, filtering, electrochemical recovery, condensation, electrolysis, and reverse osmosis are frequently used in this process.

10.5 NEW TECHNOLOGIES FOR RECYCLING E-WASTE MATERIALS

Currently, NiMH batteries, fluorescent lights, and permanent magnets are the main sources of REE recycling and/or recovery. For small consumer gadgets, recycling efforts have been minimal (Mudali et al., 2021b). Many of the industrialized nations, like Japan, the USA, the UK, Germany, and so forth, have already adapted recycling technology for the three categories to an industrial scale and are close to being fully developed. To recover metal from e-waste, numerous traditional thermochemical and biochemical techniques have undergone laboratory and pilot-scale testing. Physical methods of recovering metal are still widely used, nonetheless, because they are simpler to use and have a smaller environmental impact.

When hazardous products are recycled by converting them into secondhand goods, they ought to be environmentally friendly (Garlapati, 2016). At the national and/or regional levels, effective management of the environment necessitates the installation of gathering centers, e-waste transport, processing, storing, recovering, and dumping. These facilities should be made available by regulatory bodies, and rewards for better performance must indeed be offered. At the district, regional, and federal levels, e-waste gathering, transfer, and recycling centers must be established by manufacturers and Non-Governmental Organizations (NGOs).

To accomplish ecologically sustainable recovery of e-waste, specialist knowledge and practical training need to be required (Garlapati, 2016). To filter the dangerous and useful materials from complicated e-waste, expert employees are required. Then, distinct environmentally friendly recycling techniques must be developed for the poisonous and desirable materials individually. Obsolete technology must maintain strict environmental standards to reduce the negative environmental effects on recycling staff. The recycling plants must have effective air pollution control strategies for point source and escape emissions. Due to the fact that precious metals are included in it, the corporate sector and a number of research groups are stepping forward to support research goals today.

The subsections describe the efforts made to recycle materials from electronic waste used in biomedical applications.

10.5.1 METALS

Metal recycling from electronic waste is a profitable economic potential for small and medium-sized corporations (Mudali et al., 2021b). Nevertheless, a pyramidal structure may be seen in the e-waste recycling business, where just a few big firms are involved in the pre-processing and dismantling of e-waste, a greater percentage during pre-processing and disassembly, and an excessive amount during retrieval.

In the recovery of metal from electronic waste (industrial production reject/scrap as well as final disposal equipment), some different process routes are suggested, including biometallurgical electrometallurgical, pyrometallurgical, and hydrometallurgical methods, mechanical recycling, as well as their hybrids, are all examples of processes (Ahirwar and Tripathi, 2021; Mudali et al., 2021). The two primary methods for treating electronic trash are hydrometallurgical and pyrometallurgical procedures. For the separation and recovery of certain metals, electrometallurgical/electrochemical procedures may be applied along these paths. To determine the routes and procedures for recycling, Due to the diversity of e-waste, it is essential to categorize it in regard to categories, layout, constituents, and content. Analyzing sustainability while considering environmental, economic, and social factors is crucial for the industrial implementation of these technologies.

Physical recovery is the method for reusing e-waste that is most often applied worldwide (Mudali et al., 2021b). This is one of the most effective ways to recover metal. Before any further processing, this pre-treatment method is used to extract metallic and non-metallic materials from e-waste. Disassembling, dismantling, shredding, crushing, and other physical recycling methods are disparities of metal and nonmetals that will come next. These physical techniques mostly concentrate on nonmetal materials rather than metal materials. There have been investigations into several techniques based on the characteristics of gravity, magnetism, electrostatics, and optics.

Three methods based on electrical properties have already been employed to separate non-ferrous materials from PCBs: eddy current isolation, corona electrically charged detachment, and tribo-electrostatic dissociation (Mudali et al., 2021b). In the processing of e-waste, for the refining of basic and precious metals, the froth flotation technique is currently being researched. Low capital and operating expenses are a benefit of physical separation techniques, but a drawback is the significant loss of valuable metals (10–35%) caused by improper metal liberation.

In hydrometallurgy, the used metal must be subsequently recovered through the mixture by sedimentation, absorption, electrochemical treatment, electro-winning, or solvent evaporation (Ahirwar and Tripathi, 2021; Mudali et al., 2021). The precious metals are extracted into liquids by dissolving in the appropriate solvent or solution (acids or perhaps a base). The hydrometallurgical method can be used to extract copper, REEs, and precious metals like gold, silver, and so forth. The most effective way to recover metals, this procedure is also the most affordable, environmentally friendly, and can be carried out in under-regulated, moderate settings. High metal recoveries are provided by hydrometallurgical techniques at a comparatively cheap capital cost. Small-scale applications are a good fit for these methods.

In the past three decades, pyro-metallurgical processing has evolved into a standard technique for recovering non-ferrous and valuable metals derived from e-waste (Ahirwar and Tripathi, 2021; Mudali et al., 2021). This process includes high-temperature gas phase reactions, treating, sintering, softening, and casting in a plasma electrical furnace or exhaust gases. Because of the slag's oxide production, recovering Fe and Al is challenging. Materials that are recovered from waste PCBs are now accomplished through several pyro-metallurgical techniques. High-quality feeds are needed for these energy-intensive operations. High investment costs, environmental degradation, and non-selectivity degradation of rare earth metals are some further drawbacks. When a comprehensive pollution control mechanism is put in place, the pyrometallurgical process offers a workable method for recovering energy from electronic trash.

10.5.2 PLASTICS

The amount of electronic garbage produced globally is already growing startling at a pace of 4%–5% per year, and it has been predicted that by 2018, that number will be close to 50 million tons (Kumar et al., 2019). Plastics make up about 21% of the weight of electronic waste, coming in second to metals as the most prevalent component. Styrene-based polymers make up around 55% of the plastics in e-waste; their recycling is more complicated and difficult than that of plastics like polypropylene and polyethylene terephthalate because they contain many impurities, including flame retardants made of bromine (BFRs). Therefore, the paucity of research on utilizing plastics from e-waste is indeed not surprising.

Polymers constitute around 20% of the electronic and electrical items that are discarded (Butturi et al., 2020). This plastic fraction's recycling is a challenging problem that is significantly influenced by the presence of dangerous compounds like brominated flame retardants. From a technological and economic perspective, since the materials are mixed together and have different compositions, recycling e-waste polymers is a difficult problem, but it also raises concerns for the environment and public health because e-waste plastics may contain additives that could be dangerous. For a variety of reasons, different metals and organic materials are incorporated into the polymers in tiny amounts to improve mechanical strength and support, increase thermal and electrical conductivity, increase resistance to burning and environmental conditions, and provide other additional advantages (Ahirwar and Tripathi, 2021).

To delay ignition and stop the spread of fire in electrical equipment, brominated flame retardants (BRFs) are utilized (Butturi et al., 2020). These additions must be stable throughout the lifespan of the type of plastic and consistent with the polymers into which they are applied. The additives are permanent, bioaccumulative, and hazardous due to the needed stability (Joseph et al., 2021). This toxicity is released from the surroundings during the gathering, transport, management, storage, reworking, and waste phases. In many impoverished nations, including India, many African countries, as well as some Latin American countries, BFRs reach the ecosystem by direct diffusion belonging to treated materials under ambient temperature via photodecomposition. These countries still use inadequate recycling techniques. Additionally, unsuitable laws are used to dispose of a sizable portion of e-waste in dumps, harming the environment. In many poor nations where the tradition of open-firing WEEE is indeed widespread, the thermal breakdown of BFRs results in the release of dioxins, furans, small and large brominated species, and other pollutants into the environment. The alternative final disposal approach for e-plastics to recuperate energy is incinerating them.

Plastics are a crucial component of everyday life, from household goods to medical equipment (Joseph et al., 2021). Plastics are easily functionalized to give desired chemical or functional properties and can be produced in any shape or size. Plastics have supplanted steel, porcelain, and glass in medical equipment because they are more affordable and long-lasting. Plastics are thoroughly researched in the development of medical disposables due to the rise in diseases linked to non-disposable materials. They may be coupled with fillers as well as additives to change their surface properties or elasticity, and they are light. Care should be taken when choosing the right plastic material, as well as considering the material's durability, sterilizing capability, and chemical or thermal properties.

Emphasizing the e-plastic waste's component, many experts believe that several of the main methods of reusing e-plastics in poor countries is their reuse as recycled aggregates (also known as RPAs or recycled plastic aggregates) in the building industry (Butturi et al., 2020). Low-biodegradable plastic has become a problem for the environment. However, several of its properties, including low solidity, a flat surface, low penetrability, and good chemical resistance, make it appropriate for usage in the building sector, including general, paving, and lightweight concrete blocks. These uses contribute to lowering the over-mining as well as natural resource consumption like cement and aggregate, particularly in places where energy recovery through the burning of waste or plastic recycling is not possible due to a lack of specialized infrastructures, such as many low- and middle-income nations (LMICs). Additionally, because usage of plastic in building constructions in these nations can successfully address the demand for inexpensive construction materials brought on by the region's expanding urbanization.

Common components of e-waste materials include ABS (acrylonitrile butadiene styrene) (30%), PC/ABS (9%), PP (polypropylene) (8%), HIPS (high impact polystyrene) (25%), PC (polycarbonate) (10%), as well as other mixed materials (18%) (Butturi et al., 2020). WEEE plastics have almost 15 distinct industrial plastics. Around 3400 WEEE goods feature plastic parts, which are frequently made up of many types of polymers. Additionally, flame-retardant plastics make up around 25% of e-plastics.

Following the recovery of the precious metals, the recyclable plastics are reprocessed using either mechanical recycling or chemical recycling can produce fuel, monomeric units, or other substances (Butturi et al., 2020). The recycling process implies primary reuse, which WEEE plastics use to create new plastic items and down-cycling, and it involves using plastic materials again for different projects.

The most widely used electronic polymers, including PP, HIPS, and ABS, remain easily accessible in significant numbers and can be effectively used with the conventional mechanical technique (Butturi et al., 2020). Dismantling-based recycling, or disassembling plastic parts before shredding, is the most efficient way to recover high-quality post-consumer plastics when reprocessing a variety of materials is required (primary reuse). Plastics can be separated using optimum procedures so that various types of waste can be recycled separately, thus it is possible to identify the availability of dangerous compounds and use them as additional raw materials for reuse. The most widely used BFR is TBBPA (and its variants), which is added to thermoplastic polyesters (PET/PBT), high-impact polystyrene, polystyrene, and ABS as an additive or polymerized as just a reactive addition to polycarbonate resins (PC) and epoxy.

RoHS forbids the use of brominated flame retardants, mercury, hexavalent chromium, cadmium, and lead in electronic and electrical appliances

(Tansel, 2017). Cadmium, octabromodiphenyl ether (octa-BDE), and/or decabromodiphenyl ether (deca-BDE) are present in polymers used for both big and little home products at levels that are higher than those allowed by the WEEE and RoHS directives of the European Union. The polymers most often found in office supplies, PC displays, and computers include HIPS, ABS/PC-blend, and ABS.

The percentages of use of the different types of plastics in biomedical applications are illustrated in Figure 10.5. Examples of the main plastics used in biomedical applications are summarized in Table 10.2.

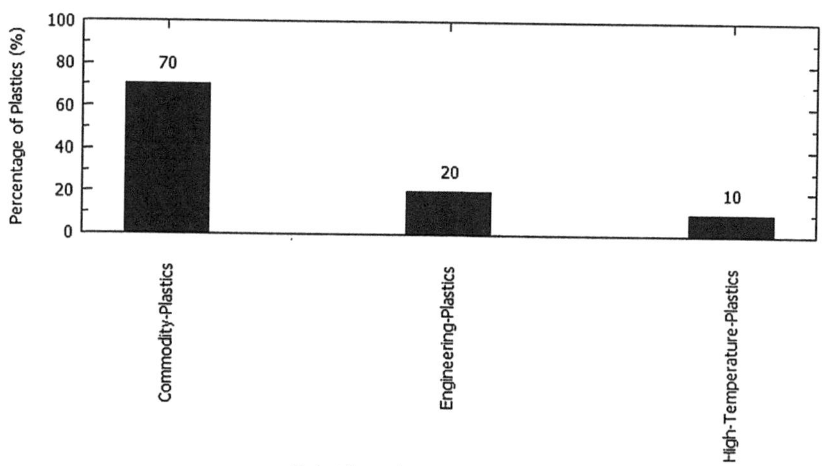

FIGURE 10.5 Percentages of use of the different types of plastics in biomedical applications.

TABLE 10.2 Main Plastics Used in Biomedical Applications

Commodity Plastics	Engineering Plastics	High-Temperature Engineering
Polyethylene (PE)	Polyamides (PA)	Polyimides (PI)
Polypropylene (PP)	Polyesters (PES)	Polyetherimides (PEI)
Polystyrene (PS)	Polycarbonates (PC)	Polysulfones (PSU)
Polyvinyl chloride (PVC)	Polyurethanes (PU)	Poly(ether-ether ketone) (PEEK)

10.5.3 RARE-EARTH ELEMENTS (REES)

Only a few businesses worldwide are actively engaged in REE recovery across all supply chains (Mudali et al., 2021b). Additionally, most of this recycling is pre-consumer discard, and well is hardly any action in the post-consumer reuse of e-waste. Japan is now pioneering the recycling of rare earth metals. The most developed method in terms of applicability at the industrial scale is the recuperation of REEs from phosphors in fluorescent lighting. For recycling rare earth elements, Honda and Hitachi are now in the lead. REEs are recovered from permanent magnets, phosphor, PCBs, NiMH batteries, and other materials using a variety of hydrometallurgy-based techniques. As a result, treating waste EEE through a hydro-metallurgical process is a feasible solution. E-waste recycling is currently done using an integrated approach to produce REEs that are effective and high purity. These low-energy environmentally friendly methods combine several diverse processes, including pyrometallurgical, hydrometallurgical, and iono/solvo metallurgical processes. In the limited cases where magnets are recycled, recycling rates are somewhat higher. Around 1% of both the rare earth metals retrieved from final disposal objects are successfully gathered and treated utilizing REE separation facilities at the collecting site. Only a few businesses worldwide are actively engaged in REE recovery across all supply chains. Furthermore, the majority of this recycling is pre-consumer trash, whereas there is hardly any action in the recovery of e-waste after consumption.

REEs are usually scattered at moderately low densities across the geological cortex and are not found in consolidated mineral deposits (Brewer et al., 2022). Although more and more extraction activities are starting to develop globally, China's mines and processing facilities predominantly provide the world's demand for REEs. REE pollution can result from mining, industrial use, and waste processing. Later decades of service, REEs have been recognized as hazardous for plants, animals, and people in high enough quantities. One possible source of considerable REE pollution may come from improperly disposed electronic trash (E-waste). Although REE densities are high, achieving 26.7 wt.% in neodymium iron boron magnets, for instance, the main composition of E-waste is quite complex. Nevertheless, most substantial E-waste may not be as consolidated. In theory, REEs may be solubilized from environmental E-waste throughout time in a way commonly found with other heavy metals. Once REEs have been initially released into the environment through leaching upon E-waste or alternative sources, their motion in native systems, and the land-water ecosystem determines how far that pollution spreads. The two main modes of REE motion in land-water systems are dispersed phase as well as colloid-assisted transport, and the distinctions between these two modes could just have a significant effect on REE's probable world contamination occurrence. As a result, environmental factors controlling pH and the presence of all other aqueous or floating groundwater components can influence REE differentiation and distribution, and are anticipated to have a significant impact on transport in soil environments. The behavior of rare earth elements is one of the potential metal contaminants that could be affected by anthropogenic or naturally occurring nanoparticles (NPs) in the environment.

Due to the following issues, there is very little REE recycling: (i) technological challenges; (ii) a lack of incentives; (iii) ineffective final disposal object accumulation; (iv) final disposal rare earth are frequently found in hard-to-disintegrate compounds and are frequently found in components buried deep inside consumer items; and (v) rare earth are frequently found in final disposal products at very low quantities (Mudali et al., 2021a). Final disposal of rare earth recovery is complicated and costly because of each of these. The main rare earth metals belonging to the group of lanthanides used in biomedical applications that generate electronic waste are outlined in Table 10.3.

TABLE 10.3 Main Rare Earth Elements (REEs) from the Lanthanide Group Used in Biomedical Applications

Rare Earth Element (REE)
Cerium (Ce)
Lanthanum (La)
Neodymium (Nd)
Praseodymium (Pr)
Dysprosium (Dy)
Gadolinium (Gd)
Samarium (Sm)
Erbium (Er)
Ytterbium (Yb)
Lutetium (Lu)

10.5.4 GLASS

Local anesthetics are kept in the world in glass bottles because they are recyclable (Gaiser et al., 2004). Recycling would lead to a decrease in hospital solid waste. Glass recycling is beneficial to the environment. Recycling glass uses less energy than making new glass. The use of fresh raw materials is not necessary. Only 2.9 million tons of the 12.9 million tons of glass produced in the US in 2000 were recycled. For unknown causes, recycling of glass

(23%) occurs far less frequently than recycling paper (45.4%). Bottles made of glass can only be recycled. Because the glass in ampoules is of lower grade, they cannot be recycled in the US. These ampoules would contaminate the recycled glass and make it unusable if they were put in the recycling bin. Glass recycling is possible in a crowded labor and delivery room. It enhances housekeeping safety and has a favorable effect on the environment. The hospital made a small financial gain by not putting the glass in the sharp container, but the gain was not substantial.

10.5.5 NANOMATERIALS

Different nanomaterials (NMs) have been developed for use in a variety of biological processes, such as medication delivery, immunization, imaging techniques, and biosensors (Elkodous et al., 2021). Toxicology continues to be the principal issue limiting their usage, nevertheless. Nanomaterials recovered from diverse waste types provide a groundbreaking way for not only preventing severe environmental impacts but also implementing circular economy principles, which are critical to attaining long-term growth. Recyclable NMs also have been employed as a risk-free, ground-breaking alternative with remarkable promise for a variety of biological applications.

Among all the possible uses for these molecules, the biological applications of NMs are of special interest and have contributed to the development of nanomedicine (Elkodous et al., 2021). Different NM types have been created, and their application has undergone a thorough investigation. When contrasted with other pharmaceutical chemicals and compounds, NMs have shown the capacity to overcome numerous significant limitations, including physiological instability, absence of water solubility, and *non-vivo* ineffectiveness. NMs are used in a variety of biological applications, including vaccines, biosensors, drug delivery systems, early disease detection, and bioimaging.

Electronic garbage, or e-waste, is a fast-expanding global issue (Elkodous et al., 2021). In European countries, 17 kg of e-waste is generated per individual each year. About 1 kg of e-waste, which contains Cu, Al, and ferrous metals, is produced annually per person in India and China. The amount of waste that reached US landfills in 2000 was 4.6 million tons, and the US Environmental Protection Agency estimates that it will more than double over the next four years.

Recycling a wide range of industrial wastes offers an inexpensive but also secure approach to producing CuO nanoparticles (NPs; Elkodous et al.,

2021). CuO NPs have also been recovered from Cu-rich industrial waste such as waste sludge, wastewater, batteries such as LIBs and NiMH batteries, and e-waste such as SIM cards and motherboards.

Nanotechnology uses on a large scale are growing quickly (Pati et al., 2016). The rising manufacturing of nanomaterials and nano-enabled goods, as well as their subsequent disposal, increases the number of nanomaterial wastes (i.e., nano-waste). Numerous nanotechnologies use scarce resources, such as rare earth elements and valuable metals, which ultimately become nano-waste. To make nanotechnology more sustainable, it is crucial to devise methods for recovering these valuable but scarce elements. The development of future nano-waste management must take the life cycle into account. Furthermore, technological advancements made to increase sustainability may come with unintended environmental costs. Identifying and analyzing these hidden effects as well as the general sustainability of a product or a process can be done using life cycle assessment (LCA), a quantifiable approach used to evaluate the cumulative environmental consequences associated with all stages of material use, from the extraction of the raw materials (cradle) through the end-of-life (grave).

Innovative techniques for recycling nano-waste might have unintended negative environmental effects that can only be found by developing thorough LCAs (Pati et al., 2016). For instance, an LCA of the synthesis of gold nanoparticles revealed that, despite appearing to be environmentally friendly on the surface, the use of benign chemicals can have considerable life cycle implications. A revolutionary process for extracting gold from effluent ash "eliminates the water demand" throughout the recovery of metals, according to recent research. The process demands temperatures of roughly 800°C for such optimum gold recovery. Because high-temperature operations frequently use high-pressure steam and a lot of fuel, they typically have large water footprints. Such indirect water uses may cancel out any decrease in the use of freshwater during recycling and recovery. It is critical to address possible environmental impacts from such a life cycle perspective while developing innovative nano-waste management technologies. By integrating novel recycling systems with LCA modeling, it may be feasible to create a new type of closed- and open-loop processes for environmental nanotechnology.

Recently, there has been a rise in research on collecting and recycling valuable metals and REEs from manmade industrial effluents, such as those produced by smartphones, hard disks, PCBs, LCD screens, discarded fluorescent lamps, and wastewater ash (e.g., seawater; Pati et al., 2016). The anticipated markets for nano-based goods suggest that nano-based applications could result in a significant consumption of expensive materials overall, even though a particular nano-enabled usage or product may only entail trace quantities of a specific metal or REE. As a result, it is critical to develop early-stage material reuse and recovery procedures that integrate with nano-manufacturing technologies, such as the one mentioned in this reference.

Even though items incorporating engineered nanoparticles have been utilized in commerce for roughly the last 21 years, disposing of them and the effects that follow are anticipated to provide a serious environmental concern in the future (Nichols, 2016). There are similar issues with electronic trash (e-waste) and a rising worry involving carbon fiber-reinforced polymers (CFRP) garbage. However, no standard method for handling nano-waste (purely designed nanomaterials, materials containing manufactured nanomaterials, or items contaminated with nanomaterials) has been developed, owing to the lack of a unified motivating factor in the nanoscience community to force standardization. The e-waste problem has lately gotten a lot of attention, and more emphasis is now being focused on CFRP trash, particularly with recycling initiatives. The lessons learned from circuitry and carbon composites can be applied to nanomaterials, which will help us understand how to cope with nano-waste.

Engineered nanoparticles are widely used in several industrial processes, which has led to both unique and exciting difficulties as well as concerns (Nichols, 2016). These include describing their life cycle, tracking their destiny and movement in the ecosystem, establishing regulatory criteria, and comprehending the risk to human health. Recent changes in the Toxic Substances Control Act (TSCA) involving nanoparticles may include a methodology for the characterization of potential nanomaterials risk. However, two of the least understood aspects of using them are dealing with the waste generated by products integrating designed nanomaterials and treating the natural nanomaterials themselves when they are discharged into the environment. Pure nanomaterials, items contaminated with nanoparticles, and things that have reached the end of their useful lives and contain nanomaterials are all considered to be nano-trash. The fact that not all nano-waste is the same complicates the situation even more. Nano-waste includes a variety of substances, including carbon nanotubes (CNTs) and quantum dots, each having unique physicochemical characteristics and toxicological profiles. These resources are not all the same, even though they are sometimes grouped under the same heading. Understanding the creation and disposal of nano-waste has emerged as a byproduct of nanotechnology-related operations. To emphasize the overall difficulty rather than go into detail on specific particle properties, the term nano-wastes will be used to refer to all components that can be found in these waste streams.

Dealing with nano-waste has three challenges (Nichols, 2016). First, 1.5 million tons of trash are predicted to be generated annually around the world. The enormous volume of waste created makes it difficult to develop suitable storage facilities and methods for sending waste to its final location securely. Second, nano-trash contains substances including silver nanoparticles, titanium dioxide, and zinc oxide that may be hazardous to both human health and the environment. Finally, no all-inclusive or standard strategy is in place to deal with the problem's full scope. Several techniques, including life cycle analysis (LCA) and movement of material analysis, may be used to calculate the volume of trash produced and the path it takes from creation to disposal. However, in the case of nanomaterials, a complete characterization of their behavior has not yet been carried out. These methodologies, in particular life cycle assessment (LCA), entirely depend on the assumption that the materials under observation behave similarly to conventional materials. In theory, a standard regulatory system for governing waste disposal, for instance, the Resource Conservation and Recovery Act, should apply to nano-waste (RCRA). However, due to toxicity studies, these models might not be sufficient to adequately control nanomaterials (such as the toxicity characteristic leaching process) be difficult to use for determining the toxicity of nanomaterials, and some nano-waste may be produced by households, making it exempt from RCRA and other existing regulations.

Even though nanoparticles are a new class of potential contaminants, some precedents can be utilized as a model for understanding why and how to address this issue as soon as possible (Nichols, 2016). Keyboards, cell phones, cathode ray tubes, and some other electrical and electronic devices that already have fulfilled their purpose are known as electronic garbage (or e-waste). Waste carbon-fiber reinforced Plastic (CFRP) cannot be remolded and is an inevitable by-product of employing carbon composites. Excellent stand-ins for electronic trash and CFRP are used to highlight the need for a consistent strategy for treating nano-waste. First, the amount generated by both forms of waste shows how quickly developing garbage types may harm or endanger the environment and human health if allowed unchecked. Second, because these waste kinds are made up of a variety of complex components, as do many manufactured nanomaterials, reducing their toxicity has been and will continue to be difficult. Third, both sorts of trash come from things that are utilized and transported all over the world, which influences the health of everyone. Finally, given that both types of waste may contain manufactured nanoparticles, particularly in more recent product categories, they can serve as a historical link to outline a foreseeable future for problems that will undoubtedly arise with nano-waste. The main materials involved in nano-waste are presented in Figure 10.6.

Rare earth element (REE)
Cerium (Ce)
Lanthanum (La)
Neodymium (Nd)
Praseodymium (Pr)
Dysprosium (Dy)
Gadolinium (Gd)
Samarium (Sm)
Erbium (Er)
Ytterbium (Yb)
Lutetium (Lu)

FIGURE 10.6 Main materials found in nano-waste.

One of the newest research areas today is nanotechnology, which draws on information from many various branches of science and technology, including those related to industry, the environment, and health (Patil and Bohara, 2020). It is developing technology for creating unique materials with entirely new features. Different physiochemical, optical, and electrical qualities of manufactured goods are used to address a range of issues that cannot be resolved by conventional technology. Nanoparticles (NPs) are essential to developing waste management, wastewater treatment, and effluent water. Because they offer ways to remove pollutants from other contaminated media as well as organic impurities in wastewater and water, NPs appear to be highly promising. Organic pollutants, medicines, phenols, anionic and cationic dyes, metalloids, and heavy metals are removed from wastewater before it is dumped into nearby bodies of water, NPs are frequently utilized when it comes to the treatment, degradation, and remediation of aquatic environmental contaminants. The benefits of NP synthesis are becoming incredibly important in many areas of research and technology that are concerned with creating an eco-friendly environment. The community and connected employees are therefore very interested in raising awareness about waste management, source reduction, waste isolation, appropriate waste transportation, and appropriate waste treatment. This will contribute to preserving ecological stability, biodiversity, and the general well-being of the world's population.

10.6 CONCLUSION

Despite advances in the handling of electronic waste, it is essential to carry on efforts to improve the environmental quality complicated due to the presence of hazardous materials that can be generated during the recovery process and the reduced percentage of recycling of the useful materials involved. The most secure method for handling electronic waste is formal recycling. It is urgent to develop new strategies at the design level to exploit the recycling of electronic waste in a more efficient way to increase the reuse of materials that have been previously used to reduce the energy required for their processing and, above all, all to take care of the planet environmentally by not depending on unprocessed raw material. Recycling of metals, polymers, nanomaterials, and rare earths will continue to be necessary as outdated biomedical equipment, and systems generate e-waste, intending to increase the drive to raise recyclability levels is the top priority for materials engineers dedicated to ensuring environmental sustainability.

KEYWORDS

- **biomedical equipment**
- **environmental sustainability**
- **e-waste management**
- **high-temperature engineering**
- **nanoparticles**
- **organic pollutants**
- **phenols**
- **recycling**

REFERENCES

Ahirwar, R., & Tripathi, A. K. (2021). E-waste management: A review of recycling process, environmental and occupational health hazards, and potential solutions. *Environmental Nanotechnology, Monitoring & Management, 15*, 100409.

Baldé, C. P., Forti, V., Gray, V., Kuehr, R., & Stegmann, P. (2017). *The Global E-Waste Monitor 2017: Quantities, Flows, and Resources* (United Nations University (UNU), International Telecommunication Union (ITU), & International Solid Waste Association (ISWA), pp. 40–50). Bonn/Geneva/Vienna.

Brewer, A., Dror, I., & Berkowitz, B. (2022). Electronic waste as a source of rare earth element pollution: Leaching, transport in porous media, and the effects of nanoparticles. *Chemosphere, 287*, 132217.

Butturi, M. A., Marinelli, S., Gamberini, R., & Rimini, B. (2020). Ecotoxicity of plastics from informal waste electric and electronic treatment and recycling. *Toxics, 8*, 99.

Elkodous, M. A., El-Husseiny, H. M., El-Sayyad, G. S., Hashem, A. M., Doghish, A. S., Elfadil, D., Radwan, Y., El-Zeiny, H. M., Bedair, H., Ikhdair, O. A., Hashim, H., Salama, A. M., Alshater, H., Ahmed, A. A., Elsayed, M. H., Nagy, M., Ali, N. Y., Elahmady, M., Kamel, A. M., Elkodous, M. A., Maallem, I., Kaml, M. B. S., Nasser, N., Nouh, A. A., Safwat, F. M., Alshal, M. M., Ahmed, S. K., Nagib, T., E-Sayed, F. M., Almahdi, M., Adla, Y., ElNashar, N. T., Hussien, A. M., Salih, A. S., Mahmoud, S. A., Magdy, S., Ahmed, D. I., Hassan, F. M. S., Edward, N. A., Milad, K. S., Halasa, S. R., Arafa, M. M., Hegazy, A., Kawamura, G., Tan, W. K., & Matsuda, A. (2021). Recent advances in waste-recycled nanomaterials for biomedical applications: Waste-to-wealth. *Nanotechnology Reviews, 10*, 1662–1739.

Gaiser, R. R., Cheek, T. G., & Gutsche, B. B. (2004). Glass recycling in the labor suite is environmentally sound and economical. *British Journal of Anaesthesia, 92*(4), 584–586.

Garlapati, V. K. (2016). E-waste in India and developed countries: Management, recycling, business and biotechnological initiatives. *Renewable and Sustainable Energy Reviews, 54*, 874–881.

Joseph, B., James, J., Kalarikkal, N., & Thomas, S. (2021). Recycling of medical plastics. *Advanced Industrial and Engineering Polymer Research, 4*, 199–208.

Kumar, U., Goonetilleke, D., Gaikwad, V., Pramudita, J. C., Joshi, R. K., Sharma, N., & Sahajwalla, V. (2019). Activated carbon from e-waste plastics as a promising anode for sodium-ion batteries. *ACS Sustainable Chemistry & Engineering, 7*(12), 10310–10322.

Lubell, I. (2019). 3 top concerns about e-waste in the healthcare industry. *Ewaste-Blog, Mayers Alloys.*

Mudali, U. K., Patil, M., Saravanabhavan, R., & Saraswat, V. K. (2021a). Review on e-waste recycling: Part I–A prospective urban mining opportunity and challenges. *Transactions of the Indian National Academy of Engineering, 6*, 547–568.

Mudali, U. K., Patil, M., Saravanabhavan, R., & Saraswat, V. K. (2021b). Review on e-waste recycling: Part II–Technologies for recovery of rare earth metals. *Transactions of the Indian National Academy of Engineering, 6*, 613–631.

Murthy, V., & Ramakrishna, S. (2022). A review on global e-waste management: Urban mining towards a sustainable future and circular economy. *Sustainability, 14*, 647.

Nichols, G. P. (2016). Exploring the need for creating a standardized approach to managing nanowaste based on similar experiences from other wastes. *Environmental Science Nano, 3*(5), 946–952.

Pati, P., McGinnis, S., & Vikesland, P. J. (2016). Waste not want not: Life cycle implications of gold recovery and recycling from nano waste. *Environmental Science Nano, 3*(5), 1133–1143.

Patil, P. M., & Bohara, R. A. (2020). Nanoparticles impact in biomedical waste management. *Waste Management & Research, 38*(11), 1189–1203.

Patil, R. A., & Ramakrishna, S. (2020). A comprehensive analysis of e-waste legislation worldwide. *Environmental Science and Pollution Research, 27*, 14412–14431.

Tansel, B. (2017). From electronic consumer products to e-wastes: Global outlook, waste quantities, recycling challenges. *Environment International, 98*, 35–45.

CHAPTER 11

Microbial Strategies for the Biodegradation of Pharmaceutical Pollutants

AMOL D. GHOLAP,* RITESH A. KHAIRNAR, SAMPADA D. SAWANT, and TEJAL D. PARAB

Department of Pharmaceutics, St. John Institute of Pharmacy and Research, Palghar, Maharashtra, India

*Corresponding author.

ABSTRACT

Pharmaceutical medicine is helping human beings with the diagnosis and treatment of ailments. One of the great challenges associated with the pharmaceutical sector is the proper disposal of expired medicines. Several methods are implemented to take care of this issue including incineration, mechanical disruption, chemical sterilization along with alkaline hydrolysis method. However, the management of pharmaceutical active pollutants is still present as a major issue and poses a serious threat to water bodies along with the terrestrial ecosystem. The self-purification and degradation ability process of the microbes can be used to counter the recalcitrant as well as therapeutic pollutants. Such microbes utilize the pollutants for their carbon source and are able to transform the contaminants with the help of the co-metabolic pathways. The degradation of the target contaminants can be done by implementing native microorganisms, bioaugmentation (BA) along biostimulation. For effective neutralization efficiency genetically, engineered microbes can be used for assistance in the pollutant's mineralization process. The detoxification of the pollutants through microbes can be able to enhance biodegradation (BD) and improve water quality standards for human benefit.

Biomedical Waste Management: Recycling and Remediation Strategies. Younis Ahmad Hajam, Javid Ahmad Malik, Anu Pandita, and Deepu Pandita (Eds.)

11.1 INTRODUCTION

Pharmaceutical medicines are used in the treatment of public health and their degradation is turning into environmental issues in the form of pollutants. The biggest global health challenge is to manage used and expired pharmaceuticals, personal care items, and over-the-counter medications. There are various ways through which pharmaceutical contaminants in the environment. Most of the time, the unmetabolized medical Excipients are disposed of in sewage for conventional or biological treatment plants. This method cannot be used for the treatment of water contaminated with traces of medication and it may be considered ineffective in removing them. Highly undesirable and toxic technologies used for waste treatment are incineration, mechanical disruption, encapsulation, chemical sterilization, and alkaline hydrolysis. Aquatic and terrestrial ecology are seriously threatened by improper disposal of contaminated used and expired pharmaceutical products. The management of hazardous, and non-hazardous, chemo waste and conservation of the environment from pharmaceutically active pollutants is a great challenge for various regulatory bodies. Even due to long-term exposure to these active pollutants, the very minimal level affects the aquatic and surface habitat. The most commonly used packaging materials for pharmaceuticals are mainly made up of plastic and metals which are difficult to dispose of and recover (Barnes et al., 2008; Teixeira et al., 2016; Woolridge and Hoboy, 2019).

For a safe environment, it is necessary to remove the pharmaceutical residues and biocides including antibiotic, analgesic, antimicrobial, anticonvulsant, beta-blockers, antihistamines, lipid regulators, antipsychotic, genotoxic waste, Infectious, pathological waste, synthetic hormones, and bulk pharmaceutical contaminants. Limited scope, high cost, and time-consuming are some major limitations of conventional treatment methods. Just like pharmaceuticals, microorganisms are also major organic matter and xenobiotics degraders which help to maintain the proper functioning of the ecosystem. After exposure of pharmaceuticals to the ecosystem, Ecotoxicity is determined by various processes such as biodegradation (BD), photodegradation, hydrolysis, and sorption into suspended solids and bed sediment. The toxic pharmaceutical compounds from the environment can be effectively removed by using processes such as photodegradation and microbial-mediated degradation. The rate of degradation depends on temperature, pH, richness of microorganisms, amount of biosolids, and Lipophilicity of pharmaceuticals. The most significant dissipation way for degradation of pharmaceuticals inside the environment is that of microbial degradation.

Microorganisms degrade pollutants by metabolic and co-metabolic pathways or enzymatic pathways as a result they are vital for the regulation and protection of the ecosystem.

Recovery of contamination is only possible if there is absence of microbial obstruction by the harmful effect of molecules (Liang et al., 2013; Singh, 2012; Wang et al., 2015; Yamamoto et al., 2009). The various microorganisms which reduce the toxic effects of pharmaceutical active contaminants by using certain co-substrate-like carbon sources include *Rhodococcus rhodochrous*, *Pseudomonas putida*, *Pseudomonas fluorescens*, *Bacillus subtilis*, *Aspergillus niger*, *Sphingomonas herbicidovorans*, white rot fungus *Phanerochaete chrysosporium* and genera Pseudomonas, Arthrobacter, and Enterobacter. The degradation of Ibuprofen and Diclofenac has been demonstrated by various studies involving a series of batch cultures of aerobic and unsaturated compounds. Exploring various advancements in microbial techniques has a wide range of benefits for the degradation of bioactive solid and liquid pharmaceutical pollutants due to their eco-friendly behavior. Some reviews largely focus on essential aspects and application of microorganisms as biodegraders to complete mineralization of pharmaceuticals active as hazardous chemicals (Monteiro and Boxall, 2010; Fang et al., 2012; Tiehm et al., 2011).

11.2 MECHANISM OF BIODEGRADATION (BD)

Hydrolysis and biological degradation are two main forms of BD. This method is further classified based on sources used, such as enzymes and bacteria. Microorganisms transformed xenobiotics through growth and cometabolic processes. For the complete mineralization of organic contaminants growth process uses organic pollutants as the source of carbon energy. For degradation of active pollutants various bacteria, fungi, yeast, and certain protozoa are involved. The primary benefit of this process is that it enhances the complete mineralization of organic compounds into carbon dioxide along with water by aerobic and anaerobic processes.

The three phases of biodegradation (BD) include:

1. Natural attenuation involves the reduction of pollutants by natural microorganisms;
2. Biostimulation is a technique to improve the biodegrading efficiency by providing nutrients and oxygen to the process; and
3. Bioaugmentation (BA), where more efficient supplementary microorganisms are added to the system in addition to natural microflora to target the specific contaminants as presented in Figure 11.1.

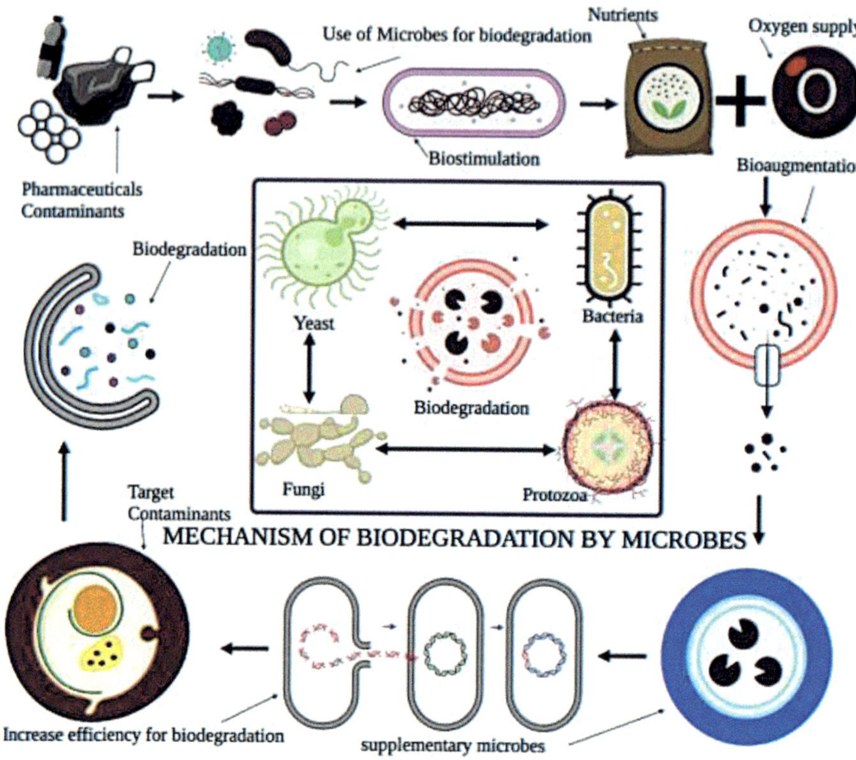

FIGURE 11.1 The mechanism of biodegradation through microbes involves biostimulation along with bioaugmentation processes. The supplementary microbes are used to be added in this process to increase the efficiency of biodegradation.

The rate along with the extent of degradation influenced by various factors such as pH, temperature, nitrogen, and phosphorus content in micro-organisms whether they use organic pollutants as substrate or cometabolite them (Tahri et al., 2013).

11.3 MICROBIAL DETOXIFICATION OF HEAVY METALS

Biodegradation (BD) is a vital process that essentially uses natural flora to transform active therapeutic pollutants like pharmaceutical and medical wastes. Microorganisms are widespread throughout the biosphere and may develop in a broad variety of environmental conditions. The area of BD pollutants the dietary and adaptability of different microorganisms has been explored. Even though modern society is inevitably undergoing industrialization and technological advancement, past few decades there have been multiple ecological and human catastrophes that have led to the industry's significant anthropogenic role for pollutants.

Industrial discharge materials have a great impact on the quality of local water because of the presence of toxic hazardous materials like lead, cadmium (Cr), and Mercury (Hg) along with toxic chemicals. Such toxic chemicals include organic chemicals such as pesticides, and dioxins along with polyaromatic hydrocarbons (PHAs), and so forth, which have a detrimental effect on human health. The enzymes along with cellular proteins react with high concentrations of metals to form complexes causing deleterious effects by the persistence of heavy metals inside the environment along with their toxic nature. The nucleic acids are also involved in it. The toxic heavy metals bioconcentrate, bio-accumulate, and then Biomagnify causing the thread to the environment. Tropic transfer of four major hazardous metals Cd, Chromium (Cr), Copper (Cu), and Mercury (Hg). Since fish is the primary source throughout the world, the resulting bioaccumulation of these hazardous heavy metals inside the fish is of major concern (Akahori et al., 1999; Ali and Khan, 2019; Liu et al., 2019; Nisha, 2015).

The toxicity in humans can be caused by consuming fish containing high levels of heavy metals. If the fish present in the environment belongs to industrial effluent-contaminated water bodies and is consumed, it may lead to a high human health risk (Javed and Usmani, 2011; Rose et al., 2015; Singh et al., 2010). The same scenario applies to heavy metals contaminated crop consumption as food major issues of risk to human health. The irrigation of agricultural fields and sewage wastewater results in the accumulation of heavy metals inside the soil which finally get transferred to food crops like cereals and vegetables along with milk. As heavy metals are transferred to biotic areas from abiotic environments, they accumulate in organisms at several tropic levels of the food chain. Significant effects on wildlife and human health can be seen by trophic transfer along with bioaccumulation. The biomagnification of such hazardous heavy metals is also playing a major role in the nutrition network.

Even at small concentrations some heavy metals there was an issue of toxicity due to their cytotoxicity along with carcinogenicity. Such cases are also involved with mutagenicity. Studies have shown that heavy metal accumulation inside human beings has induced the production of superoxide and further generation of reactive oxygen species. Such compounds are producing detrimental effects into the same as presented in Figure 11.2.

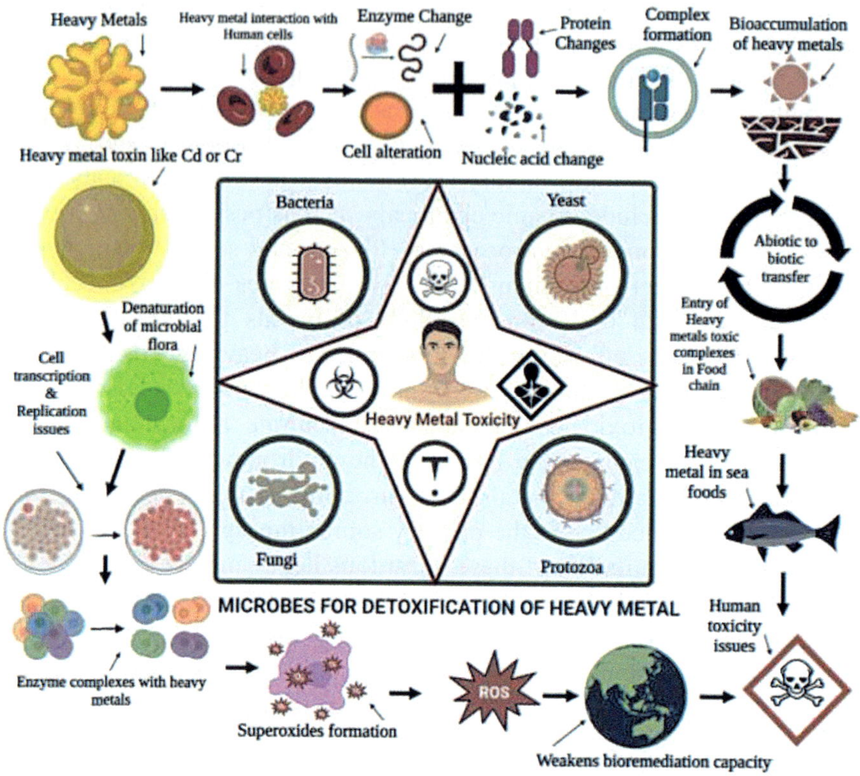

FIGURE 11.2 The contribution of microbes to the detoxification of heavy metals is required to decrease toxicity to human health induced by the formation of reactive oxygen species, which further weakens bioremediation capacity.

The major concentration of metals displaces the several enzyme-related metal ions and engages in cell activities by competing with structurally related nonmetals, the interaction of such microbes along with heavy metal functional groups inside the Biomolecules for the cells are obstructed as compared to the high concentration of single metal, metals with the combination have more detrimental effects to that of microbial diversity in the environment. The complex formation by a combination of metals generates hazardous compounds that may affect the normal functioning of organisms as per the studies done on soil bacteria the particular strains of bacteria that are resistant grow, however, short-term exposure to heavy metals can result in the selection along with growth for bacteria which are resistant type within weeks. Some metals like Cr along with Cd have the power to reduce efficiency for bioremediation. They also can denaturing action on microbial flora. It has been discovered that Cr (III) reacting along with carboxyl as well as thiol groups affects transcription and replication within the cell, affecting the enzyme structure along with the cationic Cr (III) complexes generated within.

DNA's anionic phosphate groups interact with the cell. The superoxide radicals have the ability to destroy the DNA's structure get stabilize it. Heavy metals that interact with substrates in competitive or non-competitive ways may change the enzyme configurations through allosteric effects, which could have an impact on the essential enzymatic functions. With some heavy metals like cadmium and lead, the interaction for ligands along with metal displacements from native binding sites damages the microorganism's DNA structure and cell membranes (Dixit et al., 2015; Hetzer et al., 2006; Mench et al., 2006; Tchounwou et al., 2012). There is a significant change in the nucleic acid which has an impact on the physiology of microorganisms, altering their growth, metabolism, and shape as well as disrupting cell membranes and limiting enzyme functioning. Microbial Heavy Metals Remediation Early 1980s studies of interactions between *Bacillus subtilis* and several metals, like copper, iron, and manganese revealed that discrepancies between the two parameters. One is the net anionic charge present on the bacteria while the second one is the net positive charge of metals which helps the bacterial cell to interact with the metals. The nucleation sites on the cell surfaces of the microorganisms were able to attach to the metals with the opposite charges, which ultimately led to the metals precipitating on the cell wall. The notion of bioremediation has been further investigated as a result of studies carried out by the U.S. Geological Survey (USGS) that showed interactions between specific species of bacteria along with fungi to that of metals and other harmful substances.

Genetically engineered microorganisms (GEM) are used in bioremediation to a great extent as they are able shoe degradation into the environmental toxins along with the bound metals.

The kinetic equilibrium along with metal composition plays a very important role in this process. The binding pattern of toxins is generally involving adsorption or absorption on the cellular surface binding sites. So many reactions like electrostatic interaction along with precipitation are playing a major role in it. Several other processes like ion exchange and surface complexation are also important for the same. All these factors are important for the biosorption process. Biosorption can occur when cellular fragments, dead or alive biomass, are passively absorbed onto the cell wall and other outer layers through surface complexation. The cell metabolic cycle has no bearing on this mechanism. The bioaccumulation process involves the entry of heavy metals into the plasma membrane. Once the heavy metals get into the cytoplasm through active uptake then several other processes of cells like chemical process, biological processes are used to get changed. Such processes involve changing the toxic materials into less harmful components. The interactions between bacteria and their environment are greatly influenced by their surface features. While gram-negative bacteria have layers of lipopolysaccharide, phospholipids, and peptidoglycan in their cell walls, gram-positive bacteria also have layers of peptidoglycan and teichoic acid, with the latter making up as much as 90% of the cell wall (Bissen and Frimmel, 2003; Cervantes et al., 2001; Fashola et al., 2016; Igiri et al., 2018; Rajbanshi, 2009; Sankarammal et al., 2014).

11.4 PHARMACEUTICALS IN THE ENVIRONMENT

Pharmaceuticals tend to save lives but the disposal of pharmaceutical waste and its by-products is a growing concern to the environment. The harmful effect of such compounds on habitats can be detected by using advanced analytical techniques such as GC-MS/MS, LC/MS/MS, and UPLC/MS. These techniques determine certain adverse issues of pharmaceuticals in the environment present in micro- and nano-gram quantities because of their extended infusion into the ecological matrix and destruction by various mechanisms these therapeutically active compounds are termed as pseudo persistent which leads to the creation of complex pharmaceutical pool since wastewater treatment plants are set up for handling organic products in the range of milligram which degrade quickly and medication are not entirely removed. Pharmaceutical products are differentiated from most other pollutants by:

1. Their molecular masses which is less than 500 Da;
2. Chemically complicated molecule with a variety of forms, sizes, molar masses, and functions;
3. Polar compounds which have more than one group of ionizable substances;
4. pH-dependent ionization properties;
5. Lipophilic;
6. Mild water solubility;
7. These molecules tend to adsorb and disseminate throughout a living body which biologically alters their chemical structure (Daughton, 2004; Larsson, 2014; Le-Minh et al., 2010; Lipinski et al., 2001; Stackelberg et al., 2004).

There are tons of active pharmaceutical substances in the environment, which result from "over the counter" medication impact on consumption and usage estimations. Therapeutic compounds that enter the marine environment include antibiotics, anti-inflammatory agents, analgesics, antiepileptic agents, antidepressants, lipid-lowering agents, antihistamines, beta-blockers, and so forth. When wastewater treatment and sewage treatment plants (STPs) are inefficient, microbial BD of medicines is particularly essential (Bush, 1997; Kümmerer, 2009; Küster and Adler, 2014; Rivera-Utrilla et al., 2013; Tahar et al., 2013; Ziylan and Ince, 2011).

11.5 PROPERTIES OF NATURAL MICROBIAL GROUPS IN THE BIODEGRADATION (BD) OF PHARMACEUTICALS

The natural microbial communities involving soil and water are ineffective in the various activities that monitor ecosystem integrity and control the fate of contaminants introduced into the environment; they offer ecosystem service that is known as regulation. Microorganisms help with the degradation of contaminants through metabolic and co-metabolic pathways hence they are involved in the development of ecosystem self-refinement. The elimination of several xenobiotic compounds is supported by pharmaceutical BD. This is considered to be the most crucial mechanism. Contamination recovery happens when molecule toxicity does not interfere with microbial activity. Pharmaceuticals contain numerous materials with fluctuating physical and chemical properties that are used in biochemical and environmental processes. Complex effluents from pharmaceuticals naturally enter the environment and thus a mixture of active substances is detected by natural microbial communities. This review report summarizes studies on the impact of biocides which includes various drugs like ibuprofen, gemfibrozil, and some antibiotics like sulfamethazine on natural microbial communities (Bueno et al., 2012; Gerbersdorf et al., 2011; Lahti and Oikari, 2011; Luo et al., 2014; Megharaj et al., 2011; Reed and Martiny, 2007; Rodarte-Morales et al., 2012a; Venkatesan et al., 2012).

11.6 MICROBIAL TRANSFORMATION OF PHARMACEUTICALS

Due to some factors like high surface-volume ratio, reduction in time of biomass transformation, higher metabolic rate, and easy maintenance of sterility, Microbial cells are considered to be best for the transformation of pollutants. Various factors like the incubation period of microorganisms and the chemical properties of pharmaceutical products influence the biotransformation process (Coelho and Ribeiro, 2015; Hegazy et al., 2015).

11.6.1 ANTIBIOTICS

Some naturally occurring bacteria can biodegrade the active pollutants but most of the pharmaceuticals are toxic to bacteria. Antibiotics that are used to inhibit and treat microbial infection are known as antimicrobial agents. These are secondary metabolites which are originated from bacteria and fungi or synthetic and semi-synthetic processes. The use of antibiotics and their distribution in aquatic and terrestrial ecosystems may increase their resistance to microbes. In addition, the absence of organic matter and chemical catalytic bacteria in soil and wastewater treatment facilities may be another significant issue with antibiotics. The microbial transformation of conventional antibiotics can be performed to develop novel and updated antibiotics with improved characteristics such as minimal toxicity, wide antimicrobial range, improved oral absorption, and less resistance effect. Sulfamethazine in soil can affect natural microbial population and soil respiration (Fair and Tor, 2014; Kotzerke et al., 2008; Polianciuc et al., 2020; Reis et al., 2014; Tappe et al., 2013). According to the study sulfamethazine and pyrimidine molecules can be mineralized with the use of gram-positive microbacterium species. *Microbacterium lacus* strain SDZm4 can partially mineralize sulfadiazine which is a sulfonamide antibiotic. The drugs belonging to the class sulfonamide such as sulfasalazine, sulfamethazine, sulfamethoxypyridazine, sulfapyridine, sulfadimethoxine along sulfathiazole. The concentration of sulfamethizole and sulfamethoxazole was decreased by 20% in *Rhodococcus rhodochrous*. Antimicrobial efficiency with reduced side effects can be enhanced by lankacidin-C-14-butyrate which is a bio-conversion product of lankacidin C produced from *Bacillus megaterium*. The enzyme of white rot fungus can degrade various sulfonamides, ciprofloxacin, and norfloxacin. Sulfapyridine, sulfathiazole, and ofloxacin can be eliminated by the fungi *Trametes versicolor*. The fluoroquinolone drug ofloxacin undergoes 80% degradation when treated with *Trametes Versicolor* in an Erlenmeyer flask. Transformation of enrofloxacin and other antibiotics can be performed by Basidiomycetes indigenous. The degradation of clarithromycin in 72% of cases was by *Trichoderma harzianum* by a co-metabolic process. *Labrys portucalensis* performs biotransformation through cleavage of the piperazine ring and replacement of fluorine substituent which further leads to the formation of an intermediate with less antibacterial toxicity. The biotransformation of norfloxacin to their metabolites like 8-hydroxynorfloxacin, 6-defluoro-6-hydroxynorfloxacin. The efficiency of contaminant removal in wastewater treatment plants can be improved by using these microorganisms in the bioaugmentation process (Buchicchio et al., 2016; Gros et al., 2014; Kim et al., 2011; Prieto et al., 2011; Rodríguez-Rodríguez et al., 2012; Rodarte-Morales et al., 2012a).

11.6.2 ANTICONVULSANT AND ANTIEPILEPTIC

Gabapentin and carbamazepine are difficult to metabolize in the digestive system and hence their biological excretion rate is too high. As they are highly prescribed drugs for the treatment of seizures thus it's concentration in surface and groundwater is too high. This problem has led to increased worries about aquatic ecosystems. But in the presence of glucose, *Rhodococcus rhodochrous* along with *Aspergillus niger* have reported maximum degradation of carbamazepine. Carbamazepine is biodegraded in liquid medium by *Cunninghamella elegans*. In labs, *Phanerochaete chrysosporium* was kept in a non-sterile bioreactor for 100 days about 60–80% of carbamazepine was removed. About 57% of degraded carbamazepine was removed by *Trichoderma harzianum* by using a co-metabolic process and optimal mineralization was seen by aerobic degradation done by *Trametes versicolor* in six days. With this method, the rate reached 95.6% with some major by-products such as acridine, acridone, 10,11-epoxy-carbamazepine, and 10,11-dihydro-10,11-dihydroxycarbamazepine. *Bjerkandera* species manganese peroxidase is not capable of degrading carbamazepine, but the fact changes with the addition of Mn^{2+} and glucose peptone addition in the medium. *Umbelopsis ramanniana*, *Cunninghamella elegans,* and *Pleurotus ostreatus* were also capable of degradation. About 28% of *Scenedesmus obliquus* and 35% of *Chlamydomonas mexicana* a type of microalgae had eliminated 1 mg/L of carbamazepine but at higher concentrations, the algal growth is inhibited.

A widely used drug as an anti-seizure and for bipolar disorders that is gabapentin has been detected majorly in surface water and sometimes in groundwater in concentration in tens of micrograms per liter. Gabapentin was mostly found in groundwater and soils. It was a molecular indicator for anthropogenic contamination of water bodies as its sediments while treating wastewater. It was found to accumulate from year to year in soil as studies had proved gabapentin is highly resistant to microbial degradation in soil. *C. Elegans* ATCC 9245 which is a type of fungi is capable of BD in liquid culture. *T. versicolor* and *G. lucidum*. The report says that the Bacterium *Rhodococcus rhodochrous* and the fungus *Aspergillus niger* can degrade by co-metabolism of gabapentin (100 ppm) in liquid media containing glucose. To conclude gabapentin is persistent in degradation in the environment, though, the microbial population can incompletely degrade. However, its mechanism for complete degradation is still not known. There is decreased biomass of Riverine biofilm due to degrading gabapentin reported which is harmful to aquatic animals as it causes acute toxicity (Golan-Rozen et al., 2011; Marco-Urrea et al., 2009; Xiong et al., 2016; Zhang and Geißen, 2012).

11.6.3 DEGRADATION OF ANTI-DEPRESSANTS

The conditions like anxiety, insomnia, dysthymia, and others are related to antidepressant drugs. Prozac® (fluoxetine HCl), the 1,4-benzodiazepine, Valium® (Diazepam), and their major human metabolites (norfluoxetine HCl, temazepam, and Oxazepam) have shown susceptibility to microbes in bacterial liquid culture from Sewage Sludge (SS). The studies give insights into the degradation of antidepressant drugs. *In-vitro* liquid culture experiment of SS-soil amendment (fluoxetine HCl) carried for about 60 days or extended exposure showed immunity to BD. The investigation proved that Oxazepam which is 1,4-benzodiazepine undergoes biotic transitions. About 80% loss was reported in biotic as well as abiotic medium. Prozac which is an antidepressant containing fluoxetine classified under trifluoromethylphenyl moiety is majorly prescribed and fluorinated antidepressant citalopram undergoes microbial degradation by *Bjerkandera* sp. R1, *Bjerkandera adjusta,* and *Phanerochaete chrysosporium*. The above species completely degraded citalopram of concentration 1 mg/L for 14 days treatment time but only 46% of fluoxetine was degraded. In concentration, 2 M of substrate bacterium *L. portucalensis* F11 can degrade fluoxetine giving the molar quantity of fluoride ions in the presence of a carbon source and acetate (Redshaw et al., 2008; Rodarte-Morales et al., 2012b).

11.6.4 LIPID LOWERING AGENTS (FIBRATES)

To reduce the incidents of heart diseases lipid abnormalities and controlling cholesterol levels (low density lipoproteins) are major key factors. In today's date, statins and fibrates are the group of lipids-lowering agents that are majorly prescribed. For treating coronary heart diseases Gemfibrozil, which is a cholesterol-regulating drug is prescribed. It is metabolized in the liver and excreted in urine as a glucuronide conjugate. This excreted metabolite is biodegraded by the fungus *Cunninghamella elegans* within seven days degrading about 97% of the total 10 milligrams of clofibric acid. Fungi *Ganderma lucidum* showed 47% degradation of clofibric acid by giving end products 4-chlorophenolate, 2-hydroxymethacrylic acid, and hydroxy clofibric acid. One of the most commonly prescribed medicines to decrease lipid levels is Atorvastatin (AT) but very less methods of microbial degradation are available. However, some of the researchers studied the function of white rot fungus, that is, *T. versicolor* for AT degradation. Although no studies were done to prove the laccase activities of *T. versicolor* as an important factor for the degradation of pharmaceuticals (Christen et al., 2010; Lawrence et al., 2005; Marco-Urrea et al., 2009; Rodríguez-Rodríguez et al., 2012; Veach et al., 2012).

11.6.5 ANTIHISTAMINES

Allergic diseases were treated using antihistamines for a long period. Histamine is degraded by catalyzing histamine oxidase or histamine dehydrogenase by certain microbes. Bacteria were isolated from salted fish products to degrade histamines which include *Rummeliibacillus stabekisii,* and *Agrobacterium tumefaciens*. Some other microbes are also contributing the same as *Bacillus cereus* and *Bacillus polymyxa*. The utilization of *Bacillus licheniformis*, *Bacillus amyloliquefaciens* along with *Bacillus Subtilis* and *Bacillus licheniformis* is also involved in it. The highest activity in the degradation of histamines was reported by *B. polymyxa* in liquid broth media by producing histamine dehydrogenase. Around 100% degradation was reported by *Lactobacillus plantarium* D-103. Histamine and tyramine are degraded by *Staphylococcus xylossus* (Kung et al., 2017; Lee et al., 2015).

11.6.6 BETA BLOCKER

Beta-adrenergic blockers (BAB) are the type of drugs mainly used to treat irregular heart rhythms. The most commonly prescribed drugs in treatment are atenolol, metoprolol, and propranolol. Propranolol and Atenolol are degraded by white rot fungus by different ligninolytic enzymes, peroxidase, laccases, and some intercellular enzymes. The contaminants are degraded by hydroxylation, dehalogenation, deamination, and other reactions. Fluidized bed bioreactor containing *Trametes versicolor* microbes for eight days in both sterile as well as nonsterile wastewater (Asgher et al., 2008; Cruz-Morató et al., 2013).

11.6.7 ANTISEPTIC

These are classes of drugs used to prevent the growth of microorganisms also called biocide. The drawback of antiseptic drug therapy is the development of resistance, for example, the case of cross-resistance to clinically relevant antibiotics like Triclosan present as an antiseptic or preservative in cosmetics, household cleaning products along with plastic material, paints, and so forth. As triclosan has low water solubility and high n-octanol water portion coefficient it undergoes BD in sediment and soil. Mainly two bacterial species in liquid media are found to degrade, that is, *Nitrosomonas europaea* and *betaproteobacterial Methylobacillus* sp. of concentration 0.5–2 mg/L. Studies prove triclosan BD in agricultural soils depends on 2% moisture level and types of concentration (0.05, 0.2, or 2 mg/kg). Triclosan is used as a sole carbon source for highly resistant soil bacteria *Pseudomonas putida* TriRY and *Alcaligenees xylosoxidans* subsp. *Dentrificans* TR1 (Carr et al., 2011; Chen and Rosazza, 1994; Murdoch and Hay, 2005; Quintana et al., 2005).

11.6.8 BLOOD LIPID REGULATOR

The main risk factors for cardiovascular diseases are irregularities in lipid levels. The most efficient way to maintain good cardiac health is to lower LDL concentrations of cholesterol. From 1995–2005 volume of cholesterol-lowering drugs increased significantly. Mechanism of action for such drugs are variable but two main classes of drugs were used, that is, statins and fibrates. Since the 1980s patients at high risk of coronary heart diseases have been prescribed with blood lipid regulating agent named Gemfibrozil (an acidic drug). The pharmacokinetics of gemfibrozil is supported by data that it is digested in the liver and excreted in urine by glucuronide conjugation. These conjugates are freely cleaved in the process of sewage treatment releasing the active metabolites. Fungi *Cunninghamella elegans ATCC 9245* is capable of biodegrading gemfibrozil in liquid culture by hydroxylation.

It is found in surface water as it is quite persistent in water that has a half-life of 70 to 299 days. Naturally occurring microbes inhibited growth in the presence of gemfibrozil as 100 micrograms per liter was added in river water microcosms and a negative effect was seen within 3 hr. Such concentration was also effective in wastewater treatment plants and different water toxicological tests (Christen et al., 2010; Lawrence et al., 2005; Veach et al., 2012).

KEYWORDS

- **bioaugmentation**
- **biodegradation**
- **detoxification**
- **lipid regulator**
- **microorganism**
- **pharmaceutical pollutants**

REFERENCES

Akahori, A., Gabryelak, T., Józwiak, Z., & Gondko, R. (1999). Zinc-induced damage to carp (Cyprinus carpio L.) erythrocytes *in vitro*. *IUBMB Life, 47*(1), 89–98. https://doi.org/10.1080/15216549900201083.

Ali, H., & Khan, E. (2019). Trophic transfer, bioaccumulation, and biomagnification of non-essential hazardous heavy metals and metalloids in food chains/webs—Concepts and implications for wildlife and human health. *Human and Ecological Risk Assessment: An International Journal, 25*(6), 1353–1376. https://doi.org/10.1080/10807039.2018.1469398.

Asgher, M., Bhatti, H. N., Ashraf, M., & Legge, R. L. (2008). Recent developments in biodegradation of industrial pollutants by white rot fungi and their enzyme system. *Biodegradation, 19*(6), 771–783. https://doi.org/10.1007/s10532-008-9185-3.

Barnes, K. K., Kolpin, D. W., Furlong, E. T., Zaugg, S. D., Meyer, M. T., & Barber, L. B. (2008). A national reconnaissance of pharmaceuticals and other organic wastewater contaminants in the United States—I) Groundwater. *Science of The Total Environment, 402*(2–3), 192–200. https://doi.org/10.1016/j.scitotenv.2008.04.028.

Bissen, M., & Frimmel, F. H. (2003). Arsenic—A review. Part I: Occurrence, toxicity, speciation, mobility. *Acta Hydrochimica et Hydrobiologica, 31*(1), 9–18. https://doi.org/10.1002/aheh.200390025.

Buchicchio, A., Bianco, G., Sofo, A., Masi, S., & Caniani, D. (2016). Biodegradation of carbamazepine and clarithromycin by Trichoderma harzianum and Pleurotus ostreatus investigated by liquid chromatography—high-resolution tandem mass spectrometry (FTICR MS-IRMPD). *Science of The Total Environment, 557–558*, 733–739. https://doi.org/10.1016/j.scitotenv.2016.03.119.

Bueno, M. J. M., Gomez, M. J., Herrera, S., Hernando, M. D., Agüera, A., & Fernández-Alba, A. R. (2012). Occurrence and persistence of organic emerging contaminants and priority pollutants in five sewage treatment plants of Spain: Two years pilot survey monitoring. *Environmental Pollution, 164*, 267–273. https://doi.org/10.1016/j.envpol.2012.01.038.

Bush, K. (1997). Antimicrobial agents. *Current Opinion in Chemical Biology, 1*(2), 169–175. https://doi.org/10.1016/S1367-5931(97)80006-3.

Carr, D. L., Morse, A. N., Zak, J. C., & Anderson, T. A. (2011). Biological degradation of common pharmaceuticals and personal care products in soils with high water content. *Water, Air, & Soil Pollution, 217*(1–4), 127–134. https://doi.org/10.1007/s11270-010-0573-z.

Cervantes, C., Campos-García, J., Devars, S., Gutiérrez-Corona, F., Loza-Tavera, H., Torres-Guzmán, J. C., & Moreno-Sánchez, R. (2001). Interactions of chromium with microorganisms and plants. *FEMS Microbiology Reviews, 25*(3), 335–347. https://doi.org/10.1111/j.1574-6976.2001.tb00581.x.

Chen, Y., & Rosazza, J. P. N. (1994). Microbial transformation of ibuprofen by a Nocardia species. *Applied and Environmental Microbiology, 60*(4), 1292–1296. https://doi.org/10.1128/aem.60.4.1292-1296.1994.

Christen, V., Hickmann, S., Rechenberg, B., & Fent, K. (2010). Highly active human pharmaceuticals in aquatic systems: A concept for their identification based on their mode of action. *Aquatic Toxicology, 96*(3), 167–181. https://doi.org/10.1016/j.aquatox.2009.11.021.

Coelho, M. A., & Ribeiro, B. D. (2015). *White Biotechnology for Sustainable Chemistry*. Royal Society of Chemistry. https://doi.org/10.1039/9781782624080.

Cruz-Morató, C., Ferrando-Climent, L., Rodriguez-Mozaz, S., Barceló, D., Marco-Urrea, E., Vicent, T., & Sarrà, M. (2013). Degradation of pharmaceuticals in non-sterile urban wastewater by Trametes versicolor in a fluidized bed bioreactor. *Water Research, 47*(14), 5200–5210. https://doi.org/10.1016/j.watres.2013.06.007.

Daughton, C. G. (2004). Non-regulated water contaminants: Emerging research. *Environmental Impact Assessment Review, 24*(7–8), 711–732. https://doi.org/10.1016/j.eiar.2004.06.003.

Dixit, R. C., Wasiullah, Malaviya, D., Pandiyan, K., Singh, U. B., Sahu, A., Shukla, R., Singh, B. P., Rai, J. P., Sharma, P., Lade, H., & Paul, D. (2015). Bioremediation of heavy metals from soil and aquatic environment: An overview of principles and criteria of fundamental processes. *Sustainability, 7*, 2189–2212.

Fair, R. J., & Tor, Y. (2014). Antibiotics and bacterial resistance in the 21st century. *Perspectives in Medicinal Chemistry, 6*, PMC.S14459. https://doi.org/10.4137/PMC.S14459.

Fang, Y., Karnjanapiboonwong, A., Chase, D. A., Wang, J., Morse, A. N., & Anderson, T. A. (2012). Occurrence, fate, and persistence of gemfibrozil in water and soil. *Environmental Toxicology and Chemistry, 31*(3), 550–555. https://doi.org/10.1002/etc.1725.

Fashola, M., Ngole-Jeme, V., & Babalola, O. (2016). Heavy metal pollution from gold mines: Environmental effects and bacterial strategies for resistance. *International Journal of Environmental Research and Public Health, 13*(11), 1047. https://doi.org/10.3390/ijerph13111047.

Gerbersdorf, S. U., Hollert, H., Brinkmann, M., Wieprecht, S., Schüttrumpf, H., & Manz, W. (2011). Anthropogenic pollutants affect ecosystem services of freshwater sediments: The need for a "triad plus x" approach. *Journal of Soils and Sediments, 11*(6), 1099–1114. https://doi.org/10.1007/s11368-011-0373-0.

Golan-Rozen, N., Chefetz, B., Ben-Ari, J., Geva, J., & Hadar, Y. (2011). Transformation of the recalcitrant pharmaceutical compound carbamazepine by Pleurotus ostreatus: Role of cytochrome P450 monooxygenase and manganese peroxidase. *Environmental Science & Technology, 45*(16), 6800–6805. https://doi.org/10.1021/es200298t.

Gros, M., Cruz-Morato, C., Marco-Urrea, E., Longrée, P., Singer, H., Sarrà, M., Hollender, J., Vicent, T., Rodriguez-Mozaz, S., & Barceló, D. (2014). Biodegradation of the X-ray contrast agent iopromide and the fluoroquinolone antibiotic ofloxacin by the white rot fungus Trametes versicolor in hospital wastewaters and identification of degradation products. *Water Research, 60*, 228–241. https://doi.org/10.1016/j.watres.2014.04.042.

Hegazy, M. E. F., Mohamed, T. A., ElShamy, A. I., Mohamed, A. E. H. H., Mahalel, U. A., Reda, E. H., Shaheen, A. M., Tawfik, W. A., Shahat, A. A., Shams, K. A., Abdel-Azim, N. S., & Hammouda, F. M. (2015). Microbial biotransformation as a tool for drug development based on natural products from mevalonic acid pathway: A review. *Journal of Advanced Research, 6*(1), 17–33. https://doi.org/10.1016/j.jare.2014.11.009.

Hetzer, A., Daughney, C. J., & Morgan, H. W. (2006). Cadmium ion biosorption by the thermophilic bacteria Geobacillus stearothermophilus and G. thermocatenulatus. *Applied and Environmental Microbiology, 72*(6), 4020–4027. https://doi.org/10.1128/AEM.00295-06.

Igiri, B. E., Okoduwa, S. I. R., Idoko, G. O., Akabuogu, E. P., Adeyi, A. O., & Ejiogu, I. K. (2018). Toxicity and bioremediation of heavy metals contaminated ecosystem from tannery wastewater: A review. *Journal of Toxicology, 2018*, 1–16. https://doi.org/10.1155/2018/2568038.

Javed, M., & Usmani, N. (2011). Accumulation of heavy metals in fishes: A human health concern. *International Journal of Environmental Sciences, 2*, 659–670.

Kim, D. W., Heinze, T. M., Kim, B. S., Schnackenberg, L. K., Woodling, K. A., & Sutherland, J. B. (2011). Modification of norfloxacin by a Microbacterium sp. strain isolated from a wastewater treatment plant. *Applied and Environmental Microbiology, 77*(17), 6100–6108. https://doi.org/10.1128/AEM.00545-11.

Kotzerke, A., Sharma, S., Schauss, K., Heuer, H., Thiele-Bruhn, S., Smalla, K., Wilke, B. M., & Schloter, M. (2008). Alterations in soil microbial activity and N-transformation processes due to sulfadiazine loads in pig-manure. *Environmental Pollution, 153*(2), 315–322. https://doi.org/10.1016/j.envpol.2007.08.020.

Kümmerer, K. (2009). Antibiotics in the aquatic environment–A review–Part I. *Chemosphere, 75*(4), 417–434. https://doi.org/10.1016/j.chemosphere.2008.11.086.

Kung, H. F., Lee, Y. C., Huang, Y. L., Huang, Y. R., Su, Y. C., & Tsai, Y. H. (2017). Degradation of histamine by Lactobacillus plantarum isolated from miso products. *Journal of Food Protection, 80*(10), 1682–1688. https://doi.org/10.4315/0362-028X.JFP-17-135.

Küster, A., & Adler, N. (2014). Pharmaceuticals in the environment: Scientific evidence of risks and its regulation. *Philosophical Transactions of the Royal Society B: Biological Sciences, 369*(1656), 20130587. https://doi.org/10.1098/rstb.2013.0587.

Lahti, M., & Oikari, A. (2011). Microbial transformation of pharmaceuticals naproxen, bisoprolol, and diclofenac in aerobic and anaerobic environments. *Archives of Environmental Contamination and Toxicology, 61*(2), 202–210. https://doi.org/10.1007/s00244-010-9622-2.

Larsson, D. G. J. (2014). Pollution from drug manufacturing: Review and perspectives. *Philosophical Transactions of the Royal Society B: Biological Sciences, 369*(1656), 20130571. https://doi.org/10.1098/rstb.2013.0571.

Lawrence, J. R., Swerhone, G. D., Wassenaar, L. I., & Neu, T. R. (2005). Effects of selected pharmaceuticals on riverine biofilm communities. *Canadian Journal of Microbiology, 51*(8), 655–669. https://doi.org/10.1139/w05-047.

Lee, Y. C., Lin, C. S., Liu, F. L., Huang, T. C., & Tsai, Y. H. (2015). Degradation of histamine by Bacillus polymyxa isolated from salted fish products. *Journal of Food and Drug Analysis, 23*(4), 836–844. https://doi.org/10.1016/j.jfda.2015.02.003.

Le-Minh, N., Khan, S. J., Drewes, J. E., & Stuetz, R. M. (2010). Fate of antibiotics during municipal water recycling treatment processes. *Water Research, 44*(15), 4295–4323. https://doi.org/10.1016/j.watres.2010.06.020.

Liang, X., Chen, B., Nie, X., Shi, Z., Huang, X., & Li, X. (2013). The distribution and partitioning of common antibiotics in water and sediment of the Pearl River Estuary, South China. *Chemosphere, 92*(11), 1410–1416. https://doi.org/10.1016/j.chemosphere.2013.03.044.

Lipinski, C. A., Lombardo, F., Dominy, B. W., & Feeney, P. J. (2001). Experimental and computational approaches to estimate solubility and permeability in drug discovery and development settings. *Advanced Drug Delivery Reviews, 46*(1–3), 3–26. https://doi.org/10.1016/S0169-409X(00)00129-0.

Liu, J., Cao, L., & Dou, S. (2019). Trophic transfer, biomagnification, and risk assessments of four common heavy metals in the food web of Laizhou Bay, the Bohai Sea. *Science of The Total Environment, 670*, 508–522. https://doi.org/10.1016/j.scitotenv.2019.03.140.

Luo, Y., Guo, W., Ngo, H. H., Nghiem, L. D., Hai, F. I., Zhang, J., Liang, S., & Wang, X. C. (2014). A review of the occurrence of micropollutants in the aquatic environment and their fate and removal during wastewater treatment. *Science of The Total Environment, 473–474*, 619–641. https://doi.org/10.1016/j.scitotenv.2013.12.065.

Marco-Urrea, E., Aranda, E., Caminal, G., & Guillén, F. (2009). Induction of hydroxyl radical production in Trametes versicolor to degrade recalcitrant chlorinated hydrocarbons. *Bioresource Technology, 100*(23), 5757–5762. https://doi.org/10.1016/j.biortech.2009.06.078.

Megharaj, M., Ramakrishnan, B., Venkateswarlu, K., Sethunathan, N., & Naidu, R. (2011). Bioremediation approaches for organic pollutants: A critical perspective. *Environment International, 37*(8), 1362–1375. https://doi.org/10.1016/j.envint.2011.06.003.

Mench, M., Renella, G., Gelsomino, A., Landi, L., & Nannipieri, P. (2006). Biochemical parameters and bacterial species richness in soils contaminated by sludge-borne metals and remediated with inorganic soil amendments. *Environmental Pollution, 144*(1), 24–31. https://doi.org/10.1016/j.envpol.2006.01.014.

Monteiro, S. C., & Boxall, A. B. A. (2010). Occurrence and fate of human pharmaceuticals in the environment. In *Pharmaceuticals in the Environment* (pp. 53–154). https://doi.org/10.1007/978-1-4419-1157-5_2.

Murdoch, R. W., & Hay, A. G. (2005). Formation of catechols via removal of acid side chains from ibuprofen and related aromatic acids. *Applied and Environmental Microbiology, 71*(10), 6121–6125. https://doi.org/10.1128/AEM.71.10.6121-6125.2005.

Nisha Yadav, R. K. (2015). Bioremediation of polluted soil and water obtained from the Tarai Bhavan region of Uttrakhand, India. *Journal of Bioremediation & Biodegradation, 06*(02). https://doi.org/10.4172/2155-6199.1000276.

Polianciuc, S. I., Gurzău, A. E., Kiss, B., Ştefan, M. G., & Loghin, F. (2020). Antibiotics in the environment: Causes and consequences. *Medicine and Pharmacy Reports*. https://doi.org/10.15386/mpr-1742.

Prieto, A., Möder, M., Rodil, R., Adrian, L., & Marco-Urrea, E. (2011). Degradation of the antibiotics norfloxacin and ciprofloxacin by a white-rot fungus and identification of degradation products. *Bioresource Technology, 102*(23), 10987–10995. https://doi.org/10.1016/j.biortech.2011.08.055.

Quintana, J., Weiss, S., & Reemtsma, T. (2005). Pathways and metabolites of microbial degradation of selected acidic pharmaceuticals and their occurrence in municipal wastewater treated by a membrane bioreactor. *Water Research, 39*(12), 2654–2664. https://doi.org/10.1016/j.watres.2005.04.068.

Rajbanshi, A. (2009). Study on heavy metal resistant bacteria in Guheswori sewage treatment plant. *Our Nature, 6*(1), 52–57. https://doi.org/10.3126/on.v6i1.1655.

Redshaw, C. H., Cooke, M. P., Talbot, H. M., McGrath, S., & Rowland, S. J. (2008). Low biodegradability of fluoxetine HCl, diazepam, and their human metabolites in sewage sludge-amended soil. *Journal of Soils and Sediments, 8*(4), 217–230. https://doi.org/10.1007/s11368-008-0024-2.

Reed, H. E., & Martiny, J. B. H. (2007). Testing the functional significance of microbial composition in natural communities. *FEMS Microbiology Ecology, 62*(2), 161–170. https://doi.org/10.1111/j.1574-6941.2007.00386.x.

Reis, P. J. M., Reis, A. C., Ricken, B., Kolvenbach, B. A., Manaia, C. M., Corvini, P. F. X., & Nunes, O. C. (2014). Biodegradation of sulfamethoxazole and other sulfonamides by Achromobacter denitrificans PR1. *Journal of Hazardous Materials, 280*, 741–749. https://doi.org/10.1016/j.jhazmat.2014.08.039.

Rivera-Utrilla, J., Sánchez-Polo, M., Ferro-García, M. Á., Prados-Joya, G., & Ocampo-Pérez, R. (2013). Pharmaceuticals as emerging contaminants and their removal from water: A review. *Chemosphere, 93*(7), 1268–1287. https://doi.org/10.1016/j.chemosphere.2013.07.059.

Rodarte-Morales, A. I., Feijoo, G., Moreira, M. T., & Lema, J. M. (2012a). Biotransformation of three pharmaceutical active compounds by the fungus *Phanerochaete chrysosporium* in a fed-batch stirred reactor under air and oxygen supply. *Biodegradation, 23*(1), 145–156. https://doi.org/10.1007/s10532-011-9494-9.

Rodarte-Morales, A. I., Feijoo, G., Moreira, M. T., & Lema, J. M. (2012b). Operation of stirred tank reactors (STRs) and fixed-bed reactors (FBRs) with free and immobilized *Phanerochaete chrysosporium* for the continuous removal of pharmaceutical compounds. *Biochemical Engineering Journal, 66*, 38–45. https://doi.org/10.1016/j.bej.2012.04.011.

Rodríguez-Rodríguez, C. E., Barón, E., Gago-Ferrero, P., Jelić, A., Llorca, M., Farré, M., Díaz-Cruz, M. S., Eljarrat, E., Petrović, M., Caminal, G., Barceló, D., & Vicent, T. (2012). Removal of pharmaceuticals, polybrominated flame retardants, and UV filters from sludge by the fungus *Trametes versicolor* in bioslurry reactor. *Journal of Hazardous Materials, 233–234*, 235–243. https://doi.org/10.1016/j.jhazmat.2012.07.024.

Rose, M., Fernandes, A., Mortimer, D., & Baskaran, C. (2015). Contamination of fish in UK freshwater systems: Risk assessment for human consumption. *Chemosphere, 122*, 183–189. https://doi.org/10.1016/j.chemosphere.2014.11.046.

Sankarammal, M., Thatheyus, A., & Ramya, D. (2014). Bioremoval of cadmium using *Pseudomonas fluorescens*. *Open Journal of Water Pollution and Treatment, 2014*(2), 92–100. https://doi.org/10.15764/WPT.2014.02010.

Singh, A., Sharma, R. K., Agrawal, M., & Marshall, F. M. (2010). Health risk assessment of heavy metals via dietary intake of foodstuffs from the wastewater irrigated site of a dry tropical area of India. *Food and Chemical Toxicology, 48*(2), 611–619. https://doi.org/10.1016/j.fct.2009.11.041.

Singh, S. N. (2012). *Microbial Degradation of Xenobiotics*. Springer Berlin Heidelberg. https://doi.org/10.1007/978-3-642-23789-8.

Stackelberg, P. E., Furlong, E. T., Meyer, M. T., Zaugg, S. D., Henderson, A. K., & Reissman, D. B. (2004). Persistence of pharmaceutical compounds and other organic wastewater contaminants in a conventional drinking-water-treatment plant. *Science of The Total Environment, 329*(1–3), 99–113. https://doi.org/10.1016/j.scitotenv.2004.03.015.

Tahar, A., Choubert, J. M., & Coquery, M. (2013). Xenobiotics removal by adsorption in the context of tertiary treatment: A mini-review. *Environmental Science and Pollution Research, 20*(8), 5085–5095. https://doi.org/10.1007/s11356-013-1754-2.

Tahri, N., Bahafid, W., Sayel, H., & El Ghachtouli, N. (2013). Biodegradation: Involved microorganisms and genetically engineered microorganisms. In *Biodegradation–Life of Science*. InTech. https://doi.org/10.5772/56194.

Tappe, W., Herbst, M., Hofmann, D., Koeppchen, S., Kummer, S., Thiele, B., & Groeneweg, J. (2013). Degradation of sulfadiazine by *Microbacterium lacus* strain SDZm4, isolated from lysimeters previously manured with slurry from sulfadiazine-medicated pigs. *Applied and Environmental Microbiology, 79*(8), 2572–2577. https://doi.org/10.1128/AEM.03636-12.

Tchounwou, P. B., Yedjou, C. G., Patlolla, A. K., & Sutton, D. J. (2012). Heavy metal toxicity and the environment. In *Handbook of Environmental Chemistry* (pp. 133–164). Springer. https://doi.org/10.1007/978-3-7643-8340-4_6.

Teixeira, S., Gurke, R., Eckert, H., Kühn, K., Fauler, J., & Cuniberti, G. (2016). Photocatalytic degradation of pharmaceuticals present in conventional treated wastewater by nanoparticle suspensions. *Journal of Environmental Chemical Engineering, 4*(1), 287–292. https://doi.org/10.1016/j.jece.2015.10.045.

Tiehm, A., Schmidt, N., Stieber, M., Sacher, F., Wolf, L., & Hoetzl, H. (2011). Biodegradation of pharmaceutical compounds and their occurrence in the Jordan Valley. *Water Resources Management, 25*(4), 1195–1203. https://doi.org/10.1007/s11269-010-9678-9.

Veach, A., Bernot, M. J., & Mitchell, J. K. (2012). The influence of six pharmaceuticals on freshwater sediment microbial growth was incubated at different temperatures and UV exposures. *Biodegradation, 23*(4), 497–507. https://doi.org/10.1007/s10532-011-9528-3.

Venkatesan, A. K., Pycke, B. F. G., Barber, L. B., Lee, K. E., & Halden, R. U. (2012). Occurrence of triclosan, triclocarban, and its lesser chlorinated congeners in Minnesota freshwater sediments collected near wastewater treatment plants. *Journal of Hazardous Materials, 229–230*, 29–35. https://doi.org/10.1016/j.jhazmat.2012.05.049.

Wang, C., Wang, H., & Liu, Y. (2015). Separation of aluminum and plastic by metallurgy method for recycling waste pharmaceutical blisters. *Journal of Cleaner Production, 102*, 378–383. https://doi.org/10.1016/j.jclepro.2015.04.067.

Woolridge, A., & Hoboy, S. (2019). Medical waste. In *Waste* (pp. 517–530). Elsevier. https://doi.org/10.1016/B978-0-12-815060-3.00027-X.

Xiong, J. Q., Kurade, M. B., Abou-Shanab, R. A. I., Ji, M. K., Choi, J., Kim, J. O., & Jeon, B. H. (2016). Biodegradation of carbamazepine using freshwater microalgae *Chlamydomonas mexicana* and *Scenedesmus obliquus* and the determination of its metabolic fate. *Bioresource Technology, 205*, 183–190. https://doi.org/10.1016/j.biortech.2016.01.038.

Yamamoto, H., Nakamura, Y., Moriguchi, S., Nakamura, Y., Honda, Y., Tamura, I., Hirata, Y., Hayashi, A., & Sekizawa, J. (2009). Persistence and partitioning of eight selected pharmaceuticals in the aquatic environment: Laboratory photolysis, biodegradation, and sorption experiments. *Water Research, 43*(2), 351–362. https://doi.org/10.1016/j.watres.2008.10.039.

Zhang, Y., & Geißen, S. U. (2012). Elimination of carbamazepine in a non-sterile fungal bioreactor. *Bioresource Technology, 112*, 221–227. https://doi.org/10.1016/j.biortech.2012.02.073.

Ziylan, A., & Ince, N. H. (2011). The occurrence and fate of anti-inflammatory and analgesic pharmaceuticals in sewage and freshwater: Treatability by conventional and non-conventional processes. *Journal of Hazardous Materials, 187*(1–3), 24–36. https://doi.org/10.1016/j.jhazmat.2011.01.057.

CHAPTER 12

Application and Role of Nanotechnology for the Degradation of Pharmaceutical Chemicals from the Aquatic Systems

ASHA KUMARI and RAHUL SHARMA*

Department of Chemistry, Career Point University, Bhoranj (Tikker–Kharwarian), Hamirpur, Himachal Pradesh, India

*Corresponding author.

ABSTRACT

Pharmaceuticals, a relatively new class of environmental pollutants, are becoming more problematic as contaminants in surface or groundwater in industrial and residential areas. Pharmaceuticals are discharged into aquatic habitats on a regular basis, owing to their extensive use and the difficulty of removing them in wastewater treatment facilities. Due to growing concentrations of pharmaceuticals observed in drinking and wastewater, water remediation from pharmaceutical substances is a hot topic these days. These new types of very hazardous water contaminants might be particularly harmful to the ecosystem since they can have unknown biological effects on living beings. The leakage of drugs into the aquatic environment takes place in a number of ways after they have been used. Emerging low-concentration pollutants, particularly antibiotics that are found in aquatic habitats and animals are mostly emitted by hospitals, industrial plants, residences, and livestock and fish farms. Around 4,000 distinct components make up pharmaceutical drugs. Several pollutants are detected in extremely low quantities like in a range of mg/L. As a result, degrading medicines, personal care products (PPCPs) along endocrine disrupting chemicals (EDCs) are becoming more and more difficult. Pharmaceutical-compound-containing wastewater should be treated before being discharged into sewage networks and/or surface water. Because

Biomedical Waste Management: Recycling and Remediation Strategies. Younis Ahmad Hajam, Javid Ahmad Malik, Anu Pandita, and Deepu Pandita (Eds.)

of the rapidly developing field of nanotechnology, there are several chances to effectively remove pharmaceutical compounds from water. Nanomaterials have a lot of promise for pharmaceutical product removal because of their unique nanoscale features, such as catalytic tendency in the presence of light, adsorption, filtering, and advanced oxidation processes (AOPs). The current chapter provides a summary of recent achievements in pharmaceutical product remediation utilizing nanotechnologies.

12.1 INTRODUCTION

Pharmaceuticals are a broad category of chemical substances used to treat, prevent, and cure diseases. Numerous prescription and over-the-counter medications, such as antidepressants, analgesics, antibiotics, antipyretics, blood lipid regulators, contraceptives, plus chemotherapeutic medicines, are included in the list of Pharma products (Benotti et al., 2009; Ratola et al., 2012). Figure 12.1 provides a list of common drugs that are most frequently utilized.

Pharmaceuticals are major products that are required for prescribed treatments, in-counter medicines, as well as for veterinary medications. Different health-related issues can be cured by natural or synthetic drugs, including so advantageous to society. These substances then absorbed in the blood, are typically active in minute amounts and can stay within the body to provide their complete healing outcome (Mezzelani et al., 2018). These include analgesics like acetaminophen, Ibuprofen, and naproxen; antibiotics like nitrofurantoin, levofloxacin, and chloramphenicol (CAP); antivirals like Tamiflu and zanamivir; antidepressants like alprazolam; antiepileptics like felbamate and carbamazepine; and hormonal agents like estriol and 17-b-estradiol.

Municipal wastewater, incorrect medication disposal, excreta of humans, rigorous livestock husbandry, effluents from hospitals, and the pharmaceutical industry are all foundations of pharmaceutical contamination in the environment (Mezzelani et al., 2018; Patel et al., 2019). Human feces that contain pharmaceutical components washes into freshwater bodies from the ground (Madikizela et al., 2017). The excretion of active pharmaceutical metabolites and unmetabolized pharmaceutical components endangers both aquatic life and human health (Ondarza et al., 2019). Based on the conclusions of Freyria et al. (2018) and Patel et al. (2019), pharma effluents and products for personal care are the main causes of water pollution.

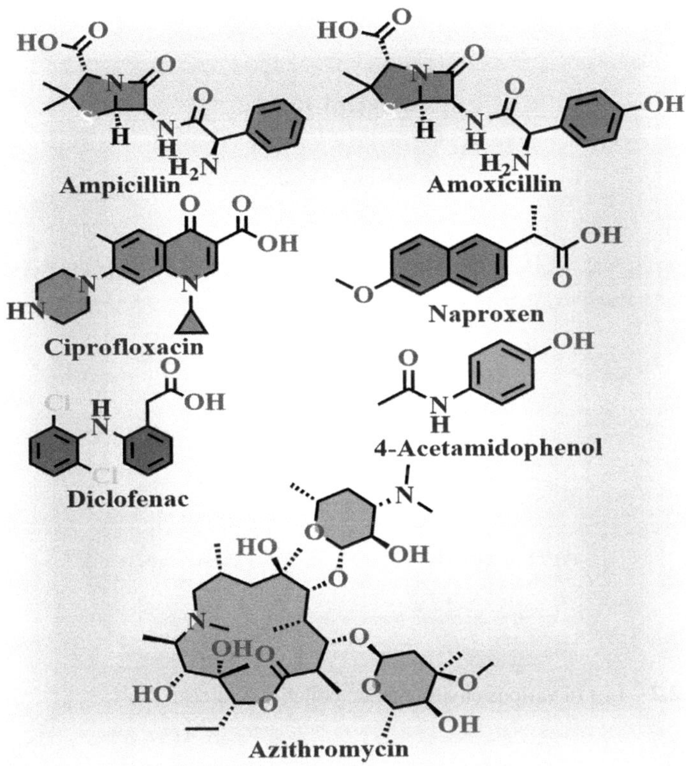

FIGURE 12.1 List of common drugs that are most frequently utilized.

Rapid expansion of pollution-causing industries like food, textile, tannery, beverage, and so forth, has caused the release of hazardous, poisonous, chemicals into the environment (Pandey, 2018; Razack et al., 2020). Furthermore, the sustainability of the environment has been hampered by the growing population's intensive use of natural resources (Mauter et al., 2018; Pandey, 2018; Razack et al., 2020). Consequently, environmental pollution results from the discharge of contaminants, such as microbial poisons, into the surrounding environment. Overuse of pesticides over the past 60 years contaminated the soil and groundwater and also caused sickness and mortality in humans including other forms of life. Figure 12.2 presents a list of various environmental pollution sources.

Different forms of water that as surface and ground waters, as well as occasionally in drinking water, pharmaceuticals, daily used personal products, and endocrine-disrupting chemicals (EDCs) that have diverse functional groups have been found. The majority of these products have high

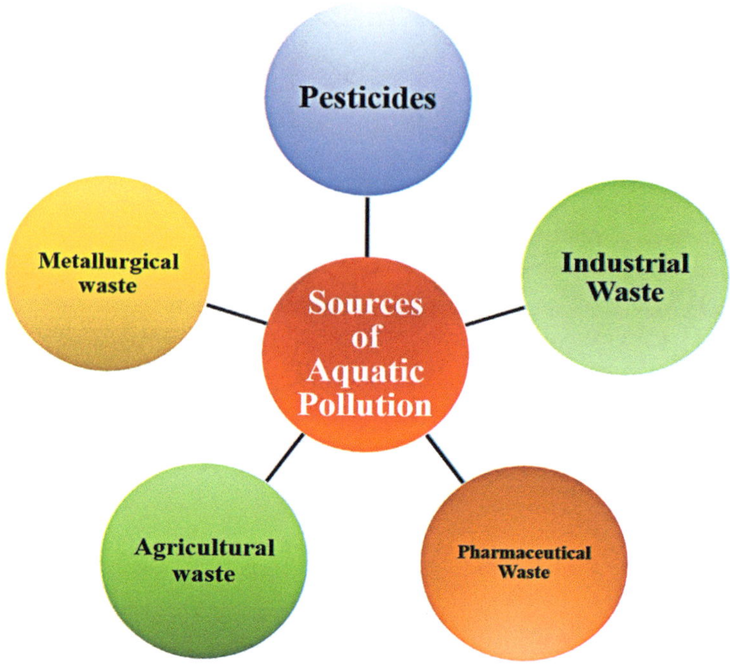

FIGURE 12.2 List of various environmental pollution sources.

polarity, low volatility, or resistance to biological breakdown mechanisms (Breton). A number of negative effects and ecological implications on the environment, including those on people and wildlife, have been connected to PPCPs and chemicals (Archer et al., 2017; Tijani et al., 2016). Health concerns have been greatly heightened due to the presence of personal care products and EDCs in water in trace amounts. There are already more than 200 widely used medications, including sulfonamide antibiotics that have been discovered in water. Around 15–25 drugs are still found in treated drinking water worldwide (Benotti et al., 2009; Jones et al., 2016). In order to prevent negative impacts on people, flora, and non-human biota, it is crucial to identify and eliminate ecologically dangerous contaminants. For the adsorption of PPCPs and EDCs in water, nanomaterials based on carbon, mainly single-wall/multi-wall carbon nanotubes (CNTs), have previously been used (Jung et al., 2015). These substances include carbamazepine, ciprofloxacin, ibuprofen, doxycycline hydrochloride, macrolides, norfloxacin, sulfonamides, quinolones (Yu et al., 2016), triclosan (Behera et al., 2010), bisphenol A (BPA; Bohdziewicz et al., 2013), tetrabromobisphenol A (Ji et al., 2012), 17-aethinyl estradiol (EE2; Joseph et al., 2011), oxytetracycline (Oleszczuk et al., 2009), tetracycline (Ji et al., 2010), and endocrine (Kumar et al., 2012).

12.1.1 COMMONLY USED TREATMENT METHODS

To eliminate organic contaminants from water, conventional wastewater treatment methods include adsorption using activated carbon, the process of ozonation, and Catalysis in the presence of UV light, these treatments are frequently used (Padhye et al., 2014). These techniques have a number of precincts, though, such as their high cost or low efficiency. It has been extensively noted that refractory pharmaceutically active compounds cannot be entirely degraded by the biological mechanisms that are most frequently used. For example, while several other commonly encountered compounds such as bezafibrate and clofibric acid can be eliminated, carbamazepine is almost completely non-biodegradable product (Castiglioni et al., 2006) In addition to this it may contain raw materials, organic solvents, substrate, intermediates, and catalysts, formulate it particularly challenging to treat the wastewater discharged by the pharmaceutical sector. Alternative therapies are therefore needed to completely eradicate PhACs.

With the recent advancement of nanotechnology, there are now greater chances to create novel methods for more effective degradation of pharmaceutical compounds during water and wastewater treatment. In general, the term "nanotechnology" describes the utilization of materials that have nanoscale dimensions between 1 and 100 nm.

12.2 ROLE OF NANOTECHNOLOGY FOR DEGRADATION OF PHARMACEUTICAL PRODUCTS

The most essential and successful technique for the removal of contaminants from water without driving resultant probable pollution at a reasonable cost is adsorption. The adsorption technique has been widely employed in wastewater treatment for many years as a quick, effective, and affordable physicochemical method.

For many years now, the creation and use of nanomaterials have been of great interest. Water treatment with nanotechnology and nanomaterials (Figure 12.3) has a significant deal of promise for high efficiency and can enhance water supply from a variety of sources. Because of the high surface-to-volume ratio along with strong reactivity, nanomaterials offer significantly better and more effective remediation for water contaminants than conventional methods like AC.

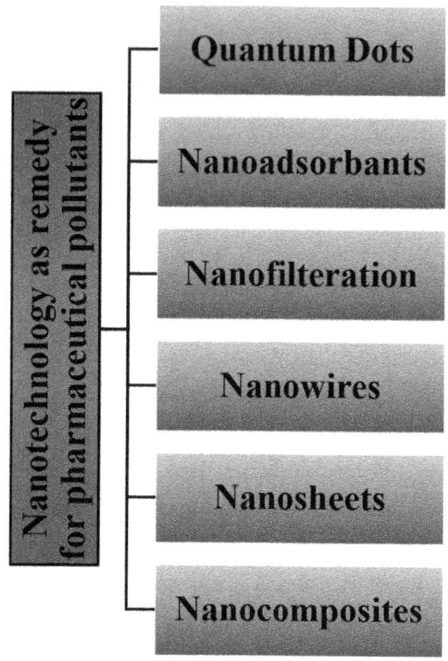

FIGURE 12.3 Different types of nanomaterials as remedies for pharmaceutical pollutants.

The porous nature of NMs and their surface characteristics, such as surface modification, have a significant impact on their adsorption capabilities. Nanoparticles, nanostructured catalytic membranes, bioactive nanoparticles, nanocatalysts, nanomembranes, or nanosorbents have all undergone considerable development, offering a different approach to the problem of global water pollution (Bhatia et al., 2023; Sharma et al., 2017). Importantly, advances in NMs and nanotechnology improve the filtration capacity and improve the water quality (Mohammad et al., 2015; Sharma et al., 2022; Wang et al., 2013). In the processes of adsorption, photocatalysis, catalytic ozonation, and filtering, nanomaterials have been widely used or studied.

12.2.1 NANOADSORBENTS FOR DEGRADATION

The creation of several nanomaterials, including nanoscale zeolites, graphene, and carbon nano-tubes, provides insight into the increased adsorption of PhACs. Great precise surface area, plentiful, selective adsorption location, little intra-particle diffusion distance, variable minute opening, simple regeneration, and reusability are just a few of the special benefits that these nano-adsorbents provide (Chauhan et al., 2016; Mestre et al., 2007).

12.2.2 CARBONACEOUS NANOMATERIALS

Different allotropes of carbon are created when carbon is hybridized to different degrees. These allotropes include CNTs, graphene, carbon nanofibers (CNFs), carbon nanoparticles, and nanobeads (CNBs; Khajeh et al., 2013). These include CNTs and graphene, which have undergone extensive research as adsorptive exclusion of PhACs from the water sources. Sp^2-hybridized carbon allotropes have also received significant consideration as promising adsorbents because of their distinctive surface properties derived from their nanostructure (Kumari et al., 2021; Laera et al., 2013; Yang et al., 2012). PhACs can bind to carbonaceous nanomaterials by five different interactions, such as hydrophobic effects, stacking, hydrogen bonds, electrostatic interactions, and covalent interactions (Sharma et al., 2021; Zhang et al., 2010). Characteristics of both carbonaceous nanomaterials, and PhACs molecules influence kinds, intensities, and contributions to the overall adsorption. Consequently, the surface of carbonaceous nanoparticles is crucial for the removal of PhACs by adsorption. The adsorbent's adsorption performance in the aqueous phase is determined by its physical characteristics, like surface area, functionalities on the surface, and pores.

12.2.3 OTHER NANO-ADSORBENTS

Some two-dimensional nanomaterials through a large specific surface area, like stacked double hydroxides, hexagonal boron nitride nanosheets, and molybdenum disulfide (MoS_2) nanosheets, have also demonstrated good adsorption. PHACs removal in water treatment procedures and have shown excellent activity against organic pollutants. Magnetic nano-adsorbents (such as Fe_3O_4) may be readily engaged and removed from indulgence water, making them appealing materials (Khin et al., 2012). The adsorption of PhACs has been reported for several magnetic characteristic-based nano-adsorbents, together with magnetic zeolites and magnetic carbon nanocomposite, and they have all demonstrated high efficiency (Attia et al., 2013; Singh et al., 2012).

12.2.4 NANOPHOTOCATALYSTS FOR DEGRADATION

For the photocatalytic oxidation of pharmaceutical pollutants, a few studies have also discussed metal oxides (Bhatia et al., 2022; Gautam et al., 2020; Gusain et al., 2019) functionalized metal-organic frameworks (MOFs), photocatalysts based on TiO_2. Numerous more reviews have also looked at various Processes or particular advanced oxidation processes (AOPs), like electro-Fenton, photoelectrocatalysis, and Fenton-like for the handling of wastewater that has pharmaceuticals and other crude compounds (Velempini et al., 2021). It has been highlighted that at the same time as previous metal oxides have lately come out as precious candidates for oxidative reactions, the majority of papers from the past exclusively cover TiO_2 for heterogeneous photocatalytic degradation of pharmaceuticals.

12.2.5 OZONATION (ORGANIC ELIMINATION)

Organic elimination technique ozonation is an established method that has been used extensively in dealing with drinking water and wastewater. Organic contaminants are typically oxidized through indirect O_3 degradation by producing OH or direct molecular ozone (O_3) processes (Farhadian et al., 2019; Lei et al., 2019). While cephalexin, penicillin G, is mostly converted by OH, many pharmaceutically active compounds, such as quinolones, sulfonamides, and tetracyclines, are primarily destroyed by oxidation. PhACs' ozonation kinetics, however, are rather slow. Additionally, the ozonation process frequently results in the formation of carboxylic acids and aldehydes, neither of which can interact with O_3. As a result of its enormous potential to efficiently degrade PhACs such as CNTs (Yin et al., 2021) and metal oxides (Wang et al., 2020) has drawn increased attention in recent years. However, there is still a dearth of information regarding the catalytic ozonation of PhACs utilizing nanotechnology.

12.2.6 FILTRATION

Filtration using nanotechniques nanofiltration (NF), and ultrafiltration (UF), reverse osmosis are just a few of the filtration technologies that have been explored and used extensively to eliminate contaminants from wastewater. The main drawbacks of the filtering methods are membrane fouling and excessive energy usage (Hu et al., 2020) with the advancement of nanotechnology, it is now possible to add useful nanoparticles to membranes to improve their permeability, stability, fouling resistance, or self-cleaning capabilities. This makes filtration more useful for engineering applications (Bhatia et al., 2021; Zhang et al., 2020) Surface coating and usage of a materials composite are the modification methods that are most frequently utilized (CNTs, nanofibers, zeolites, and so forth).

12.3 DIFFERENT TYPE OF NANOMATERIALS USED FOR DEGRADATION

12.3.1 METAL NANOPARTICLES FOR DEGRADATION

Metal nanoparticles of different metals like silver, iron, zinc, and cobalt were used for the treatment of wastewater by degrading organic pollutants. Physical and chemical processes have historically been used to prepare metal nanoparticles. Physical methods for creating nanoparticles are very expensive, energy-intensive, and space-intensive. Due to the attachment of dangerous substances to the surface of chemically produced nanoparticles, they are not suitable for medical applications. Additionally, the pollutants produced through the chemical pathway are harmful to the environment. Consequently, there is a growing demand for the environmentally friendly method of synthesizing metal nanoparticles for use in biomedical applications along with other applications. In literature under the influence of visible light, the photocatalytic activity of AgNPs against norfloxacin was tested. The significant absorption at 340 nm with 10 mg of norfloxacin was used to explain the photocatalytic degradation behavior. The superior lowering ability after 80 minutes of visible light exposure was studied in the degradation of norfloxacin (Kanagamani et al., 2019). The pharmaceutical's reductive breakdown is promoted by the iron layer's blockage of the active sites and subsequent reticence of electron transfer on the surface, which also increases the absorption of the pharmaceutical and its by-products. The interaction of iron and pharmaceuticals plays a significant part in the elimination process since the amount of contact relies mainly on the charge of the compound and the iron exterior (Lumbaque et al., 2019). A significant source of severe water pollution worldwide, the fabric industry and related wastewater have been growing rapidly. It has been believed that all kinds of life are adversely affected by water that contains various sorts of dyes. The capacity of manufactured copper nanoparticles to degrade dyes or organic pollutants catalytically was investigated. These copper nanoparticles' distinctive shape, high surface area, exceptional mechanical and thermal stability, and optimal optical, magnetic, and catalytic capabilities have been credited with their broad use (Sharma et al., 2018).

12.3.2 METAL OXIDE NANOPARTICLES FOR DEGRADATION

Metal oxide nanoparticles were used in literature for the degradation of antibiotics. The elimination of amoxicillin (AMX) by the greenly produced iron oxide nanoparticles employing plant extract was accomplished. The antibiotic was removed by adsorption and degradation processes for metal oxide nanoparticles (Demirezen et al., 2019). Another study reported the employment of iron oxide nanoparticles for removal and mineralization of diclofenac was about 65% and 48%, respectively, under nearly ideal operating conditions. In the tests, there was very little iron leaching, which supports the efficient encapsulation and durability of the catalytic nanoparticles in the ceramic porous matrix, enabling multiple membrane usage. Future research directions for technique optimization are suggested, with a focus on membrane modification for better organic pollutant adsorption and their oxidation inside the catalytic membrane matrix (Yang et al., 2022). Many persistent organic pollutants have been suggested to be degraded by copper oxide nanoparticles (CuO-NPs), which work well as catalysts. Enrofloxacin, a fluoroquinolone, a significant class of antibiotic medications, was used for the first time in humans in the 1980s and for veterinary purposes a decade later. The degradation of enrofloxacin was investigated in the current work. An organofluorine chemical called enrofloxacin is used as a feed supplement to speed up animal growth and display antibacterial action against Gram-positive and negative bacteria. Numerous water sources have been reported to contain enrofloxacin. It was proposed that the reaction route in this instance involved the production of Reactive Oxygen Species and their subsequent interaction with Enrofloxacin (Dror et al., 2020). In another study, Pharmaceuticals are polar micropollutants that are typically not dealt with well by conventional wastewater treatment facilities. Antibiotics should be eliminated before they are released into ecosystems. To further boost single ozonation's effectiveness, catalytic ozonation is applied by using zinc oxide nanoparticles. In the current investigation, semi-batch tests were carried out to determine how well single ozonation and catalytic ozonation degraded oxytetracycline (OTC), a pollutant that poses environmental risks, in aqueous solutions. Ozone (ZnO/O_3) was utilized as a catalyst in conjunction with zinc oxide nanoparticles (ZnO). Investigated was the impact of operational factors on the ZnO/O_3 process, including pH from range 3–11, starting oxy-tetracycline (10–100 mg/l), temperature (15–35°C), and ozone generation rate and catalyst dosage. Ozone dose was discovered to have the apparent (Moshin et al., 2021).

This is preferable to release organic-containing aquaculture wastewater into the environment after action with more than 98% detoxification. The treatment of wastewater using customary remediation methods veterinary pharmaceutical residues were typically only partially removed after the use of veterinary medications. Aquaculture produces a lot of waste that is released into the environment. Some of this waste is dangerous and difficult to break down. In this study, nano-ZnO was successfully created to speed up the breakdown of AMX in an ultrasound/H_2O_2 system from aquaculture effluent (Ayanda et al., 2021).

12.3.4 MAGNETIC/NANOCOMPOSITE/NANOPARTICLES

Nanocomposites and magnetic nanoparticles show great adsorption tendency. This section focuses on some nanocomposite compositions and magnetic nanoparticles playing a role in the degradation of pharmaceuticals. In a previously reported study, a straightforward one-pot approach was used to successfully create a composite catalyst called CuO@C, which is made up of bare-core CuO with a carbon shell. This composite shows an exceptional catalytic performance upon its utilization as a heterogeneous catalyst to make active peroxymonosulfate (PMS). Ultimately showed a positive mineralization effect on tetracycline (TC), with a good mineralization rate of 75.3% in just 60 minutes, according to total organic carbon (TOC) measurements. Researchers looked for the effect of pH, co-existing ions, PMS dosage, TC concentration, and catalyst annealing temperature. It was discovered that TC elimination was facilitated by an initial alkaline pH, yet the reaction for degradation was efficient throughout a wide pH range (Yang et al., 2021). Another study involves the use of natural phosphate ore to create zinc oxide-hydroxyapatite nanocomposites, which were then tested for their ability to photodegrade two antibiotics in solution. The synthesis of homogeneously dispersed photocatalytic nanocrystals in the apatite phase was made possible through in situ growth of ZnO and heat treatment. When compared to the individual phases, nanocomposites showed higher rates of ofloxacin and ciprofloxacin sorption. Depending on the antibiotic being evaluated, the performances of the nanocomposites in terms of photodegradation under UV irradiation were equal to or better than those of the photocatalytic particles alone at high ZnO loadings. This dependence reveals diverse degradation processes that were linked to various antibiotic-nanocomposite-particle affinities. The approach outlined here seems quite promising for development because of its simplicity and adaptability (El Bekkali et al., 2018).

TiO_2 has large band gap energy. As a result, the photocatalytic activity of titanium dioxide stimulated by UV light is minimal. It is possible to increase the activity of titanium dioxide. By splitting apart pairs of electron-holes. In this, hexadecyl trimethylammonium bromide was used as a template to produce TiO_2–WO_3 nanocomposites among varying percentages of WO_3. Numerous characterization techniques were used to test the photocatalytic, physical, and structural characteristics of titanium dioxide and its nanocomposites. Using visible light, the effectiveness of these nanocomposites was evaluated for degradation of paracetamol. Nanocomposites perform very well in the photocatalytic breakdown of paracetamol. TiO_2–WO_3—3 wt.% photocatalyst, 1.5 g/l of TiO_2–WO_3—3 wt.% photocatalyst, 30 minutes of irradiation period, and 100% paracetamol photocatalytic degradation were used to achieve this. TiO_2–WO_3 nanocomposites perform photocatalytic degradation of paracetamol better than TiO_2 does (Namshah et al., 2018). Under UV light, photocatalytic degradation of AMX by using titanium dioxide nanoparticles placed on graphene oxide (GO/TiO_2) was tested. The findings of the experiments demonstrated that important variables such as the beginning pH, GO/TiO_2 dose, UV intensity, and initial AMX concentration all had a substantial impact on the degradation of AMX. The AMX degradation efficiency was found to be greater than 99% under the conditions tested compared to the photolysis and adsorption procedures (Balarak et al., 2021). The simultaneous gelation of titanium alkoxide and precipitation of a naturally occurring phosphate mineral in ammonia was used to create TiO_2-hydroxyapatite nanocomposites, which were then used to test for drug degradation in solution under UV light. On the apatite-rich composites, pharmaceuticals were more readily adsorbed than on the other materials. In the phases that were TiO_2-rich, photodegradation worked better. When compared to TiO_2 related to the antibiotic affinity, photocatalytic degradation of nanocomposites performs better at high TiO_2 content. Ciprofloxacin and ofloxacin were completely photodegraded in 15 and 120 minutes, respectively, using nanocomposites as the photocatalyst. For the improved adsorption and photocatalytic degradation of ciprofloxacin and ofloxacin in water, the suggested nanocomposites are effective catalysts (Bouyarmane et al., 2021).

The capacity of magnetic nanocomposite with graphene-resistant nanocomposites of iron oxide, zinc oxide, and tin oxides to photocatalytically breakdown azithromycin in an aqueous environment is examined in other investigations. The $GO@Fe_3O_4/ZnO/SnO_2$ nanocomposite was examined using XRD, FT-IR, TEM, and VSM. In a batch system, the effects of pH, contact duration, catalyst content, and azithromycin starting concentration were investigated. Optimal conditions were pH 1–4, 120 min with 1 g/L of $GO@Fe_3O_4/ZnO/SnO_2$, and UV-C irradiation, 90.06% of the 30 mg/L azithromycin was destroyed. According to the findings, it can be inferred that the compound $GO@Fe_3O_4/ZnO/SnO_2$ can be employed as an efficient catalyst for the degradation of antibiotics from aquatic environments because of its strong ability to remove azithromycin (Sayadi et al., 2019). Antibiotic ciprofloxacin is frequently used to treat infections. It is primarily excreted in its unmetabolized form before finally entering the water through sewage and wastewater discharge. The purpose of this study is to create $ZnFe_2O_4@CMC$ and analyze how well it removes CIP during the photocatalytic reaction. The $ZnFe_2O_4@CMC$ nanobiocomposite was successfully made using our hydrothermal technique. They may be used as an excellent photocatalysts with high efficiency, cheap cost, simple preparation, and easy operation. These bionanocomposites have tremendous potential for the removal of antibiotics like ciprofloxacin from aqueous solutions (Malakootian et al., 2019). Hazardous chemicals that can cause cancer in humans are found in pharmaceutical waste. Therefore, one of the top priorities in environmental engineering is the removal of these residues from wastewater. In this investigation, we assessed the efficacy of a nanocomposite of nickel ferrites $FeNi_3$, chitosan, and bismuth as a photocatalyst for the degradation of metronidazole (MTZ) in the presence of simulated sunlight (Arghavan et al., 2021). In the current work, a hybrid Fe_3O_4/ZnO nanostructured photoactive material has been produced and tested for wastewater treatment applications. Four antibiotics were chosen as models for persistent organic molecules in water, and degradation tests were carried out to examine the photocatalytic performance of each antibiotic. Because of their electrostatic properties, antibiotic adsorption onto the nanocomposite surface was crucial for lowering the initial concentration of antibiotics (Fernández et al., 2019).

Combining hydrothermal synthesis with microwave heating led to the effective synthesis of F-Pdx co-doped TiO_2 nanocomposites. The addition of F and Pd to the TiO_2 crystal structure creates additional energy levels between the valence and conductive bands, increasing TiO_2's photocatalytic absorption to the visible and near-infrared light spectrums. A new phase begins to occur if the amount of Pd dopant exceeds a certain threshold, in this case, 5%. Thus, by adding 10% palladium to the TiO_2 lattice, the optical band gap of the F-Pdx codoped TiO_2 was reduced from 3.26 eV to 0.54 eV. However, when this nanocomposite was tested for sulfamethoxazole degradation (30 ppm), up to 98.4% of sulfamethoxazole was removed in under 40 minutes. While fluorine's insertion had a substantial impact on the photodegraded product, the addition of Pd to TiO_2 had an exceptional effect on the degradation of the medication sulfamethoxazole (Jahdi et al., 2020).

The fabrication of GO/CuBDC-Fe_3O_4 ternary nanocomposite utilizing a green solvothermal process was described as a unique approach. The goal of this nanocomposite was to create a peroxymonosulfate (PMS) activator to swiftly break down the antibiotic ciprofloxacin (CIP). Graphene oxide (GO) nanosheets were used as a suitable platform for the formation of Fe_3O_4 and metal-organic frameworks (MOF). According to the findings, Cu/Fe species and C-O groups in ternary nanocomposite catalyzed PMS in a way that enabled the production of hydroxyl and sulfate radicals, which were subsequently employed to degrade CIP. The ternary nanocomposite has also shown excellent recyclability, enabling easy degradation of pharmaceutics (Alamgholiloo et al., 2021).

12.4 MECHANISM FOR DEGRADATION OF PHARMACEUTICAL DRUGS

Degradation of pharmaceutical chemicals from aquatic systems became very important. In this chapter, we have focused on different compositions of nanoparticles, and nanocomposites for the degradation of these chemicals. Various properties of nanomaterials are helpful for degradation. These include a high surface-to-volume ratio of the nanomaterials, stability, cost-effectiveness, and so forth. Photocatalytic degradation and the adsorption tendency of nanomaterials play important roles in promoting degradation. The catalytic process that is induced, which conducts a redox reaction to further oxidize or decrease the organic molecule, is a part of the heterogeneous process known as photo-catalysis. Bhatia et al. (2022) discussed the removal of cefuroxime axetil an antibiotic using a photo-catalysis process.

To remove non-degradable organic contaminants from groundwater, drinking water preparation, process water, or other water sources, adsorption is most typically used as a tertiary cleaning step after biological water purification. Adsorption is the process by which liquid molecules cling to the surface of a solid. Biochar (BC) can help the photocatalyst remove a variety of organic pollutants from aqueous solutions. BSPs are more effective than single nanoparticles at photodegrading organic pollutants due to their high absorption capacity. BC, a member of the carbon family, performs magnificently as a photocatalyst when combined with other nanomaterial catalysts, producing a novel hybrid nanomaterial known as biochar aided photocatalyst (BSPs). These BSPs have amazing physio-chemical properties and are highly efficient at deteriorating water pollutants. When compared to pure nanophotocatalyst, BC-supported catalysts have a larger surface area, are extremely porous, and have greater catalytic efficiency as well as stability and recoverability (Ahmaruzzaman, 2021). Figure 12.4 represents a possible mechanism for degradation.

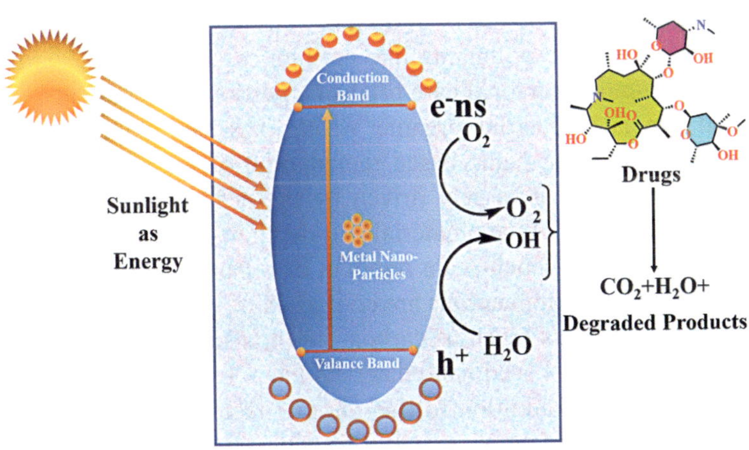

FIGURE 12.4 Mechanism for degradation.

12.5 ELEMENTS THAT IMPACT THE ACCURACY OF DRUGS' DEGRADATION KINETIC DATA

Numerous factors, including pH, temperature, the research medium (solvent, dielectric constant, and ionic strength), the quantity of light exposure, and excipients, can affect how quickly pharmaceuticals degrade. The quality of the data produced from the kinetic analysis of pharmaceutical degradation can be impacted by each of these elements acting alone or in combination. Studying each variable's unique impact on the rate of degradation is crucial for producing high-quality kinetic study data. The interpretation of kinetic data may become more challenging if certain variables have an overlapping influence. The rate and the order of the degradation response can both be altered by simultaneous effects of factors, which can further provide false positive results. One of the most significant elements that can have a direct impact on the rate of deterioration among all of these parameters is the temperature employed for the degradation kinetic research (Bhangare et al., 2022).

12.6 CONCLUSION

Pharmaceutical active compounds are ingested in ever-increasing amounts and varieties, and they are regularly found in surface water, groundwater, and even drinking water. However, their removal using traditional sewage treatment techniques is hardly ever sufficient. The advancement of nanotechnology presents prospects for developing cutting-edge methods for effective elimination. The use of several cutting-edge nanomaterials for the remediation of organic pollutants like antibiotics, antipyretics, medicines, endocrine disruptors, personal care products, pesticides, detergents, organic dyes, and typical industrial organic wastes has been summarized and covered in this chapter. Nano-based remediation of organic and inorganic pollutants from water is largely driven by the adsorption, oxidation, and degradation capability of nanomaterials. In-depth research has been done on nanomaterials for the bench-scale removal of organic pollutants. The bulk of applied complex nanomaterials are composed of various combinations of inorganic nanomaterials (such as carbon- and transition metal-based NMs) and organic molecular-based nanomaterials (such as nano-membranes, and organic polymers). In addition to this various factors that are responsible for affecting the degradation capacity of nanomaterials are also discussed. Photocatalysis is the most common technique for degradation. Photocatalysis has demonstrated to have tremendous promise for the elimination of pollutants in water and wastewater treatment due to its low cost, ease of use, and environmentally friendly characteristics.

KEYWORDS

- **adsorption**
- **advanced oxidation processes**
- **drug degradation**
- **nanomaterials**
- **nanotechnology**
- **photocatalysis**

REFERENCES

Ahmaruzzaman, M. (2021). Biochar-based nanocomposites for photocatalytic degradation of emerging organic pollutants from water and wastewater. *Materials Research Bulletin, 140*, 111262.

Alamgholiloo, H., Pesyan, N. N., Mohammadi, R., Rostamnia, S., & Shokouhimehr, M. (2021). Synergistic advanced oxidation process for the fast degradation of ciprofloxacin antibiotics using a GO/CuMOF-magnetic ternary nanocomposite. *Journal of Environmental Chemical Engineering, 9*(4), 105486.

Archer, E., Petrie, B., Kasprzyk-Hordern, B., & Wolfaardt, G. M. (2017). The fate of pharmaceuticals and personal care products (PPCPs), endocrine disrupting contaminants (EDCs), metabolites, and illicit drugs in WWTW and environmental waters. *Chemosphere, 174*, 437–446.

Arghavan, F. S., Al-Musawi, T. J., Rumman, G. A., Pelalak, R., Khataee, A., & Nasseh, N. (2021). The photocatalytic performance of a nickel ferrite/chitosan/bismuth (III) oxyiodide nanocomposite for metronidazole degradation under simulated sunlight illumination. *Journal of Environmental Chemical Engineering, 9*(4), 105619.

Attia, T. M. S., & Hu, X. L. (2013). Synthesized magnetic nanoparticles coated zeolite for the adsorption of pharmaceutical compounds from aqueous solution using batch and column studies. *Chemosphere, 93*(9), 2076–2085.

Ayanda, O. S., Aremu, O. H., Akintayo, C. O., Sodeinde, K. O., Igboama, W. N., Oseghe, E. O., & Nelana, S. M. (2021). Sonocatalytic degradation of amoxicillin from aquaculture effluent by zinc oxide nanoparticles. *Environmental Nanotechnology, Monitoring & Management, 16*, 100513.

Balarak, D., Mengelizadeh, N., Rajiv, P., & Chandrika, K. (2021). Photocatalytic degradation of amoxicillin from aqueous solutions by titanium dioxide nanoparticles loaded on graphene oxide. *Environmental Science and Pollution Research, 28*(36), 49743–49754.

Behera, S. K., Oh, S. Y., & Park, H. S. (2010). Sorption of triclosan onto activated carbon, kaolinite, and montmorillonite: Effects of pH, ionic strength, and humic acid. *Journal of Hazardous Materials, 179*(1–3), 684–691.

Benotti, M. J., Trenholm, R. A., Vanderford, B. J., Holady, J. C., Stanford, B. D., & Snyder, S. A. (2009). Pharmaceuticals and endocrine disrupting compounds in US drinking water. *Environmental Science & Technology, 43*(3), 597–603.

Bhangare, D., Rajput, N., Jadav, T., Sahu, A. K., Tekade, R. K., & Sengupta, P. (2022). Systematic strategies for degradation kinetic study of pharmaceuticals: An issue of utmost importance concerning current stability analysis practices. *Journal of Analytical Science and Technology, 13*(1), 1–13.

Bhatia, N., Kumari, A., Chauhan, N., Thakur, N., & Sharma, R. (2023). *Duchesnea indica* plant extract mediated synthesis of copper oxide nanomaterials for antimicrobial activity and free-radical scavenging assay. *Biocatalysis and Agricultural Biotechnology, 47*, 102574.

Bhatia, N., Kumari, A., Sharma, A., & Sharma, R. (2021). Bio-inspired green formulation of various metal nanoparticles: A review. *Journal of Materials Science Research and Reviews, 8*, 1–38.

Bhatia, N., Kumari, A., Thakur, N., Sharma, G., Singh, R. R., & Sharma, R. (2022). Phytochemically stabilized chitosan encapsulated Cu and Ag nanocomposites to remove cefuroxime axetil and pathogens from the environment. *International Journal of Biological Macromolecules, 212,* 451–464.

Bohdziewicz, J., & Kamińska, G. (2013). Kinetics and equilibrium of the sorption of bisphenol A by carbon nanotubes from wastewater. *Water Science and Technology, 68*(6), 1306–1314.

Bouyarmane, H., El Bekkali, C., Labrag, J., Es-Saidi, I., Bouhnik, O., Abdelmoumen, H., & Robert, D. (2021). Photocatalytic degradation of emerging antibiotic pollutants in waters by TiO_2/Hydroxyapatite nanocomposite materials. *Surfaces and Interfaces, 24*, 101155.

Castiglioni, S., Bagnati, R., Fanelli, R., Pomati, F., Calamari, D., & Zuccato, E. (2006). Removal of pharmaceuticals in sewage treatment plants in Italy. *Environmental Science & Technology, 40*(1), 357–363.

Chauhan, K., Sharma, R., Dharela, R., Chauhan, G. S., & Singhal, R. K. (2016). Chitosan-shimmer stabilized silver nanocomposites for antimicrobial and antioxidant applications. *RSC Advances, 6*(79), 75453–75464.

Demirezen, D. A., Yıldız, Y. Ş., & Yılmaz, D. D. (2019). Amoxicillin degradation using green synthesized iron oxide nanoparticles: Kinetics and mechanism analysis. *Environmental Nanotechnology, Monitoring & Management, 11*, 100219.

Dror, I., Fink, L., Weiner, L., & Berkowitz, B. (2020). Elucidating the catalytic degradation of enrofloxacin by copper oxide nanoparticles through the identification of the reactive oxygen species. *Chemosphere, 258*, 127266.

El Bekkali, C., Bouyarmane, H., El Karbane, M., Masse, S., Saoiabi, A., Coradin, T., & Laghzizil, A. (2018). Zinc oxide-hydroxyapatite nanocomposite photocatalysts for the degradation of ciprofloxacin and ofloxacin antibiotics. *Colloids and Surfaces A: Physicochemical and Engineering Aspects, 539*, 364–370.

Farhadian, N., Akbarzadeh, R., Pirsaheb, M., Jen, T. C., Fakhri, Y., & Asadi, A. (2019). Chitosan modified N, S-doped TiO_2 and N, S-doped ZnO for visible light photocatalytic degradation of tetracycline. *International Journal of Biological Macromolecules, 132*, 360–373.

Fernández, L., Gamallo, M., González-Gómez, M. A., Vázquez-Vázquez, C., Rivas, J., Pintado, M., & Moreira, M. T. (2019). Insight into antibiotics removal: Exploring the photocatalytic performance of a Fe_3O_4/ZnO nanocomposite in a novel magnetic sequential batch reactor. *Journal of Environmental Management, 237*, 595–608.

Freyria, F. S., Geobaldo, F., & Bonelli, B. (2018). Nanomaterials for the abatement of pharmaceuticals and personal care products from wastewater. *Applied Sciences, 8*(2), 170.

Gautam, S., Agrawal, H., Thakur, M., Akbari, A., Sharda, H., Kaur, R., & Amini, M. (2020). Metal oxides and metal-organic frameworks for the photocatalytic degradation: A review. *Journal of Environmental Chemical Engineering, 8*(3), 103726.

Gusain, R., Gupta, K., Joshi, P., & Khatri, O. P. (2019). Adsorptive removal and photocatalytic degradation of organic pollutants using metal oxides and their composites: A comprehensive review. *Advances in Colloid and Interface Science, 272*, 102009.

Hu, K., Li, R., Ye, C., Wang, A., Wei, W., Hu, D., & Yan, K. (2020). Facile synthesis of Z-scheme composite of TiO2 nanorod/g-C3N4 nanosheet efficient for photocatalytic degradation of ciprofloxacin. *Journal of Cleaner Production, 253*, 120055.

Jahdi, M., Mishra, S. B., Nxumalo, E. N., Mhlanga, S. D., & Mishra, A. K. (2020). Smart pathways for the photocatalytic degradation of sulfamethoxazole drug using F-Pd co-doped TiO_2 nanocomposites. *Applied Catalysis B: Environmental, 267*, 118716.

Ji, L., Chen, W., Bi, J., Zheng, S., Xu, Z., Zhu, D., & Alvarez, P. J. (2010). Adsorption of tetracycline on single-walled and multi-walled carbon nanotubes as affected by aqueous solution chemistry. *Environmental Toxicology and Chemistry, 29*(12), 2713–2719.

Ji, L., Zhou, L., Bai, X., Shao, Y., Zhao, G., Qu, Y., & Li, Y. (2012). Facile synthesis of multiwall carbon nanotubes/iron oxides for removal of tetrabromobisphenol A and Pb (II). *Journal of Materials Chemistry, 22*(31), 15853–15862.

Jones, O. A., Lester, J. N., & Voulvoulis, N. (2005). Pharmaceuticals: A threat to drinking water? *Trends in Biotechnology, 23*(4), 163–167.

Joseph, L., Heo, J., Park, Y. G., Flora, J. R., & Yoon, Y. (2011). Adsorption of bisphenol A and 17α-ethinyl estradiol on single-walled carbon nanotubes from seawater and brackish water. *Desalination, 281*, 68–74.

Jung, C., Son, A., Her, N., Zoh, K. D., Cho, J., & Yoon, Y. (2015). Removal of endocrine disrupting compounds, pharmaceuticals, and personal care products in water using carbon nanotubes: A review. *Journal of Industrial and Engineering Chemistry, 27*, 1–11.

Kanagamani, K., Muthukrishnan, P., Shankar, K., Kathiresan, A., Barabadi, H., & Saravanan, M. (2019). Antimicrobial, cytotoxicity, and photocatalytic degradation of norfloxacin using *Kleinia grandiflora* mediated silver nanoparticles. *Journal of Cluster Science, 30*(6), 1415–1424.

Khajeh, M., Laurent, S., & Dastafkan, K. (2013). Nanoadsorbents: Classification, preparation, and applications (with emphasis on aqueous media). *Chemical Reviews, 113*(10), 7728–7768.

Khin, M. M., Nair, A. S., Babu, V. J., Murugan, R., & Ramakrishna, S. (2012). A review on nanomaterials for environmental remediation. *Energy & Environmental Science, 5*(8), 8075–8109.

Kumar, A. K., & Mohan, S. V. (2012). Removal of natural and synthetic endocrine disrupting estrogens by multi-walled carbon nanotubes (MWCNT) as adsorbent: Kinetic and mechanistic evaluation. *Separation and Purification Technology, 87*, 22–30.

Kumari, A., Sharma, A., Sharma, R., Malairaman, U., & Singh, R. R. (2021). Biocompatible and fluorescent water-based NIR emitting CdTe quantum dot probes for biomedical applications. *Spectrochimica Acta Part A: Molecular and Biomolecular Spectroscopy, 248*, 119206.

Laera, G., Lens, P. N. L., Virkutyte, J., Jegatheesan, V., Kim, S. H., & AlAbed, S. (2013). Nanotechnology for water and wastewater treatment: Potentials and limitations. In *Nanotechnology for Water and Wastewater Treatment* (pp. 1–26). IWA Publishing.

Lei, J., Chen, B., Lv, W., Zhou, L., Wang, L., Liu, Y., & Zhang, J. (2019). An inverse opal TiO_2/g-C_3N_4 composite with a heterojunction for enhanced visible-light-driven photocatalytic activity. *Dalton Transactions, 48*(10), 3486–3495.

Lumbaque, E. C., Tiburtius, E. R. L., Barreto-Rodrigues, M., & Sirtori, C. (2019). Current trends in the use of zero-valent iron (Fe^0) for degradation of pharmaceuticals present in different water matrices. *Trends in Environmental Analytical Chemistry, 24*, e00069.

Madikizela, L. M., Tavengwa, N. T., & Chimuka, L. (2017). Status of pharmaceuticals in African water bodies: Occurrence, removal, and analytical methods. *Journal of Environmental Management, 193*, 211–220.

Malakootian, M., Nasiri, A., Asadipour, A., & Kargar, E. (2019). Facile and green synthesis of $ZnFe_2O_4$@ CMC as a new magnetic nanophotocatalyst for ciprofloxacin degradation from aqueous media. *Process Safety and Environmental Protection, 129*, 138–151.

Mauter, M. S., Zucker, I., Perreault, F., Werber, J. R., Kim, J. H., & Elimelech, M. (2018). The role of nanotechnology in tackling global water challenges. *Nature Sustainability, 1*(4), 166–175.

Mestre, A. S., Pires, J., Nogueira, J. M. F., & Carvalho, A. P. (2007). Activated carbons for the adsorption of ibuprofen. *Carbon, 45*(10), 1979–1988.

Mezzelani, M., Gorbi, S., & Regoli, F. (2018). Pharmaceuticals in the aquatic environments: Evidence of emerged threat and future challenges for marine organisms. *Marine Environmental Research, 140*, 41–60.

Mohammad, A. W., Teow, Y. H., Ang, W. L., Chung, Y. T., Oatley-Radcliffe, D. L., & Hilal, N. (2015). Nanofiltration membranes review: Recent advances and prospects. *Desalination, 356*, 226–254.

Mohsin, M. K., & Mohammed, A. A. (2021). Catalytic ozonation for removal of antibiotic oxy-tetracycline using zinc oxide nanoparticles. *Applied Water Science, 11*(1), 1–9.

Namshah, K. S., & Mohamed, R. M. (2018). WO_3–TiO_2 nanocomposites for paracetamol degradation under visible light. *Applied Nanoscience, 8*(8), 2021–2030.

Oleszczuk, P., Pan, B., & Xing, B. (2009). Adsorption and desorption of oxytetracycline and carbamazepine by multi-walled carbon nanotubes. *Environmental Science & Technology, 43*(24), 9167–9173.

Ondarza, P. M., Haddad, S. P., Avigliano, E., Miglioranza, K. S., & Brooks, B. W. (2019). Pharmaceuticals, illicit drugs, and their metabolites in fish from Argentina: Implications for protected areas influenced by urbanization. *Science of the Total Environment, 649*, 1029–1037.

Padhye, L. P., Yao, H., Kung'u, F. T., & Huang, C. H. (2014). Year-long evaluation on the occurrence and fate of pharmaceuticals, personal care products, and endocrine-disrupting chemicals in an urban drinking water treatment plant. *Water Research, 51*, 266–276.

Pandey, G. (2018). Challenges and prospects of agri-nanotechnology for sustainable agriculture in India. *Environmental Technology & Innovation, 11*, 299–307.

Patel, M., Kumar, R., Kishor, K., Mlsna, T., Pittman Jr, C. U., & Mohan, D. (2019). Pharmaceuticals of emerging concern in aquatic systems: Chemistry, occurrence, effects, and removal methods. *Chemical Reviews, 119*(6), 3510–3673.

Ratola, N., Cincinelli, A., Alves, A., & Katsoyiannis, A. (2012). Occurrence of organic microcontaminants in the wastewater treatment process: A mini-review. *Journal of Hazardous Materials, 239*, 1–18.

Razack, S. A., & Duraiarasan, S. (2020). A way to create a sustainable environment: Green nanotechnology–With an emphasis on noble metals. In *The ELSI Handbook of Nanotechnology: Risk, Safety, ELSI and Commercialization* (pp. 359–425). Springer.

Sayadi, M. H., Sobhani, S., & Shekari, H. (2019). Photocatalytic degradation of azithromycin using GO@ Fe_3O_4/ZnO/SnO_2 nanocomposites. *Journal of Cleaner Production, 232*, 127–136.

Sharma, A., Sharma, R., Bhatia, N., & Kumari, A. (2021). Review on synthesis, characterization, and applications of silver sulfide quantum dots. *Journal of Materials Science Research and Reviews, 42*, 42–58.

Sharma, P., Pant, S., Poonia, P., Kumari, S., Dave, V., & Sharma, S. (2018). Green synthesis of colloidal copper nanoparticles capped with *Tinospora cordifolia* and its application in catalytic degradation in textile dye: An ecologically sound approach. *Journal of Inorganic and Organometallic Polymers and Materials, 28*(6), 2463–2472.

Sharma, R., & Kumari, A. (2022). Potential applications of biorenewable nanocomposite materials for electrocatalysis, energy storage, and wastewater treatment. In *Biorenewable Nanocomposite Materials, Vol. 1: Electrocatalysts and Energy Storage* (pp. 25–46). American Chemical Society.

Sharma, R., Singh, P., Dharela, R., Chauhan, G. S., & Chauhan, K. (2017). Thiourea functionalized β-cyclodextrin as a green reducing and stabilizing agent for silver nanocomposites with enhanced antimicrobial and antioxidant properties. *New Journal of Chemistry, 41*(21), 12645–12654.

Singh, K. P., Singh, A. K., Singh, U. V., & Verma, P. (2012). Optimizing the removal of ibuprofen from water by magnetic nanocomposite using Box–Behnken design. *Environmental Science and Pollution Research, 19*(3), 724–738.

Tijani, J. O., Fatoba, O. O., Babajide, O. O., & Petrik, L. F. (2016). Pharmaceuticals, endocrine disruptors, personal care products, nanomaterials, and perfluorinated pollutants: A review. *Environmental Chemistry Letters, 14*(1), 27–49.

Velempini, T., Prabakaran, E., & Pillay, K. (2021). Recent developments in the use of metal oxides for photocatalytic degradation of pharmaceutical pollutants in water—A review. *Materials Today Chemistry, 19*, 100380.

Wang, S., Sun, H., Ang, H. M., & Tadé, M. O. (2013). Adsorptive remediation of environmental pollutants using novel graphene-based nanomaterials. *Chemical Engineering Journal, 226*, 336–347.

Wang, Y., Zhu, C., Zuo, G., Guo, Y., Xiao, W., Dai, Y., & Xian, Q. (2020). 0D/2D Co_3O_4/TiO_2 Z-Scheme heterojunction for boosted photocatalytic degradation and mechanism investigation. *Applied Catalysis B: Environmental, 278*, 119298.

Yang, D., Hong, P., Hu, Y., Li, Y., Wang, C., He, J., & Liu, J. (2021). Carbon framework-encapsulated copper oxide particles activate peroxymonosulfate for the efficient degradation of tetracycline. *Applied Surface Science, 552*, 149424.

Yang, L., Xiong, Z., Li, J., Wu, Z., Zhao, X., Zhao, C., & Lai, B. (2022). Iron active sites encapsulated in N-doped graphite for efficiently selective degradation of emerging contaminants via peroxymonosulfate (PMS) activation: Inherent roles of adsorption and electron-transfer dominated nonradical mechanisms. *Chemical Engineering Journal, 444*, 136623.

Yang, W., Lu, Y., Zheng, F., Xue, X., Li, N., & Liu, D. (2012). Adsorption behavior and mechanisms of norfloxacin onto porous resins and carbon nanotube. *Chemical Engineering Journal, 179*, 112–118.

Yin, H., Cao, Y., Fan, T., Qiu, B., Zhang, M., Yao, J., & Chen, S. (2020). Construction of carbon bridged TiO_2/CdS tandem Z-scheme heterojunctions toward efficient photocatalytic antibiotic degradation and Cr (VI) reduction. *Journal of Alloys and Compounds, 824*, 153915.

Yu, F., Li, Y., Han, S., & Ma, J. (2016). Adsorptive removal of antibiotics from aqueous solution using carbon materials. *Chemosphere, 153*, 365–385.

Zhang, S., Shao, T., Kose, H. S., & Karanfil, T. (2010). Adsorption of aromatic compounds by carbonaceous adsorbents: A comparative study on granular activated carbon, activated carbon fiber, and carbon nanotubes. *Environmental Science & Technology, 44*(16), 6377–6383.

Zhang, X., Wang, J., Dong, X. X., & Lv, Y. K. (2020). Functionalized metal-organic frameworks for photocatalytic degradation of organic pollutants in the environment. *Chemosphere, 242*, 125144.

CHAPTER 13

Role of Genetically Modified Microbial Consortium in the Management of Biomedical Degradable Waste

PIYUSH SHUKLA,[1*] APEKSHA PANDEY,[2] DEEPALI SHUKLA,[2] YASMEEN BANO,[1] MONIKA BHADAURIA,[3] PRAMOD KUMAR TIWARI,[1] and SATENDRA KUMAR NIRALA[4]

[1]*Department of Molecular Human Genetics, Jiwaji University, Gwalior, Madhya Pradesh, India*

[2]*IPS Institute of Technology and Management, Indore; Amity Institute of Law, Gwalior, Madhya Pradesh, India*

[3]*Department of Rural Technology and Social Development, Guru Ghasidas Vishwavidyalaya, Bilaspur, Chhattisgarh, India*

[4]*Department of Zoology, Guru Ghasidas Vishwavidyalaya, Bilaspur, Chhattisgarh, India*

*Corresponding author.

ABSTRACT

There are some restrictions on how microorganisms can be used to treat trash that contains xenobiotics, heavy metals, aromatic compounds, and organic matter that is decaying. Microbiologists and environmentalists are interested in a novel approach to genetically engineered microorganisms that could overcome such challenges. Numerous techniques, including directed evolution and rational designing, are now applied on a global scale thanks to the development of micro-biotechnology. Therefore, genetically modified microorganisms (GMM) offer a great replacement for wild strains that digest waste components slowly or not at all. When compared to conventional

Biomedical Waste Management: Recycling and Remediation Strategies. Younis Ahmad Hajam, Javid Ahmad Malik, Anu Pandita, and Deepu Pandita (Eds.)

methods, the use of genetically engineered microorganisms for waste management is both economical and environmentally friendly.

13.1 INTRODUCTION

Restrictions abound regarding the use of microorganisms in treating waste laden with xenobiotics, heavy metals, aromatic compounds, and decaying organic matter. Scientists and stewards of the environment seek a groundbreaking strategy involving genetically engineered microorganisms, aiming to surmount these formidable challenges. The ability of the microbes to degrade materials was improved by the addition of genetically altered microorganisms created by the insertion of particular unique genes. An environment that is safer and cleaner is produced by genetically altered bacteria that exhibit desired degrading traits. The microbial communities seen in contaminated locations have acquired special traits in their DNA that enable them to withstand challenging circumstances. It is now possible to create microorganisms with a particular gene combination that does not occur naturally thanks to advanced technology and cutting-edge procedures. Numerous techniques, including directed evolution and rational designing, are now applied on a global scale thanks to the development of microbiotechnology. Because of this, GMMs provide a fantastic alternate approach for wild strains that break down waste components slowly or not at all. When compared to conventional methods, the use of genetically engineered microorganisms for waste management is both economical and environmentally friendly. To create genetically altered microorganisms, a range of molecular techniques are used, including horizontal gene transfer, CRISPR-Cas system, and molecular cloning. The use of genetically modified microbes in the processing of solid waste is covered in this chapter. Synthetic biologists have developed bacteria over the past two decades to carry out specialized functions for a variety of applications, including biosensing, biocomputing, and biomanufacturing. An increasing toolset of biological parts and technologies has helped and contributed to this advancement.

Genetically modified bacteria have been developed by altering the genome to introduce desired genes or eliminate undesirable ones. Most applications have included crops, where genes are inserted into the plants to produce an insecticide or confer herbicide tolerance. Due to disagreements among various governments over how to define the approval standards, several apps have not yet been commercialized. For instance, standards in the European Union are construed more narrowly than in the US and Canada when it comes to the concept of significant equivalence. However, GMO research is anticipated to continue in order to address concerns about food security and adaptation to climate change (Motarjemi et al., 2013).

13.2 MICROBIAL CONSORTIA

Design complexity is rising in several microbial synthetic biology applications, including biocomputing and metabolic engineering. Large genetic circuits can be taxing and challenging to optimize, making it difficult to implement complex tasks in single populations. Microbial consortia can be created to share difficult tasks among a number of populations in order to get around these restrictions. Programming microbial consortia for both fundamental understanding and possible applications has advanced significantly in recent studies. Microbial consortia have been created using a variety of techniques, such as programming mutualistic relationships, utilizing programmed population control to stop specific populations from overgrowing and employing spatial segregation to lessen competition. The advancements in metabolic engineering, biofilm generation for engineered living materials, biocomputing, and biosensing are highlighted here as examples of the contribution of microbial consortia.

Microbes are required for a vital alternative method of issue-solving. Microorganisms may thrive anywhere in the biosphere and can live in a wide range of environmental situations thanks to their incredible metabolic activity. Microorganisms are excellent for bioremediation of environmental toxins because they may eat a variety of different foods. Bioremediation is extensively involved in the destruction, eradication, immobilization, or detoxification of numerous chemical wastes and physically harmful substances from the environment through the all-encompassing active action of microorganisms. The primary concept is to degrade and transform pollutants, such as hydrocarbons, oil, heavy metals, pesticides, dyes, and others. This is carried out enzymatically through metabolizing, and as a result, it significantly affects the solution of many environmental problems. There are two distinct types of elements that influence the rate of degradation: biotic conditions and abiotic conditions. Around the world, many methods and strategies are now being used in the field. These include biostimulation, bioaugmentation (BA), bioventing, biopiles, and bioattenuation, to name a few typical ones. Its bioremediation technology has its benefits and drawbacks because it has a particular use.

13.2.1 ROLE OF MICROORGANISMS

One of the main issues in the current situation is environmental preservation and sustainability. The greatest threat to humans directly in the medium and long term is the continuous emission of toxic waste and toxins owing to faulty industrialization and unrestrained urbanization, which also puts an enormous strain on the environment. The ecology and environment are now in the hands of sustainability and are at risk due to unscientific and poor management of urban and industrial wastes and toxins. It is currently necessary to repair and clean up the contaminated NR in order to build a healthy and habitable ecological for future generations. It is well known in the West and in developing nations like India that waste generation and economic development are positively correlated. Urbanization has a big impact on garbage generation in India, which displays a variety of tendencies. About 42 million tons of solid trash are produced annually in India. For the treatment of hazardous waste, greener, and more effective approaches are required. Microorganisms may have a future in this situation. Microorganisms' specific traits can be effectively used to restore the environment (Table 13.1). For biodegradation (BD) and the restoration of polluted habitats, microorganisms can be exploited as "miracle cures." Today, radioactive waste is successfully removed using a technique called nano-bioremediation that successfully combines microorganisms with nanotechnology (Mondal et al., 2019).

13.2.1.1 MICROALGAE REMOVES PHARMACEUTICAL CONTENTS FROM WATER

Due to their significant ecotoxicities and related health problems, the rise in the amount of pharmaceutical pollutants in water around the world has become a new environmental concern. Since microalgal bioremediation is a solar-powered, ecologically thorough, and sustainable reclamation technique, microalgal bioremediation of pharmaceutical pollutants has recently attracted scientific interest. We thoroughly outline the present research on the potential uses of microalgae for eliminating pharmaceutical pollutants from aqueous media in this review. Building microbial consortia, acclimation, and cometabolism for better removal of pharmaceutical pollutants by microalgae would be a few unique ways that would increase the practical viability of these technologies. Also suggested are some cutting-edge ideas

TABLE 13.1 Effect of Biological (Microbial) Pre-Treatment of Solid Waste on Biogas Yield

Sl. No.	Type of Waste	Microbe	Advantage
1.	Wheat straw	*Polyporus brumalis* Microbial consortium TC-5 Microbial consortium	• 52% higher methane yield • 36.6% higher methane yield • 80.34% higher methane yield
2.	Rice straw	*Pleurotus ostreatus* *Trichoderma reesei* *Bacillus* sp. Rumen fluid	• 120% higher methane yield • 78.3% higher methane yield • 76% higher biogas production • 82.6% higher methane yield
3.	Corn silage	*Trametes versicolor* MC: Cellulose degrading bacteria	• Increase in methane yield • 38% higher methane yield
4.	Corn straw	Microbial consortium *Bacillus subtilis* MC: Yeast, cellulolytic bacteria, lactic acid bacteria	• % higher biogas production
5.	Brewery spent grain	*Pseudobutyrivibrio xylanivorans* Mz 5T	• 17.8% higher biogas production
6.	Sweet corn processing residues	MC: Predominated by genus *Clostridium*	• 15% higher biogas production
7.	Cassava residues	MC: Beta proteobacterium HMD444 + *Thermoanaerobacterium thermosaccharolyticum* strain M18 + *Thermanaerovibrio acidaminovorans* DSM	• 96.63% higher biogas production

for utilizing integrated techniques and genetic changes to realize algal-based bioremediation technology (Xiong et al., 2018).

13.3 WASTE MANAGEMENT IN INDIA

Increased solid waste in India is mostly caused by the country's population growth and unplanned urbanization. India's population has increased from 1,028 million in 2001 to 1,366 million today. Management of solid waste (SWM) is still in its infancy in India, and outdated methods of rubbish collection and disposal are still infrequently used (Table 13.2). Technology is also advised.

TABLE 13.2 Comparison of Major MSW Management Technology

Technology	Advantages	Disadvantages
Incineration	Incineration is used to reduce 90% of solid waste volume. Ash generated can be used in the roads and construction industry. The heat produced during the process can be utilized for electricity or steam generation. Air emissions can be controlled.	High operational and maintenance costs. High-skilled workers are required. Significant cost required for installing air emission control devices. Negative public perception due to the release of harmful gases in the air.
Landfilling	More economical. Required less-skilled workers for its implementation and maintenance. It can be used to generate energy in the form of biogas for its utility in household and small-scale industries.	Day-to-day operation for leachate collection and treatment. Large area required for its implementation. Generates secondary pollutants that can degrade nearby groundwater, air, and soil.
Composting	Helps in the generation of high-quality organic fertilizers by utilizing biodegradable organic matter. Appreciable volumetric reduction of waste can occur.	Large space requirement for its implementation. Requires regular operational and maintenance costs with lesser environmental or economic costs. Generation of odor creating public inconvenience.
Anaerobic digestion/ biomethanation	Reduction in waste volume. The end product is high-quality fertilizer and gaseous fuel. Greenhouse gas emissions to the environment are bypassed because all the gases produced are in an enclosed system.	Large space requirement for its implementation. Requires regular operational and maintenance costs are lesser.
Bioethanol production	Bioethanol from MSW cannot create problems like food security. It may help in the mitigation of climate change. Reduction in GHGs.	The cost of technology is higher. The production process is tedious.

According to the Press Information Bureau, India produces 62 million tons (MT) of rubbish annually. Only 69% of waste is collected, and only 19% of that is treated; the other 50% is just dumped in landfills. The basic issue with trash management in metropolitan India is that plastic, commercial and industrial waste and e-waste are all thrown together and rarely separated from solid debris (Table 13.3). Large cities produce an average of 0.62 kg of MSW per person per day, compared to 0.17 kg per person per day in small towns.

A scientific study claims that the majority of Indians are ignorant of trash management and entrust it to the authorities. Therefore, raising knowledge about trash management is the biggest difficulty.

TABLE 13.3 Biomedical Waste and Role of Microbes

Types of Waste	Microbes/Microbial Consortium	Impact
Food waste	*Dysgonomonas* sp., *Pseudomonas caeni* strain, *Aeribacillus pallidus* strain, *Pseudomonas* sp., *Lactobacillus salivarius* strain, *Bacillus thuringiensis* strain, and *Bacillus cereus* strain	Antiacidification increases organic acid degradation.
Food scraps and dry leaves	Lactic acid bacteria, photosynthetic bacteria, and yeast	–
Sugarcane leaves and dairy manure	*B. licheniformis* (TA65), *A. nidulans* (GXU-1), and *A. oryzae* (GXU-11)	Thermophilic and degradative enzymes were produced by microbes.
Food waste (FW) and maize straw	*Pseudomonas fragi* (KY283110), *Pseudomonas simiae* (KY283111), *Clostridium vincentii* (KY283112), *Pseudomonas jessenii* (KY283113), and *Iodobacter fluviatilis* (KY283114)	Cold adapted
Wheat straw	A cellulolytic consortium of *Trichoderma* sp., *P. chrysosporium*, and *A. oryzae*	–
Fruit wastes, vegetable wastes, leaves, hay, newspaper, wheat straw, and rice husks	*Bacillus subtilis* and *Pseudomonas* sp.	Increased compost maturity
Food waste	Yeast strain *Pichia kudriavzevii*	Accelerated the composting process
OFMSW including vegetables, food, garden, and office waste	*Phanerochaete chrysosporium* (MTCC787), *Trichoderma viride* (MTCC793), and *Pseudomonas aeruginosa* (MTCC2295)	Enzymatic activity
MSW	*Phanerochaete chrysosporium* and *Trichoderma reesei*	Cellulolytic microbial consortium increases the speed of composting.
MSW	Mixed culture (*Nitrobacter* and *Thiobacillus*, lignin decomposition composite, and fungi	–
Organic fraction of municipal solid waste (OFMSW)	*Trichoderma viride*, *Aspergillus niger*, and *Aspergillus flavus*	High degradation of organic matter and early maturity
Kitchen-waste	*Bacillus thermoamylovorans*, mixed *Bacillus* species (such as *B. brevis*, *B. coagulans*, and *B. licheniformis*)	Composting process efficiency increased.

TABLE 13.3 *(Continued)*

Types of Waste	Microbes/Microbial Consortium	Impact
MSW	Cellulolytic Consortium of *Clostridia*	Remove petroleum hydrocarbons
Petroleum oily sludge	*Acinetobacter radioresistens* strain KA2 and *Enterobacter hormaechei* strain KA3	–
Paper cup waste	*Bacillus anthracis*, *B. endophyticus*, *B. funiculus*, *B. thuringiensis*, *B. cereus*, *B. toyonensis*, *Virgibacillus chiquenigi*, *Acinetobacter baumanni*, and *Lactobacillus pantheries*	Fasten the process (19 to 12 weeks)
Fly ash and wastepaper.	*P. fluorescens*	Phosphorus enriched
Municipal green waste	*P. chrysosporium* and *A. chroococcum*	Nutrient-rich

SWM regulation and administration in India were primarily the responsibility of a number of government organizations, including the State Pollution Control Board (SPCB), National Environmental Engineering Research Institute (NEERI), Ministry of Environment and Forests and Climate Change (MoEF), Central Pollution Control Board (CPCB), and Ministry of Urban Development (MoUD). ULBs, or Urban Local Bodies, are responsible for carrying out the plan locally. Better MSW management facilities were implemented for the first time by the Government of India (GoI) during the fourth five-year plan (1969–1974) by providing loans and grants to the state.

Because of bad management, inappropriate technology, and inadequate planning, these facilities are currently not in use. Nevertheless, a number of non-governmental organizations, neighborhood associations, volunteer organizations, and people have recently revived a number of decentralized and centralized organizations.

The Ministry of Environment and Forests (MoEF) formed the National Waste Management Council (NWMC) in 1990, emphasizing municipal solid waste management (MSWM). The objective of NWMC was to estimate the amount of recyclable waste and determine the best way to handle it. In 1993, the NWMC also established the National Plastic Waste Management Task Force to raise awareness of the damaging environmental effects of plastic recycling and human health. Due to inappropriate urban SWM, an epidemic broke out in Surat in 1994, which prompted the creation of the Dr. Bajaj organization, a powerful organization for urban solid waste management (USWM). The Indian government mandated garbage segregation in 2017 through a two-bin national campaign, but little changed on the ground. The six major metropolises, with production rates. Burning of crop residue, often called "on-farm" crop burning, was used to dispose of a significant amount of crop residue (8–80%). Among other Indian states, Uttar Pradesh has reported the most significant crop-burning incidents on farms, after Haryana and Punjab in April and afterward in December.

13.4 COMPOST PRODUCTION FROM DIFFERENT SOLID WASTES BY ADDING MICROBES AND MICROBIAL CONSORTIUM

13.4.1 ENZYMATIC HYDROLYSIS

An extracellular enzyme transforms complicated organic stuff (protein, lipid, and carbohydrates) into molecules that are easier to dissolve (amino acids, fatty acids, and sugars) during this step. Microorganisms include *Cellulomonas* sp., *Clostridium* sp., *Bacillus* sp., *Thermomonospora* sp., *Ruminococcus* sp., *Baceriodes* sp., *Erwinia* sp., *Acetovibrio* sp., and *Microbispora* can hydrolyze complex organic materials into soluble monomeric or dimeric substrates (Table 13.4).

TABLE 13.4 Important Fermentive Bacteria in Anaerobic Digestion

Sl. No.	Fermentation Types	Genera	Important Products
1.	Acetate fermentation	*Acetobacterium, Clostridium, Sporomusa*	Acetate, CO_2
2.	Alcohol fermentation	*Saccharomyces*	Ethanol, CO_2
3.	Butyrate fermentation	*Butyribacterium, Clostridium*	Butyrate, butanol, isopropanol, ethanol, CO_2
4.	Lactate fermentation	*Lactobacillus, Streptococcus*	Lactic acid, CO_2
5.	Propionate fermentation	*Clostridium*	Propionate, acetate, CO_2

13.4.1.1 FERMENTATION

By the presence of fermentative bacteria, reduced end products from hydrolysis are transformed into a combination of short-chain volatile fatty acids (VFAs) and other products such as CO_2, hydrogen, and acetic acid during fermentation. Table 13.4 demonstrates the importance of the primary fermentative bacteria in AD, which transforms the hydrolyzed soluble end products into different intermediates like VFAs, CO_2, alcohols, and hydrogen gas.

13.5 SOLID WASTE MANAGEMENT TECHNIQUES

The two main categories of MSW management techniques are thermochemical and biochemical techniques. Examples of thermochemical processes include combustion, pyrolysis, and gasification; examples of biochemical processes include aerobic and anaerobic composting, landfills, and bioethanol (Table 13.5). In contrast to pyrolysis and gasification, incinerating waste is the thermochemical practice used for SWM that is most frequently used in industrialized countries. Conversely, the most used biochemical approach is aerobic composting (ASMEWE, 2008; Denison et al., 1996; Themelis et al., 2003; UNEP, 1996).

TABLE 13.5 Sources and Types of Solid Waste

Sl. No.	Sources	Origin of Waste	Solid Wastes
1.	Residential	Single and multifamily residency	Plastics, paper, cardboard, metals, glass, food wastes, e-waste, and other household hazardous wastes
2.	Commercial	Office buildings, markets place, restaurants, storehouses, hotels	Wood, plastic, paper, cardboard, metals, and hazardous waste
3.	Institutional	Hospitals, prisons, schools, government center	Wood, plastic, paper, cardboard, metals, hazardous waste, and biomedical waste
4.	Municipal utility	Recreational areas, land-scaping, parks, beaches	Used polyethylene, tissues, napkins, organic wastes, wood trash, paper, cards, and plastic
5.	Industrial	Light and heavy manufacturing, fabrication, refineries, power and chemical plants, mineral extraction, and processing	Housekeeping wastes, packaging, food wastes, hazardous wastes, ashes, special wastes, scrap material, slay, tailings
6.	Construction and demolition	Construction sites, road repair, renovation sites, demolition of buildings	Wood, metal, concrete, dirt, and so forth.
7.	Agriculture	Crops, orchards, vineyards, dairies, feedlots, farms	Spoiled food wastes, agricultural wastes, hazardous wastes (e.g., pesticides)

KEYWORDS

- **agricultural wastes**
- **biomedical waste**
- **degradable waste**
- **demolition**
- **genetically modified microbes**
- **hazardous waste**
- **microbial consortium**
- **pesticides**

REFERENCES

American Society of Mechanical Engineers. (2008). *Waste-to-Energy: A Renewable Energy Source from Municipal Solid Waste*. White paper submitted to Congress. New York.

Denison, R. A. (1996). Environmental life-cycle comparison of recycling, landfilling, and incineration: A review of recent studies. *Annual Review of Energy and the Environment, 21*, 191–237.

Mondal, S., & Palit, D. (2019). Effective role of microorganisms in waste management and environmental sustainability. In *Sustainable Agriculture, Forest, and Environmental Management* (pp. 485–515).

Motarjemi, Y., Moy, G., & Todd, E. (2013). *Encyclopedia of Food Safety*. Academic Press.

Themelis, N. J. (2003). An overview of the global waste-to-energy industry. *Waste Management World, 3*, 40–46.

United Nations Environment Program. (1996). *International Source Book on Environmentally Sound Technologies for Municipal Solid Waste Management*. United Nations Environment Programme. Osaka/Shiga.

Index

B

D

E

F

G

Q

R

S

T

U

9781774919286